Advances in Research for Legume Genomics, Genetics, and Breeding

Advances in Research for Legume Genomics, Genetics, and Breeding

Guest Editors

Naoufal Lakhssassi
Adnane Boualem
Gunvant Patil

MDPI

Basel • Beijing • Wuhan • Barcelona • Belgrade • Novi Sad • Cluj • Manchester

Guest Editors

Naoufal Lakhssassi
Department of
Biological Sciences
Hampton University
Hampton, VA
USA

Adnane Boualem
Institute of Plant Sciences
Paris-Saclay (IPS2)
Universite Paris-Saclay
Orsay
France

Gunvant Patil
Department of Plant and
Soil Sciences
Texas Tech University
Lubbock, TX
USA

Editorial Office
MDPI AG
Grosspeteranlage 5
4052 Basel, Switzerland

This is a reprint of the Special Issue, published open access by the journal *International Journal of Molecular Sciences* (ISSN 1422-0067), freely accessible at: https://www.mdpi.com/journal/ijms/special_issues/Legume_Genetics.

For citation purposes, cite each article independently as indicated on the article page online and as indicated below:

Lastname, A.A.; Lastname, B.B. Article Title. *Journal Name* **Year**, *Volume Number*, Page Range.

ISBN 978-3-7258-6153-8 (Hbk)
ISBN 978-3-7258-6154-5 (PDF)
https://doi.org/10.3390/books978-3-7258-6154-5

Contents

About the Editors

Naoufal Lakhssassi

Naoufal Lakhssassi is a plant molecular biologist and geneticist whose research spans gene discovery, functional genomics, and crop improvement. Dr. Lakhssassi began his academic career at the University of Malaga in Spain, where he gained foundational expertise in plant genetics and biotechnology. He then continued his research at Southern Illinois University (SIU), focusing on advanced molecular breeding and gene function analysis in economically important crops. Dr. Lakhssassi was awarded the 2022 Inventor of the Year award in recognition of his innovations, contributions, and esteemed research (Southern Illinois University system). This prestigious honor reflects his exceptional work and impact in his field of study. Dr. Lakhssassi currently serves as an Assistant Professor in the Department of Biological Sciences at Hampton University, where he was granted the HU 2025 Faculty Grant Award. He has authored numerous peer-reviewed publications, contributed to book chapters, and secured multiple patents and research grants. Dr. Lakhssassi's work is distinguished by his innovative use of mutational breeding, genomics, and molecular tools to enhance crop resistance to pests and diseases, as well as to improve seed composition. In addition to his research, Dr. Lakhssassi is dedicated to mentoring students and fostering the next generation of plant scientists through teaching and collaborative projects.

Adnane Boualem

Adnane Boualem is a Research Director at the French National Institute for Agriculture, Food, and Environment (INRAE) in Paris. Renowned for his expertise in plant genetics and molecular biology, Dr. Boualem has made significant contributions to the understanding of genetic regulation in crop species, particularly in the fields of plant development, sex determination, and reproductive biology. With over 5,000 citations to his work, Dr. Boualem is recognized internationally for his impactful research, often integrating advanced genomics and biotechnological approaches to address key challenges in agriculture and food security. His research has not only furthered scientific knowledge but also provided practical insights for crop improvement and sustainable agriculture. Dr. Boualem continues to mentor young scientists and collaborate with international partners, strengthening INRAE's position as a leader in agricultural research.

Gunvant Patil

Gunvant Patil, an expert in genetic engineering, quantitative genetics, and translational genomics, is an Assistant Professor at the Institute of Genomics for Crop Abiotic Stress Tolerance and Department of Plant and Soil Sciences at Texas Tech University. Dr. Patil received his B.S. degree from North Maharashtra University, Jalgaon, India, and his M.S. and Ph.D. degrees from the University of Pune and National Center for Plant Biotechnology, New Delhi, India. Prior to joining the Tech faculty, Patil served as a research scientist at the University of Minnesota (St. Paul). He also worked as a postdoctoral associate at the University of Missouri (Columbia), and a visiting researcher at the Swedish University of Agricultural Sciences in Uppsala, Sweden.

International Journal of
Molecular Sciences

MDPI

Article

Deep Learning Model for Classifying and Evaluating Soybean Leaf Disease Damage

Sandeep Goshika [1], Khalid Meksem [2], Khaled R. Ahmed [1] and Naoufal Lakhssassi [2,*]

[1] School of Computing, Southern Illinois University, Carbondale, IL 62901, USA; sandeep.goshika@siu.edu (S.G.); khaled.ahmed@siu.edu (K.R.A.)
[2] School of Agricultural Sciences, Southern Illinois University, Carbondale, IL 62901, USA; meksem@siu.edu
* Correspondence: naoufal.lakhssassi@siu.edu

Abstract: Soybean (*Glycine max* (L.) Merr.) is a major source of oil and protein for human food and animal feed; however, soybean crops face diverse factors causing damage, including pathogen infections, environmental shifts, poor fertilization, and incorrect pesticide use, leading to reduced yields. Identifying the level of leaf damage aids yield projections, pesticide, and fertilizer decisions. Deep learning models (DLMs) and neural networks mastering tasks from abundant data have been used for binary healthy/unhealthy leaf classification. However, no DLM predicts and categorizes soybean leaf damage severity (five levels) for tailored pesticide use and yield forecasts. This paper introduces a novel DLM for accurate damage prediction and classification, trained on 2930 near-field soybean leaf images. The model quantifies damage severity, distinguishing healthy/unhealthy leaves and offering a comprehensive solution. Performance metrics include accuracy, precision, recall, and F1-score. This research presents a robust DLM for soybean damage assessment, supporting informed agricultural decisions based on specific damage levels and enhancing crop management and productivity.

Keywords: deep neural networks; soybean leaf damage detection; automatic labeling; computer vision

1. Introduction

The imperative of sustainable food production, coupled with escalating environmental challenges and soil pollution, underscores the need to optimize farmland resources. Soybean is one of the best sources of protein (~40%), oil (~20%), and carbohydrates (~30%) in livestock diets and confers nutritional benefits contributing to diabetes and heart disease prevention [1–4]. In 2021/2022, U.S. soybean meal prices ranged from USD 300 to USD 400 per ton, from which 21 million metric tons of soymeal was used by poultry, 5.8 million metric tons by swine, 4.9 million metric tons by dairy, and 1.8 million metric tons by beef. As a vital protein source globally, soybeans are uniquely positioned to address future food security by 2050. Predominantly cultivated in major producing countries such as the USA, Brazil, Argentina, China, and India, they face challenges in meeting domestic demand for soybeans due to pathogen infections, diseases, and suboptimal agronomic practices [5,6]. While primarily used for animal feed (70%), soy finds applications in biofuels, lubricants, and other industries. Disease-related annual yield loss in the U.S. has reached nearly 11%. Diseases pose significant economic threats, exacerbated by poor cultivation practices and pathogen diversity. Recognizing diverse symptoms like anthracnose, bacterial blight, and rust is pivotal, as these diseases contribute to a 14% reduction in global food output. Production loss due to infections caused by soybean cyst nematodes (SCN) alone is estimated at more than USD 1.5 billion of dollars in the U.S. Due to the importance of this trait, several studies have investigated the SCN resistance mechanism since 1960 [7–12]. Nematodes are not the only soybean pathogens; several other significant pathogens exist across fungi, bacteria, and oomycetes, causing millions of dollars related to soybean yield loss [13–16]. Therefore, early disease detection is crucial for appropriate and fast interventions.

Traditionally, cultural practices and limited pesticide use have mitigated diseases. Manual identification by visual examination remains challenging, requiring expert intervention. The overuse of pesticides has negatively impacted the environment. Early, accurate disease detection and classification are crucial for sustainable agroecosystems. Manual assessment, while effective, needs more precision, demanding reliable technology for damage detection and evaluation. The intersection between artificial intelligence and agriculture to efficiently utilize data, improve resource management, and integrate new approaches and technologies is deemed essential to improve U.S. and global food and agriculture areas. Machine learning (ML), deep learning (DL), and computer vision (CV) have enabled rapid and accurate classification of soybean leaf disease damage severity. In this context, this research focuses on developing a system to precisely identify the extent of damage caused by soybean leaf diseases, aiding in optimal pesticide selection and crop yield enhancement. By leveraging advanced technologies, this study contributes to disease prevention, reduced pesticide use, increased product quality, and enhanced yield.

The current study aims to pioneer leaf damage assessment using computer vision and drone-mounted cameras in agriculture. While prior research has tackled healthy/unhealthy leaf detection, this study addresses the gap in quantifying damage severity. The main objectives include the following: (1) deep learning model development—create a novel deep learning model to gauge leaf damage accurately, working with images of 600 × 400 pixels or lower; (2) multiclass damage classification—categorize soybean leaf damage into five levels within a single image and optimizing computational efficiency; (3) dataset curation and training—compile a diverse dataset from Southern Illinois University farms and training the deep learning algorithm to estimate damage percentage; and (4) comprehensive evaluation—evaluate the framework's accuracy, speed, efficiency, and offering insights into its applicability in agricultural settings. By achieving these objectives, the study aims to advance leaf damage assessment techniques, with potential implications for enhanced agricultural practices and yield management.

Traditional plant disease identification relies on visual inspection, which can be limited by observer experience and early-stage invisibility [17]. Tools like magnifying glasses can be inadequate for precise identification due to subtle variations in shape, color, and light reflections [18]. Black lights can identify some diseases, but not all, while soil fluorescence under UV light adds complexity [19]. This approach demands significant effort and expertise and may not be suitable for novices. Computer vision techniques are employed to detect soybean leaf defects, but several challenges persist. Low-quality images with noise, blur, or distortion affect algorithm performance [20]. Lighting variations introduce shadows, impacting accuracy. Background objects may lead to false positives. Image preprocessing removes noise but introduces artifacts with limited adaptability to the growth stage or leaf location.

Machine learning shows promising solutions to the above challenges but still presents some limitations. Traditional ML struggles mainly with raw data processing. Invasive techniques benefit from speed and accuracy while conventional ML requires feature expertise. Conventional ML depends on variable patterns and feature extraction, requiring repeated training. Despite successes, to the best of our knowledge, previous work has not classified damage levels [21]. Deep learning has shown promising results in various applications in agriculture, such as weed control [22–26], soil classification [27,28], and soil quality assessment [29–31]. In addition, DL demonstrates progress in identifying soybean leaf defects. DL models learn intricate patterns, aiding complex data analysis. DL eliminates manual feature engineering and scaling for large datasets and complex models [32], while transfer learning accelerates training by leveraging previous knowledge [33]. The current research employs DL for detecting soybean diseases via aerial imagery. The hybrid DL model [34] uses an optimization algorithm for soybean health in addition to detecting defect classification but not damage severity.

To our knowledge, there is a gap in deep learning-based classification of soybean leaf damage levels. This research aims to fill this void by introducing a method that collaborates deep learning and object detection, categorizing the phenotypical damage into five levels.

2. Results and Discussion

This section focuses on detecting and classifying damage levels in soybean leaves to enhance productivity and optimize pesticide use. The experimental setup and visualization of performance metrics for trained models on a custom dataset are outlined. Model performance, indicated by inference time, precision, and recall, is visually presented. The challenges in soybean leaf damage classification and their solutions are elaborated and discussed.

2.1. Environment Origin, Dataset, and Hyperparameter Tuning

The YOLOv5s model was developed using an NVIDIA Tesla V100 (32 GB) GPU, an Intel i9 11th generation CPU, 32 GB of RAM, and a 1 TB SSD to classify the level of damage in soybean leaves. The experimental setup employed the Anaconda environment with Python 3.10.9 and libraries such as Pytorch, OpenCV2, NumPy, Pillow, Matplotlib, and Pandas to train the model. The Coco data loader facilitated dataset loading and partitioning into training and validation sets. The dataset comprised 2930 images containing healthy and unhealthy soybean leaves. These images were sourced from a soybean farm at the Horticulture Research Center (Southern Illinois University) and captured using an iPhone 13 Pro camera with black and grey mats as backgrounds. Annotations followed the YOLO format, and the dataset was split into a 70% training dataset, a 20% validation dataset, and a 10% test dataset. Uniform resizing to 600×400 pixels was applied to all images. The different class distributions reveal that ClassOne contains nearly 3500 instances, indicating that the dataset has the highest proportion of small defects (Figure 1).

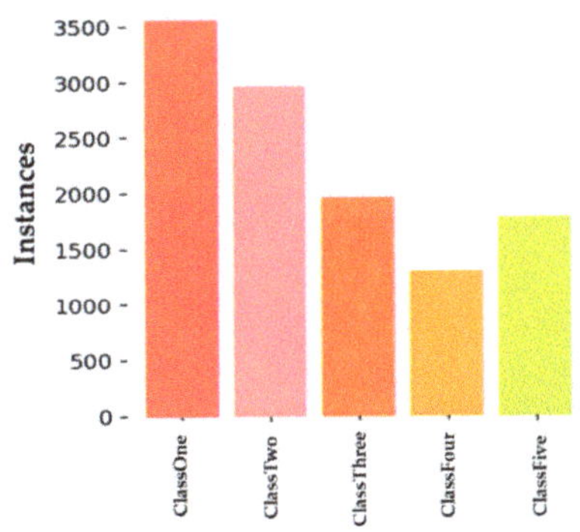

Figure 1. Graphical representation of instances of each class corresponding to the collected soybean leaf damage.

Hyperparameter tuning plays a pivotal role in optimizing model performance. Table 1 lists some hyperparameters that significantly affect optimization, such as learning rate, batch size, epochs, optimizer, input image size, regularization techniques, and anchor box dimensions. Genetic Algorithm (GA) was employed for hyperparameter optimization, given the complexity of around 30 parameters. YOLOv5's built-in 'evolve' function facilitated parameter optimization. All models underwent training up to 500 epochs, utilizing the Stochastic Gradient Descent (SGD) approach for optimization [35]. The learning rate starts with lr0 and ends with lrf to improve the generalization in the YOLOv5 model (Table 1). A batch size of 16 was employed for YOLOv5 to manage memory allocation and prevent assertion errors. The Anchor_t parameter, also known as the anchor-multiple thresholds, is used to determine the maximum adjustment that can be made to the anchor boxes during training. Neurons within the neural network process input signals using the sigmoid activation function to calculate weights and facilitate information propagation. Applying the YOLOv5's evolve function helps in finetuning and optimizing the training hyperparameters, subsequently enhancing the training outcomes.

Table 1. Values of the hyperparameters.

Hyperparameter	Value
Learning rate	lr0: 0.00334–lrf: 0.15135
Weight decay	0.00025
Batch size	16
Epochs	500
Optimizer	SGD
Input image size	600 × 800
Anchor_t	4

2.2. Practical Analysis and Metric Comparisons

The performance evaluation of YOLOv5s (Y_s) model encompasses various metrics, including inference speed, precision, recall, mean Average Precision (mAP) at different IoU thresholds (0.5 and 0.5–0.95 with 0.05 increments), the area curves of F1, precision (P), and recall (R). The evaluation metrics for model performance are explained in depth in Section 3.7. Table 2 presents the performance metrics of the generated soybean leaf detection and classification model. Precision, recall, average precision, and mAP values are determined using Equations (3), (4), (7), and (8), respectively (Section 3.7). Inference speed is a crucial factor for damage detection efficiency, as highlighted in Table 2. The Y_s model can classify the damage levels with an mAP of 92% while exhibiting the highest inference speed of 8.1 ms. Additionally, the results show that the Y_s model achieved 76% mAP@0.5–0.95, 88% F1 score, 95% P curve, 97% recall, and 93% PR area under the curve. Notably, the Y_s model achieves the highest recall and mAP at an intersection over the IoU threshold of 0.5.

Table 2. Performance metrics of soybean leaf damage detection and classification model.

Performance Metric	Model (Y_s)
Inference Speed (ms)	8.1
Precision (%)	88
mAP@0.5 (%)	92
mAP@0.5–0.95 (%)	76
F_1 Curve (%)	88
P Curve (%)	95
R Curve (%)	97
PR Curve (%)	93

The Y_s model's confusion matrix is demonstrated in Figure 2. Generally, a confusion matrix is utilized to show the performance of classifying the soybean leaf damage levels, in the tested Y_s model. The center diagonal line in the confusion matrix shows the prediction results giving the soybean leave damage levels classes. In contrast, the vertical and horizontal lines show background false negative and false positive, respectively. The values on the center line range from 0 to 1, where 1 shows 100% prediction accuracy. The model performed well in predicting the soybean leaf damage classes, as shown by the values in the center diagonal line (Figure 2). These values represent the percentage of correct predictions for each class. The model achieved 89% accuracy for ClassOne, 87% for ClassTwo, 83% for ClassThree, 82% for ClassFour, and 93% for ClassFive. These results indicate that the model performed well even when the leaves had significant damage. The model was able to accurately predict the damage level for each leaf based on its features.

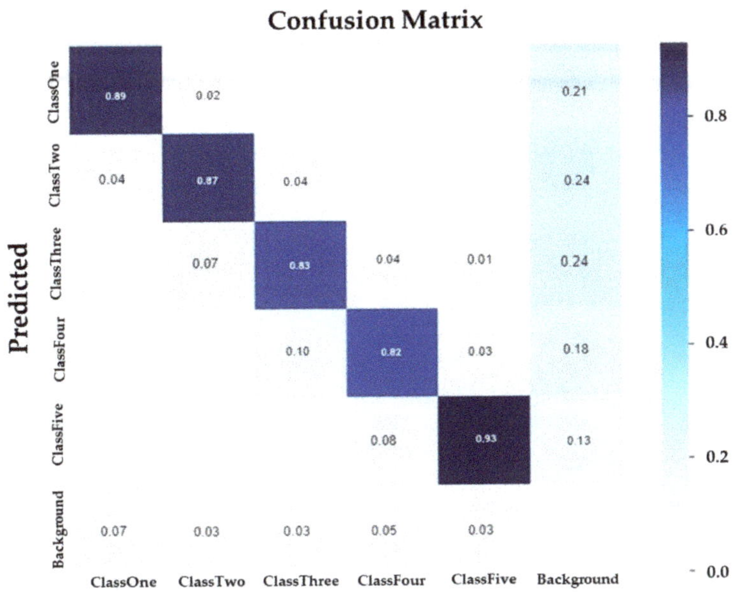

Figure 2. Confusion matrix of the YOLOv5 model in the test set (10% of the data). Percentages of correctly classified soybean damage classes fall on the diagonal. However, percentages that fall off the diagonal correspond to incorrectly classified classes.

During the training of the Y_s model, the classification loss was calculated as a function of epochs (Figure 3a). Figure 3a indicates that the Y_s model was trained successfully, reaching a low loss value after 40 epochs and achieving low classification loss as it stabilizes after 450 epochs. Figure 3b shows the F1 score curve, which combines the precision and recall in one metric, varying with the confidence threshold score for Y_s model. The performance of the Y_s model for different soybean damage levels and confidence scores is shown in Figure 3b. The model achieved the best F1 score of 0.88 when the confidence threshold was set to 0.648, which indicates a high accuracy in classifying the soybean damage levels.

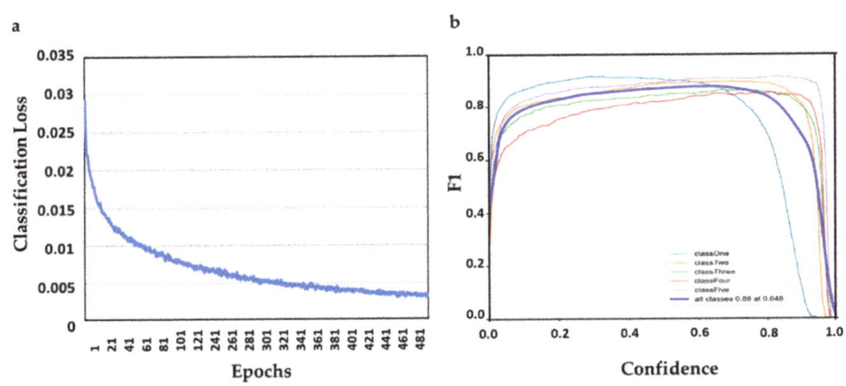

Figure 3. (**a**) represents the classification loss of the YOLOv5 model and (**b**) illustrates the YOLOv5s model's F1 score curves corresponding to the five damage levels classes.

2.3. Testing and Validation of Trained Y_s Model

The test dataset (~300 images) has been used to evaluate the trained model. Figure 4 shows the classification confidence results for each class of soybean leaf damage. As shown by the images, the model successfully detected and classified the damage severity for different types of leaf damage (Figure 4). The model's performance in detecting different damage classes of soybean leaf damage is illustrated in Figure 4, where the images show how the model can identify and classify the severity of the damage.

Figure 4. Model detection of damage to leaves. (**a**) Model detection of unhealthy leaf, (**b**) model detection of unhealthy leaf, and (**c**) model detection of healthy leaf.

Figure 4 shows that the generated model effectively determines whether the leaf is damaged (Figure 4a,b) or healthy (Figure 4c). Most importantly, Figure 4a,b depict that the model succeeds in detecting and classifying the damages in nonoverlapped soybean leaves, where bounding boxes and confidence scores indicate the detected instances of damage. Despite that the generated model successfully detected the level of soybean leaf damage from the five classes (mAP of 92%), detecting soybean leaf damage levels in an uncontrolled environment (field) is challenging due to leaves overlapping and different backgrounds, which still need to be elucidated.

3. Material and Methods

3.1. Object Detection and Machine Learning

The methodology revolves around identifying the optimal object detection model for damage classification. The cascaded detector, an early real-time detection approach with high accuracy, is widely used for face, pedestrian, and car detection [36]. This architecture implements sliding window detectors and has two main research streams to enhance speed: fast feature extraction and cascade learning. However, its limitation lies in constructing multiclass detectors within this design. YOLO (You Look Only Once) presents a noteworthy alternative, producing object detections within a 7×7 grid [22]. Despite a slight loss in detection precision, YOLO operates at approximately 40 frames per second. Combining losses on intermediate network layers enhances object identification.

Machine learning's prowess lies in supervised learning, notably classification and regression. Learning involves improving task performance with experience "E" and performance measure "P". Machine learning systems are categorized by factors such as human supervision, handling large data (online and batch learning), and prediction model development [37]. Supervised learning takes precedence as images are labeled to create the dataset. The proposed methodology involves labeling images to establish a labeled dataset. Consequently, this study emphasizes supervised learning techniques to determine the most effective object detection model for damage classification.

3.2. Dataset Creation and Annotation

Datasets are pivotal in machine learning, serving as examples for algorithms to learn from. They consist of labeled examples that guide predictions toward success or failure. Datasets are typically divided into training, validation, and test sets, training the algorithm to recognize patterns and generalize to new data. Deep learning advancements leverage data augmentation, customization, and annotation to improve model performance. This study's dataset comprises 2930 soybean leaf images, sized at 600 × 800 pixels. It encompasses 2430 images depicting various degrees of damage and 500 healthy images sourced from Kaggle and SIU's soybean farm. The wild-type Forrest seeds from Southern Illinois University Carbondale Agricultural Research Center were planted in the greenhouse and then transplanted in the field during summer until the end of vegetative growth as described earlier in [38]. Insect-damaged and infected leaves were collected at the V5 stage, and then pictures were taken under controlled lighting conditions using an iPhone 13 Pro camera. The images exhibit different background colors for enhanced segmentation.

Damage types include bacterial blight and defoliation. Bacterial blight lesions caused by the bacterium *Pseudomonas savastanoi* pv. *Glycinea* are angular and reddish-brown, surrounded by yellow halos [39]. As the disease progresses, lesions often grow together to produce large, irregularly shaped areas of dead tissue. The centers of older lesions frequently fall out, causing leaves to appear tattered [39]. Early symptoms of bacterial blight can be hard to distinguish from symptoms of several other diseases. It can be commonly confused with septoria brown spot, bacterial pustule, downy mildew, soybean rust, soybean vein necrosis, target spot, wildfire caused by other bacterial, oomycetes, or fungal diseases, in addition to insect and mite injury diseases. Tattered leaves can help distinguish bacterial blight, in addition to the symptoms observed in the upper canopy [39]. Defoliation refers to leaf loss due to pests, diseases, or natural causes. Notably, the dataset does not differentiate between these two symptoms, treating them as a unified condition. Figure 5 illustrates healthy leaves and diverse images with distinct damages and backgrounds of bacterial blight and defoliation.

Figure 5. Sample images from the dataset: (**a**) healthy leaf with brown color as background, (**b**) defoliation of leaf with brown color as background, (**c**) healthy leaf with black color as background, and (**d**) yellow disease leaf with black color as background.

Each image's leaf defects are categorized into five damage levels (ClassOne to ClassFive) (Table 3), determined through a custom image processing script [40]. This dataset paves the way for model training and enhances model accuracy through specific photo labeling, a cornerstone of effective crowdsourcing strategies.

Table 3. Classification and the percentage damage in soybean leaf.

Classification	Percentage Damage of Each Defect in Leaf (D_p)
ClassOne	Greater than 0 and less than 1.1
ClassTwo	Greater than 1.1 and less than 2.4
ClassThree	Greater than 2.4 and less than 4.1
ClassFour	Greater than 4.1 and less than 6.6
ClassFive	Greater than 6.7

3.3. Automated Annotation for Image Labeling

Manual image labeling poses challenges such as time consumption, errors, and discrepancies. A custom script was developed in this research for automated labeling of leaf damage using computer vision, which drastically reduced labeling time to 50%. This script inputs leaf images and generates annotations, including bounding box coordinates, image size, total damage, image path, name, and defect name in XML format. The automated labeling process employs libraries like OpenCV, NumPy, and Matplotlib to extract image features, context, and edges. The level of damage detection relies on parameters such as leaf and damage area. For instance, to find the leaf area (Figure 6a), the edges are detected and enclosed to calculate its area (Figure 6b). Inner damages are found by identifying edges within the leaf's outer area (Figure 6c), using morphological operations like erosion and dilation for noise reduction and edge enhancement. Automatic thresholding techniques like OTSU separate the leaf from the background. The area of a contour is calculated using the Shoelace formula (Equation (1)), considering its vertices' coordinates. For multiple contours representing different damage areas, the percentage of damage is calculated relative to the total leaf area. The script establishes a PASCAL VOC XML format containing these annotations and calculations. Automated labeling proves efficient and consistent, mitigating human-related errors and enhancing dataset quality.

$$A = 0.5 * |(x_1 * y_2 + x_2 * y_3 + \ldots + x_n * y_1) - (y_1 * x_2 + y_2 * x_3 + \ldots + y_n * x_1)| \quad (1)$$

where n is the number of vertices in the polygon, and $x_1, y_1, x_2, y_2, \ldots, x_n, y_n$ are the coordinates of the vertices in order. The average area of the n leaves in the dataset is given by

$$A_T = \frac{A_{T_1} + A_{T_2} + A_{T_3} \ldots\ldots\ldots + A_{T_n}}{n}$$

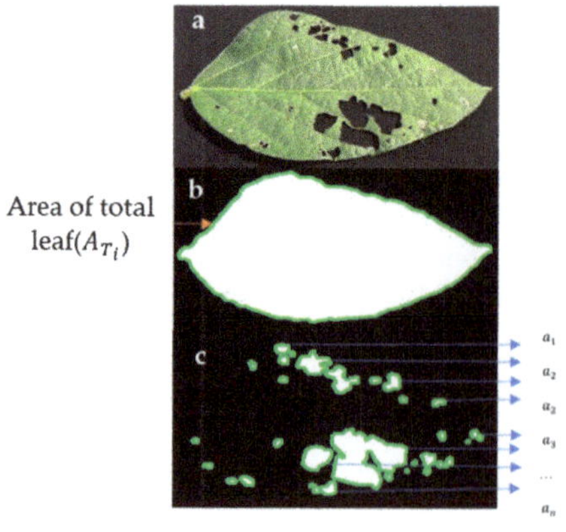

Figure 6. Original leaf images and converted binary images were used to draw contours. (**a**) Original image with damage, (**b**) total leaf area with contours, and (**c**) inner damage of leaf with contours.

The percentage damage of each defect in the leaf is given by

$$D_p = \frac{A_{T_i}}{A_T} \times c \times 10$$

where c is the constant used to scale up the value, A_{T_i} is the area of defect i-th in the leaf.

3.4. Enhancing Dataset through Data Augmentation

Data augmentation techniques can mitigate challenges related to dataset size, resolution, and ground truth boxes in the collection process. Data augmentation, a vital deep learning strategy, expands the training dataset by generating new instances from existing data. This technique bolsters deep learning model performance by curbing overfitting, enhancing generalization, and augmenting data diversity [41]. YOLOv5 employs a specific augmentation technique called Mosaic augmentation (Figure 7). This method fuses four distinct images into a mosaic image, subsequently used to train the object detection model. Mosaic augmentation enhances the model's capability to identify objects within intricate settings with multiple objects and backgrounds.

Figure 7. An example of Mosaic augmentation generated by YOLOv5 for the soybean dataset. It includes more diverse soybean leaves and backgrounds.

Additionally, as mosaic images replicate a range of scenarios, it diminishes the requirement for extensive training data. However, this technique can be computationally intensive, demanding the creation of numerous images and labels per training batch. Proper hyperparameter adjustments, like crop patch size and overlap, are essential to ensure effective learning from mosaic images. To bolster deep learning model efficacy in scenarios with limited or imbalanced data, several strategies like geometric transformations, color space transformations, kernel filters, and meta-learning can be implemented for image augmentation [41]. Data augmentation generates new data akin to the original dataset but with slight variations [42]. Techniques include cropping, flipping, rotating, blurring, scaling, translating, color perturbations, noise addition, and more. By embracing these augmentation strategies, the model learns from an enriched dataset, surmounting data scarcity and imbalance constraints.

3.5. Model Architecture and Training

Challenges related to dataset size, resolution, and ground truth boxes in the collection process can be mitigated using data augmentation techniques. Data augmentation, a vital deep learning strategy, expands the training dataset by generating new instances from existing data. This technique bolsters deep learning model performance by curbing overfitting, enhancing generalization, and augmenting data diversity [41].

The YOLOv5, a convolutional neural network (CNN)-based object detector, offers advantages like high accuracy, rapid detection speed, and lightweight nature [43]. Object detectors of this kind are classified into different categories based on their attributes and features. These detectors primarily consist of two components: a CNN-based backbone for image feature extraction and a detection head to predict object classes and bounding boxes. The YOLOv5 object detector incorporates intermediate layers, the neck of the detector, positioned between the backbone and the head, as depicted in Figure 8.

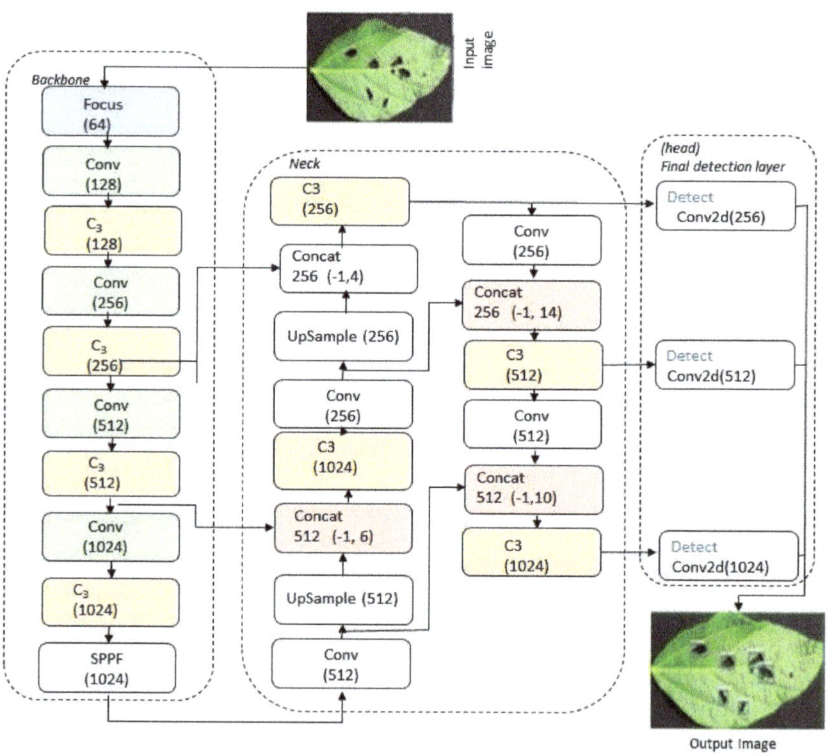

Figure 8. Architecture of the YOLOv5 neural network that includes 10 backbone layers, 14 neck layers, and 3 Conv2d layers in the head.

Backbone: YOLOv5 employs a CSPDarknet53 backbone, a modified version of Darknet53, known for its efficiency in various computer vision tasks. Cross-stage partial connections (CSP) are integrated into CSPDarknet53 to streamline information flow and reduce parameter count, enhancing its efficiency compared to the original Darknet53 [44]. The essential component of this backbone is the CBS module, which consists of a Convolution, BatchNorm, and SiLu activation function [45,46]. They stacked together and formed the C3 module. Additionally, the YOLOv5 model includes supplementary layers like Spatial Pyramid Pooling Fast (SPP) (https://github.com/ultralytics/yolov5, accessed on 12 June 2023) to boost its performance further [47]. The SSPF module includes several maxpooling layers and a CBS module. These modules fuse the feature maps of different receptive fields and enrich the expression ability of feature maps. Each Concat layer is used to slice the previous layer. This backbone demonstrates robust feature extraction capabilities for classification and other tasks, often fine tuned for versatility across activities.

Neck: The neck, a pivotal component in the object detection framework, enhances the utilization of features extracted by the backbone. It optimally processes and employs the feature maps extracted by the backbone across various stages. The YOLOv5 neck also incorporates up-sampling and down-sampling stages, combining CSPDarknet53 feature maps using C3 to create the pyramid. Additionally, lateral connections enable feature propagation across scales. The neck outputs feature maps of $80 \times 80 \times 256$, $40 \times 40 \times 512$, and $20 \times 20 \times 1024$ that correspond to target objects of a small, medium, and large scale.

Head: The head encompasses the final 3 Conv2D layers of the network, which are responsible for generating model output. While the backbone focuses on classification, the head manages object positioning tasks, determining object likelihood within bounding boxes and their respective categories. The head employs feature maps extracted from the backbone to estimate bounding box offsets relative to predefined anchor boxes, enhancing model accuracy [47].

3.6. YOLOv5 Architecture and Analysis

YOLOv5 is a one-stage object detector, structured with 1 focus layer, 10 backbone layers, and 14 neck layers. For instance, "Conv (256)" is the third convolutional layer, taking 128 pixels as input and producing 256 pixels as output for the next layer. The backbone extracts feature information, directing it to specialized convolutional layers responsible for object prediction and detection. SPPF generates fixed-size windows, independent of input image dimensions, enhancing damage detection in soybean leaves. Max pooling with kernel lengths 5, 9, and 13 contributes to this process. The last convolutional layer in the backbone bridges the head and backbone, creating feature pyramids for object scaling. The model architecture includes customizable particular convolutional layers. For the level of damage detection in soybean leaves, layers 17, 20, and 23 are selected with 256, 512, and 1024 pixels, respectively, and used to determine object classes with kernel stride.

The YOLOv5 model builds upon its predecessors, combining elements from YOLO to YOLOv4. It divides the image into $S * S$ grids, each responsible for detecting the presence of objects (soybeans), including leaf damage. β bounding boxes predict the extent of leaf damage, each associated with a confidence score. Each grid cell supports multiple bounding box predictions characterized by x, y, w, h, and c attributes. The model's output computation for a single image involves $S * S * \beta * 5$ calculations. The model determines the level of damage in the soybean leaf depending on confidence scores. Speed and accuracy are prioritized, influencing activation optimizations and regularization loss values. Model loss is a summation of bounding box loss and classification loss.

Activation layers employ the SiLU (Sigmoid Linear Unit) activation function, while SGD (Stochastic Gradient Descent) serves as the optimization function. Pytorch is the model design and detection platform, with binary cross-entropy as the loss function. Anchor boxes are applied to features, generating output vectors containing class probabilities, object scores, and bounding boxes.

3.7. Evaluation Metrics for Model Performance

Evaluation metrics are crucial in assessing a deep learning model's performance on a specific task. These metrics provide insights into the quality of the model's predictions and guide improvements. Statistical methods, such as confusion matrix, F1 curve, precision, recall, accuracy, and GIoU (Generalized Intersection over Union), are commonly employed for model evaluation.

Parameters used in detection and evaluation include T_P: True Positive, T_N: True Negative, F_N: False Negative, and F_P: False Positive.

A correct prediction of the object's level of damage to the soybean leaf within the bounding box is categorized as a true positive. Specifically, if the Intersection over Union (IoU) between the predicted and actual bounding boxes is greater than or equal to a defined threshold (denoted by α), it is considered a true positive. Conversely, a frame is labeled as a false positive when the object is absent from the bounding box and IoU is less than α. If the bounding box fails to capture the target object, it results in a false negative. True negative is assigned when the object is correctly absent or not predicted in the frame.

Model accuracy hinges on the correctness of its detections in unseen data. The following evaluation metrics are utilized in this research to assess the model's performance:

IoU: Intersection of Union is a term used to define the extent of overlap of two bounding boxes. The greater the region of overlap, the greater the IOU. It is given by the ratio area of the intersection of two boxes to the area of the union of two boxes.

$$\mathrm{IoU} = \frac{\mathrm{A} \cap \mathrm{B}}{\mathrm{AUB}} \tag{2}$$

where A, B predicted and ground truth boxes.

GIoU: Generalized IoU (GIoU) is an extension of the Intersection over Union (IoU). This metric explains the size and location of the predicted bounding box and the ground truth bounding box, where C is the smallest convex shape that encloses both boxes A and B. We calculate a ratio between the area occupied by C excluding A and B and divide by the total area occupied by C. Finally, GIoU is calculated by subtracting this ratio from the IoU value as shown in Equation (3) [48].

$$\mathrm{GIoU} = \mathrm{IoU} - \frac{|\mathrm{C} \backslash (\mathrm{A} \cup \mathrm{B})|}{|\mathrm{C}|} \tag{3}$$

Recall score is calculated as the ratio of true positives (T_P) to the sum of true positives and false negatives (F_N).

$$\mathrm{R} = \frac{T_P}{T_P + F_N} \tag{4}$$

Precision is the ratio of true positives to the sum of true positives and false positives.

$$\mathrm{P} = \frac{T_P}{T_P + F_P} \tag{5}$$

Accuracy is the ratio of the number of correct predictions (true positives and true negatives) to the total number of predictions T_P, T_N, F_P, and F_N.

$$\mathrm{A} = \frac{T_P + T_N}{T_P + T_N + F_P + F_N} \tag{6}$$

F_1 Curve is the harmonic mean of precision and recall, and it is a single number that summarizes the performance of the model across all possible classification thresholds.

$$F_1 = \frac{2 * \mathrm{P} * \mathrm{R}}{\mathrm{P} + \mathrm{R}} \tag{7}$$

The area under the precision–recall curve is calculated using numerical integration, which gives the AP score.

$$\text{AP@}\alpha = \int_0^1 p(r)dr \tag{8}$$

Mean Average Precision is the average of the AP@α calculated for all the classes over IoU Threshold α depending on the detection challenge.

$$\text{mAP@}\alpha = \frac{1}{N}\sum_{i=1}^{N} \text{AP}_i \tag{9}$$

The loss function is used to measure the difference between the predicted output of a model and the true output, and it is given by the summation of coordination loss, classification loss, and object loss, as follows.

$$\lambda_{coord}\sum_{i=0}^{S^2}\sum_{j=0}^{\beta} l_{ij}^{obj}(x_i-\hat{x}_i)^2+(y_i-\hat{y}_i)^2 \tag{10}$$

Equation (9) is used to find the loss of the bounding box of position (x, y) and the actual position $(\hat{x}, \hat{y})$ from the training data. This function computes a sum over each bounding box predictor ($j = 0\ldots\beta$) of each grid cell ($i = 0\ldots S^2$)l_{ij}^{obj} where l_{ij}^{obj} implies that the object appears in cell i and indicates that jth bounding box predictor in cell i is responsible for that prediction [48].

The coordination error calculated by the prediction box w.r.t width/height is given below in Equation (11).

$$\lambda_{coord}\sum_{i=0}^{S^2}\sum_{j=0}^{\beta} l_{ij}^{obj}\left[\left(\sqrt{w_i}-\sqrt{\hat{W}_i}\right)^2+\left(\sqrt{h_i}-\sqrt{\hat{h}_i}\right)^2\right] \tag{11}$$

The classification loss based on the confidence score for each bounding box class is given by Equation (11).

$$\lambda_{coord}\sum_{i=0}^{S^2}\sum_{j=0}^{\beta} l_{ij}^{obj}(C_i-\hat{C}_i)^2+\lambda_{noobj}\sum_{i=0}^{S^2}\sum_{j=0}^{\beta} l_{ij}^{noobj}(C_i-\hat{C}_i)^2 \tag{12}$$

The Equation (11) is used to initialize the λ value for the presence of an object differently to gain model stability or the presence of an object λ_{coord} = 5, else λ_{noord} = 0.5.

4. Conclusions and Future Work

The necessity of accurately assessing the degree of damage in soybean leaves for enhanced production and reduced disease susceptibility has prompted the utilization of neural networks. Training a neural network from the ground up aimed to ascertain the extent of leaf damage and subsequently compare outcomes against a compiled dataset. Various augmentation techniques were implemented to amplify the data size, encompassing image blurring, rotation, and resolution adjustments. Experimental findings highlight the Y_s model's commendable performance in terms of mAP@0.5. In extending this research, several avenues for improvement emerge. Firstly, there is a demand to expand the dataset by capturing a broader range of samples with overlapped leaves, encompassing diverse forms of soybean leaf damage. Furthermore, transitioning from solely detecting leaf damage to developing a neural network capable of automatically identifying overall plant defects emerges as a promising avenue for further exploration.

Author Contributions: Conceptualization, S.G., N.L. and K.R.A.; methodology, S.G. and K.R.A.; software, S.G. and K.R.A.; validation, S.G., N.L. and K.R.A.; investigation, S.G., N.L. and K.R.A.; resources, N.L., K.M. and K.R.A.; data curation, S.G., N.L. and K.R.A.; writing—original draft preparation, S.G., K.M., N.L. and K.R.A.; writing—review and editing. All authors have read and agreed to the published version of the manuscript.

Funding: This research received no external funding.

Institutional Review Board Statement: Not applicable.

Informed Consent Statement: Not applicable.

Data Availability Statement: The data presented in this study are available on request from the corresponding author.

Conflicts of Interest: The authors declare no conflicts of interest.

References

1. Zhou, Z.; Lakhssassi, N.; Cullen, M.A.; El Baz, A.; Vuong, T.D.; Nguyen, H.T.; Meksem, K. Assessment of Phenotypic Variations and Correlation among Seed Composition Traits in Mutagenized Soybean Populations. *Genes* **2019**, *10*, 975. [CrossRef] [PubMed]
2. Zhou, Z.; Lakhssassi, N.; Knizia, D.; Cullen, M.A.; El Baz, A.; Embaby, M.G.; Liu, S.; Badad, O.; Vuong, T.D.; AbuGhazaleh, A.; et al. Genome-wide identification and analysis of soybean acyl-ACP thioesterase gene family reveals the role of GmFAT to improve fatty acid composition in soybean seed. *Theor. Appl. Genet.* **2021**, *134*, 3611–3623. [CrossRef] [PubMed]
3. Lakhssassi, N.; Lopes-Caitar, V.S.; Knizia, D.; Cullen, M.A.; Badad, O.; El Baze, A.; Zhou, Z.; Embaby, M.G.; Meksem, J.; Lakhssassi, A.; et al. TILLING-by-Sequencing(+) Reveals the Role of Novel Fatty Acid Desaturases (GmFAD2-2s) in Increasing Soybean Seed Oleic Acid Content. *Cells* **2021**, *10*, 1245. [CrossRef] [PubMed]
4. Lakhssassi, N.; Zhou, Z.; Cullen, M.A.; Badad, O.; El Baze, A.; Chetto, O.; Embaby, M.G.; Knizia, D.; Liu, S.; Neves, L.G.; et al. TILLING-by-Sequencing(+) to Decipher Oil Biosynthesis Pathway in Soybeans: A New and Effective Platform for High-Throughput Gene Functional Analysis. *Int. J. Mol. Sci.* **2021**, *22*, 4219. [CrossRef] [PubMed]
5. Chang, W.S.; Lee, H.I.; Hungria, M. *Soybean Production in the Americas*; Springer International Publishing: Cham, Switzerland, 2015.
6. Di Matteo, F.; Otsuki, K.; Schoneveld, G. Soya Bean Expansion in Mozambique: Exploring the Inclusiveness and Viability of Soya Business Models as an Alternative to the Land Grab. *The Public Sphere*. 2016, pp. 61–86. Available online: https://hdl.handle.net/10568/94497 (accessed on 17 October 2023).
7. Lakhssassi, N.; Knizia, D.; El Baze, A.; Lakhssassi, A.; Meksem, J.; Meksem, K. Proteomic, Transcriptomic, Mutational, and Functional Assays Reveal the Involvement of Both THF and PLP Sites at the GmSHMT08 in Resistance to Soybean Cyst Nematode. *Int. J. Mol. Sci.* **2022**, *23*, 11278. [CrossRef] [PubMed]
8. Piya, S.; Pantalone, V.; Zadegan, S.B.; Shipp, S.; Lakhssassi, N.; Knizia, D.; Krishnan, H.B.; Meksem, K.; Hewezi, T. Soybean gene co-expression network analysis identifies two co-regulated gene modules associated with nodule formation and development. *Mol. Plant Pathol.* **2023**, *24*, 628–636. [CrossRef] [PubMed]
9. Cook, D.E.; Lee, T.G.; Guo, X.; Melito, S.; Wang, K.; Bayless, A.M.; Wang, J.; Hughes, T.J.; Willis, D.K.; Clemente, T.E.; et al. Copy number variation of multiple genes at Rhg1 mediates nematode resistance in soybean. *Science* **2012**, *338*, 1206–1209. [CrossRef] [PubMed]
10. Bayless, A.M.; Zapotocny, R.W.; Grunwald, D.J.; Amundson, K.K.; Diers, B.W.; Bent, A.F. An atypical N-ethylmaleimide sensitive factor enables the viability of nematode-resistant Rhg1 soybeans. *Proc. Natl. Acad. Sci. USA* **2018**, *115*, E4512–E4521. [CrossRef]
11. Bayless, A.M.; Smith, J.M.; Song, J.; McMinn, P.H.; Teillet, A.; August, B.K.; Bent, A.F. Disease resistance through impairment of α-SNAP-NSF interaction and vesicular trafficking by soybean Rhg1. *Proc. Natl. Acad. Sci. USA* **2016**, *113*, E7375–E7382. [CrossRef]
12. Bent, A.F. Exploring Soybean Resistance to Soybean Cyst Nematode. *Annu. Rev. Phytopathol.* **2022**, *60*, 379–409. [CrossRef]
13. Hosseini, B.; Voegele, R.T.; Link, T.I. Diagnosis of Soybean Diseases Caused by Fungal and Oomycete Pathogens: Existing Methods and New Developments. *J. Fungi* **2023**, *9*, 587. [CrossRef] [PubMed]
14. Escamilla, D.; Rosso, M.L.; Zhang, B. Identification of fungi associated with soybeans and effective seed disinfection treatments. *Food Sci. Nutr.* **2019**, *7*, 3194–3205. [CrossRef] [PubMed]
15. Huynh, T.V.; Dahlbeck, D.; Staskawicz, B.J. Bacterial blight of soybean: Regulation of a pathogen gene determining host cultivar specificity. *Science* **1989**, *245*, 1374–1377. [CrossRef] [PubMed]
16. Million, C.R.; Wijeratne, S.; Karhoff, S.; Cassone, B.J.; McHale, L.K.; Dorrance, A.E. Molecular mechanisms underpinning quantitative resistance to Phytophthora sojae in *Glycine max* using a systems genomics approach. *Front. Plant Sci.* **2023**, *14*, 1277585. [CrossRef] [PubMed]
17. Liu, J.; Wang, X. Plant diseases and pests detection based on deep learning: A review. *Plant Methods* **2021**, *17*, 22. [CrossRef] [PubMed]
18. Mignoni, M.; Aislan, H.; Kunst, R.; Righi, R.; Massuquetti, A. Soybean Images Dataset for Caterpillar and Diabrotica speciosa pest detection and classification. *Data Brief* **2021**, *40*, 107756. [CrossRef] [PubMed]
19. Bevers, N.; Sikora, E.J.; Hardy, N.B. Soybean disease identification using original field images and transfer learning with convolutional neural networks. *Comput. Electron. Agric.* **2022**, *203*, 107449. [CrossRef]
20. Zhang, K.; Wu, Q.; Chen, Y. Detecting soybean leaf disease from synthetic image using multi-feature fusion faster R-CNN. *Comput. Electron. Agric.* **2021**, *183*, 106064. [CrossRef]
21. Krishna, R.; Prema, K. Soybean crop disease classification using machine learning techniques. In Proceedings of the IEEE International Conference on Distributed Computing, VLSI, Electrical Circuits and Robotics (DISCOVER), Udupi, India, 30–31 October 2020.

22. Almalky, A.M.; Ahmed, K.R. Deep Learning for Detecting and Classifying the Growth Stages of Consolida regalis Weeds on Fields. *Agronomy* **2023**, *13*, 934. [CrossRef]
23. Razfar, N.; True, J.; Bassiouny, R.; Venkatesh, V.; Kashef, R. Weed detection in soybean crops using custom lightweight deep learning models. *J. Agric. Food Res.* **2022**, *8*, 100308. [CrossRef]
24. Hu, K.; Wang, Z.; Coleman, G.; Bender, A.; Yao, T.; Zeng, S.; Song, D.; Schumann, A.; Walsh, M. Deep learning techniques for in-crop weed recognition in large-scale grain production systems: A review. *Precis. Agric.* **2023**. [CrossRef]
25. Almalky, A.M.; Ahmed, K.R.; Guzel, M.; Turan, B. An Efficient Deep Learning Technique for Detecting and Classifying the Growth of Weeds on Fields. In Proceedings of the Future Technologies Conference (FTC) 2022, Vancouver, BC, Canada, 20–21 October 2022; Arai, K., Ed.; Springer International Publishing: Cham, Switzerland, 2023; Volume 2, pp. 818–835.
26. Almalky, A.; Khaled, A. Real Time Deep Learning Algorithm for Counting Weeds Growth Stages. In Proceedings of the 15th International Symposium on Autonomous Decentralized System (ISADS2023), Mexico City, Mexico, 15–17 March 2023.
27. Srivastava, P.; Shukla, A.; Bansal, A. A comprehensive review on soil classification using deep learning and computer vision techniques. *Multimed. Tools Appl.* **2021**, *80*, 14887–14914. [CrossRef]
28. Li, X.; Fan, P.; Li, Z.; Chen, G.; Qiu, H.; Hou, G. Soil Classification Based on Deep Learning Algorithm and Visible Near-Infrared Spectroscopy. *J. Spectrosc.* **2021**, *2021*, 1508267. [CrossRef]
29. Sumathi, P.; Karthikeyan, V.V.; Kavitha, M.S.; Karthik, S. Improved Soil Quality Prediction Model Using Deep Learning for Smart Agriculture Systems. *Comput. Syst. Sci. Eng.* **2023**, *45*, 1545–1559. [CrossRef]
30. Padarian, J.; Minasny, B.; McBratney, A.B. Using deep learning to predict soil properties from regional spectral data. *Geoderma Reg.* **2019**, *16*, e00198. [CrossRef]
31. Nyakuri, J.P.; Bizimana, J.; Bigirabagabo, A.; Kalisa, J.B.; Gafirita, J.; Munyaneza, M.A.; Nzemerimana, J.P. IoT and AI Based Smart Soil Quality Assessment for Data-Driven Irrigation and Fertilization. *Am. J. Comput. Eng.* **2022**, *5*, 1–14. [CrossRef]
32. Wu, Q.; Zhang, K.; Meng, J. Identification of Soybean Leaf Diseases via Deep Learning. *J. Inst. Eng. India Ser. A* **2019**, *100*, 659–666. [CrossRef]
33. Zamani, A.S.; Anand, L.; Rane, K.P.; Prabhu, P.; Buttar, A.M.; Pallathadka, H.; Raghuvanshi, A.; Dugbakie, B.N. Performance of Machine Learning and Image Processing in Plant Leaf Disease Detection. *J. Food Qual.* **2022**, *2022*, 1598796. [CrossRef]
34. Annrose, J.; Rufus, N.; Anantha, H.; Rex, C.; Selva, R.; Immanuel, E.; Godwin, D. Cloud-Based Platform for Soybean Plant Disease Classification Using Archimedes Optimization Based Hybrid Deep Learning Model. *Wirel. Pers. Commun.* **2022**, *122*, 2995–3017. [CrossRef]
35. Ketkar, N. Stochastic Gradient Descent. In *Deep Learning with Python: A Hands-On Introduction*; Ketkar, N., Ed.; Apress: Berkeley, CA, USA, 2017; pp. 113–132.
36. Viola, P.; Jones, M.J. Robust Real-Time Face Detection. *Int. J. Comput. Vis.* **2004**, *57*, 137–154. [CrossRef]
37. Hartman, G. *Compendium of Soybean Diseases and Pests*; APS Publications: St. Paul, MN, USA, 2016.
38. Lakhssassi, N.; Piya, S.; Knizia, D.; El Baze, A.; Cullen, M.A.; Meksem, J.; Lakhssassi, A.; Hewezi, T.; Meksem, K. Mutations at the Serine Hydroxymethyltransferase Impact Its Interaction with a Soluble NSF Attachment Protein and a Pathogenesis-Related Protein in Soybean. *Vaccines* **2020**, *8*, 349. [CrossRef] [PubMed]
39. Mueller, D.; Wise, K.; Sisson, A.; Smith, D.; Sikora, E.; Bradley, C.; Robertson, A. *A Farmer's Guide to Soybean Diseases*; APS Press: St. Paul, MN, USA, 2016.
40. Anantrasirichai, N.; Hannuna, S.L.; Canagarajah, C.N. Automatic Leaf Extraction from Outdoor Images. *arXiv* **2017**. [CrossRef]
41. Shorten, C.; Khoshgoftaar, T.M. A survey on Image Data Augmentation for Deep Learning. *J. Big Data* **2019**, *6*, 1–48. [CrossRef]
42. Perez, L.; Wang, J. The Effectiveness of Data Augmentation in Image Classification using Deep Learning. *arXiv* **2017**. [CrossRef]
43. Wang, H.; Zhang, S.; Zhao, S.; Wang, Q.; Li, D.; Zhao, R. Real-time detection and tracking of fish abnormal behavior based on improved YOLOV5 and SiamRPN++. *Comput. Electron. Agric.* **2022**, *192*, 106512. [CrossRef]
44. Wang, C.-Y.; Liao, H.-Y.M.; Wu, Y.-H. Cspnet: A new backbone that can enhance learning capability of CNN. In Proceedings of the IEEE/CVF Conference on Computer Vision and Pattern Recognition Workshops (CVPRW), Seattle, WA, USA, 14–19 June 2020.
45. Ioffe, S.; Szegedy, C. Batch Normalization: Accelerating Deep Network Training by Reducing Internal Covariate Shift. In Proceedings of the 32nd International Conference on Machine Learning, Lille, France, 6–11 July 2015; Francis, B., David, B., Eds.; Volume 37, pp. 448–456.
46. Elfwing, S.; Uchibe, E.; Doya, K. Sigmoid-weighted linear units for neural network function approximation in reinforcement learning. *Neural Netw.* **2018**, *107*, 3–11. [CrossRef]
47. Tan, M.; Pang, R.; Le, Q.V. EfficientDet: Scalable and Efficient Object Detection. In Proceedings of the 2020 IEEE/CVF Conference on Computer Vision and Pattern Recognition (CVPR), Online, 13–19 June 2020; pp. 10778–10787.
48. Ahmed, K.R. Smart Pothole Detection Using Deep Learning Based on Dilated Convolution. *Sensors* **2021**, *21*, 8406. [CrossRef]

International Journal of
Molecular Sciences

MDPI

Article

Proteomic, Transcriptomic, Mutational, and Functional Assays Reveal the Involvement of Both THF and PLP Sites at the GmSHMT08 in Resistance to Soybean Cyst Nematode

Naoufal Lakhssassi [1], Dounya Knizia [1], Abdelhalim El Baze [1], Aicha Lakhssassi [2], Jonas Meksem [3] and Khalid Meksem [1,*]

1 Department of Plant, Soil and Agricultural Systems, Southern Illinois University, Carbondale, IL 62901, USA
2 Faculté des Sciences et Techniques, Université Abdelmalek Essaâdi, Tanger 90000, Morocco
3 Trinity College of Arts and Sciences, Duke University, Durham, NC 27708, USA
* Correspondence: meksem@siu.edu

Abstract: The serine hydroxymethyltransferase (SHMT; E.C. 2.1.2.1) is involved in the interconversion of serine/glycine and tetrahydrofolate (THF)/5,10-methylene THF, playing a key role in one-carbon metabolism, the de novo purine pathway, cellular methylation reactions, redox homeostasis maintenance, and methionine and thymidylate synthesis. *GmSHMT08* is the soybean gene underlying soybean cyst nematode (SCN) resistance at the *Rhg4* locus. GmSHMT08 protein contains four tetrahydrofolate (THF) cofactor binding sites (L129, L135, F284, N374) and six pyridoxal phosphate (PLP) cofactor binding/catalysis sites (Y59, G106, G107, H134, S190A, H218). In the current study, proteomic analysis of a data set of protein complex immunoprecipitated using GmSHMT08 antibodies under SCN infected soybean roots reveals the presence of enriched pathways that mainly use glycine/serine as a substrate (glyoxylate cycle, redox homeostasis, glycolysis, and heme biosynthesis). Root and leaf transcriptomic analysis of differentially expressed genes under SCN infection supported the proteomic data, pointing directly to the involvement of the interconversion reaction carried out by the serine hydroxymethyltransferase enzyme. Direct site mutagenesis revealed that all mutated THF and PLP sites at the GmSHMT08 resulted in increased SCN resistance. We have shown the involvement of PLP sites in SCN resistance. Specially, the effect of the two Y59 and S190 PLP sites was more drastic than the tested THF sites. This unprecedented finding will help us to identify the biological outcomes of THF and PLP residues at the GmSHMT08 and to understand SCN resistance mechanisms.

Keywords: PLP; THF; SCN resistance; SHMT; soybean; mutational analysis; composite hairy root transformation

1. Introduction

Soybeans are the largest source of proteins and the second-largest source of oil worldwide. The production value of soybeans in the United States amounted to USD 46.06 billion in 2020 [1]. Soybean production, however, is affected by the presence of a microscopic parasitic roundworm, soybean cyst mematode (SCN), which contributes dramatically to increased yield loss in soybean crops nationwide, causing an estimated USD 1.5 billion in damage [2]. Emerging SCN populations have adapted to the resistance found in certain varieties of soybean, rendering the plant susceptible to infection. In fact, more than 95% of cultivated soybeans in the U.S. use SCN-resistant varieties based on the PI 88788 source of resistance, and 3% of varieties carry resistance from Peking. Due to SCN adaptation, a reduction in the effectiveness of resistant cultivars is taking place [3]. The shift in virulence of the pathogen resulted in 80% of fields in Midwest having SCN that can reproduce on PI 88788 [4]. Peking-type of resistance presents a sustainable alternative to breed for soybean lines with broad resistance to SCN. Cloning novel genes and understanding Peking-type resistance is essential and key to creating soybean varieties.

Peking-type reaction has been reported to be bigenic, requiring both the *rhg1* and the *Rhg4* loci [5]. The gene underlying resistance to SCN at the *Rhg4* locus, the *GmSHMT08*, has been identified and functionally characterized [2,6,7]. Additionally, copy numbers of the *Rhg4* were shown to play an essential role in broad resistance to SCN against five nematode races [8]. Although the soybean genome encodes at least 13 GmSHMT members, only the cytosolic GmSHMT08c was shown to play a role in SCN resistance, with the absence of functional redundancy by the other GmSHMT members, including the other cytosol-targeted GmSHMT05, the four nucleic-targeted GmSHMTs, the two plastidial-targeted GmSHMTs, and the five mitochondrial-targeted GmSHMTs [7].

The serine hydroxymethyltransferase (SHMT) is commonly present in plant and animal species. SHMT plays essential role in methionine synthesis, one-carbon metabolism, and the maintenance of redox homeostasis during photorespiration [9–12]. The SHMT is involved in the interconversion of serine/glycine and tetrahydrofolate (THF)/5,10-methyleneTHF through a transaldimination reaction [9]. The enzymatic co-factor THF is involved in the biosynthesis of various biologically important molecules including purine and pyrimidine nucleotides [13]. On the other hand, PLP acts as a coenzyme in all transamination reactions and in certain decarboxylation, deamination, and racemization reactions of amino acids. PLP, the active form of vitamin B6, is required for hundreds of different reactions in human metabolism, primarily for the synthesis of amino acids and amino acid metabolites and for the synthesis and/or catabolism of certain neurotransmitters and degradation pathways [14]. The SHMT enzyme is therefore essential to directing one-carbon units to the folate-mediated one-carbon metabolism that is required for nucleotide biosynthesis, methyl group biogenesis, and vitamin and amino acid metabolism during glycine biosynthesis [15]. During serine biosynthesis, SHMT plays a major role in the photorespiration metabolic reaction and is therefore essential for C3 plants. Through the glyoxylate cycle, SHMT plays a role in the maintenance of redox homeostasis, involving the gluthatione synthase, peroxidases, and other related genes. It is known that mutations at the mitochondrial *AtSHMT1* cause a photorespiratory deficiency in the plant model *Arabidopsis thaliana* [16]. Mutations in the human SHMT protein were shown to cause cancers and cardiovascular diseases [17–19].

SCN resistance in Forrest is derived from Peking (PI 548402) and is considered to be a promising cultivar that confers resistance to SCN that differs from SCN resistance in PI 88788. Two naturally occurring mutations, P130R and N358Y, distinguish the Forrest *GmSHMT08* allele from the susceptible soybean alleles contained in Essex and Williams 82 [2]. The GmSHMT enzyme contains several PLP and THF binding and catalysis sites that are essential to carrying out the transaldimination reaction [7]. Recently, the two Forrest-specific polymorphic substitutions (P130R and N358Y) that differ from the susceptible Essex have been reported to impact the mobility of a loop near the entrance of the (6S)-tetrahydrofolate binding site [20]. Both ligand binding and kinetic studies indicate a severe reduced affinity for folate, which dramatically impaired enzyme activity in Forrest GmSHMT08 [20].

In the current study, we performed proteomic analysis of a set of protein complexes that was immunoprecipitated using GmSHMT08 antibodies under SCN infected soybean roots. Although soybean cyst nematodes infect soybean roots, leaves play an important role by supplementing nematodes with most of the nutrients that they use to grow and complete their life cycle. In the current study, root and leaf transcriptomic analysis of differentially expressed genes under SCN infection supported the data from LC-MS. In fact, integration of proteomic and transcriptomic data pointed to the involvement of several proteins that belong mainly to pathways that use glycine/serine as a substrate/precursor. Therefore, the obtained data pointed to the involvement of the interconversion reaction carried out by the serine hydroxymethyltransferase protein. Most importantly, site-directed mutagenesis combined with composite hairy root transformations in addition to mutational analysis of the previously identified 18 EMS *Gmshmt08* Tilling mutants derived from Forrest [2,6,7] uncovered the impact of the four THF cofactor binding sites, the four PLP cofactor binding sites, and the two PLP cofactor catalysis sites at the GmSHMT08 protein on SCN resistance. This study reveals for the first time the large effect of both PLP cofactor binding and PLP cofactor catalysis sites on SCN resistance when compared to THF cofactor binding sites.

2. Results

2.1. Mass Spectrometry Identifies the Presence of Proteins That Use Serine and Glycine as Substrates/Precursors

To identify components related to the SCN resistance mechanism, we analyzed the mass spectrometry data from the immunoprecipitated protein complex that was conducted using anti-GmSHMT08 antibodies immobilized to beads in a chromatography column. Several peptides related to SCN infection were present when comparing non-infected and SCN-infected root-eluted fractions. Under non-SCN-infected conditions, only the GmSHMT08 was present in the analyzed proteomic root fractions of Forrest and Essex soybean (Supplemental Table S1). For SCN-infected soybean roots, mass spectrometry analysis showed that the obtained fragmented peptides belong to 37 proteins in the resistant genotype "Forrest" only, while three proteins were identified in the susceptible genotype "Essex" only (Figure 1). Additionally, 23 proteins were common between Forrest and Essex (Figure 1). In addition to translation, growth, cell differentiation, response to stress, flower development, and carbohydrate metabolic processes that were found in both cultivars, many other additional categories (biological processes), such as cellular metabolic process, transport, biosynthetic process, and signal transduction, were contained in the resistant cultivar only (Figure 1). When comparing the biological processes that differentiate the SCN resistance reaction from susceptibility, two categories were mainly dominant in the susceptible cultivar Essex. A total of 75% of the genes that correlate with the presence of the nematode were linked to translation versus 26% in the resistance line. Surprisingly, the growth process occupied 25% in Essex versus 5% in Forrest. The presence of these two main processes in Essex is coherent with the development and growth of plant root cells in the susceptible lines to form syncytial feeding structures.

In silico analysis of the fragmented peptides obtained from the LC-MS analysis (Supplemental Table S1) identified 62 genes that belong to the 37 proteins that were identified in Forrest only, 3 genes that belong to the 3 proteins that were identified in Essex only, and 74 candidate genes that belong to the 23 members that were common between Essex and Forrest (Figure 2A). As expected, mass spectrometry analysis revealed the presence of the GmSHMT08, GmSNAP18, and GmPR08-BetVI proteins, which is coherent with the previously reported physically interacting GmSHMT08/GmSNAP18/GmPR08-BetVI protein complex in resistance to SCN [21]. Interestingly, we were able to identify several proteins (i.e., glycine decarboxylase (GLDC), glycine dehydrogenase (GlyDH), serine/glycine hydroxymethyltransferase (SHMT/GHMT), etc.) that belong mainly to cycles that use glycine/serine as substrate/precursor and therefore that are directly related to the interconversion reaction carried out by the serine hydroxymethyltransferase protein (Supplemental Table S2).

2.2. Identification of Induced Gene Expression in Response to SCN Infection

Although a nematode interacts mostly during its life cycle with soybean roots, most of the nutrients that the nematode uses to grow and complete its life cycle are transported from the leaves. Therefore, to understand and identify the biological pathways that are linked to the presence of nematodes, the current study explored the expression of genes in both leaves and roots in response to SCN infections. Using an integrated approach combining the mass spectrometry data of 136 genes (identified in Essex MS and Forrest MS infected roots) and RNAseq data of 1538 DEG (SCN-infected Forrest root) and 8282 DEG (SCN-infected Forrest leaves), we identified several genes that were differentially expressed under SCN infection in soybean roots (Figure 2C). Transcripts of these identified genes were induced up to 3.9 Log2FoldChange in the SCN-infected soybean roots and up to 10.42 Log2FoldChange in the SCN-infected soybean leaves. Transcripts were downregulated up to 5.92 Log2FoldChange in the SCN-infected soybean roots and up to 11.41 Log2FoldChange in the SCN-infected soybean leaves (Figure 2D). These data reveal the importance of soybean leaves during root SCN infections in the resistant reaction. Of 136 genes identified in Forrest MS, 78 were differentially expressed under SCN infection in the resistant Forrest, including root and leave tissues. Out of the 78 genes, 56 were differentially expressed in Forrest leaves, 8 were differentially expressed in Forrest roots, and 14 were differentially expressed in both Forrest roots and leaves. Most of the identified genes belong mainly to 30 gene families, of which most were found to be related to redox hemeostasis, serine/glycine conversion, glyoxylate cycle, glycolysis, succinyl-CoA, heme biosynthesis related enzymes, cytoskeleton-related enzymes, and ATP mitochondrial related genes (Supplemental Table S2, Figure 3).

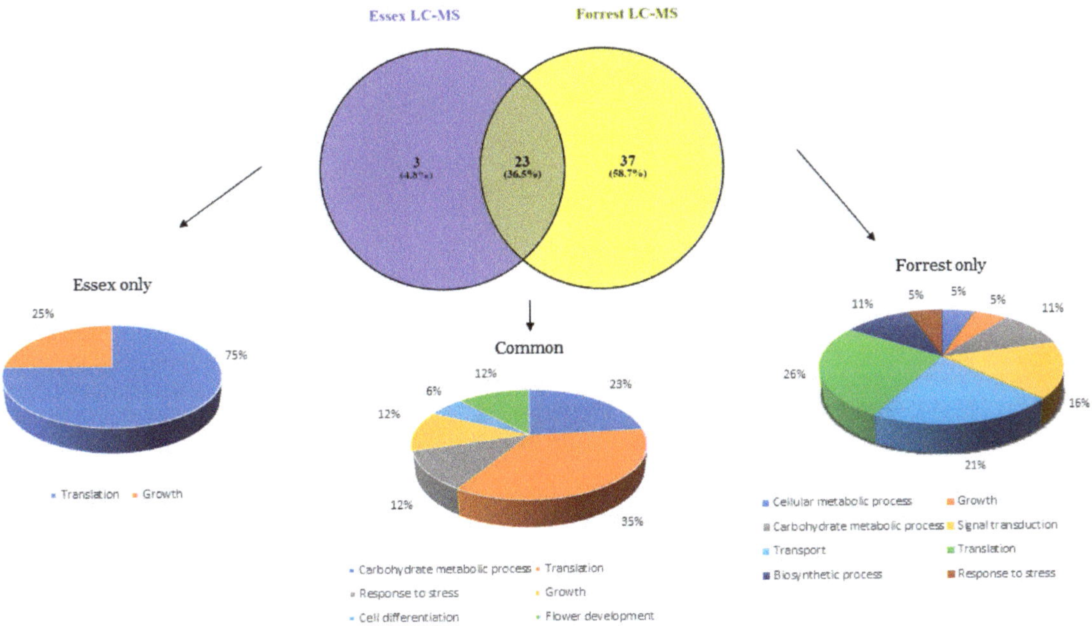

Figure 1. Venn diagram showing relationships between different biological processes obtained using mass spectrometry from Forrest and Essex roots under SCN infection (3 DAI). Thirty-seven proteins were found in the resistant genotype "Forrest" only and correspond to 8 different biological processes, while 3 proteins were identified in the susceptible genotype "Essex" and correspond to two biological processes. Twenty-three proteins were common between Forrest and Essex and correspond to six different biological processes.

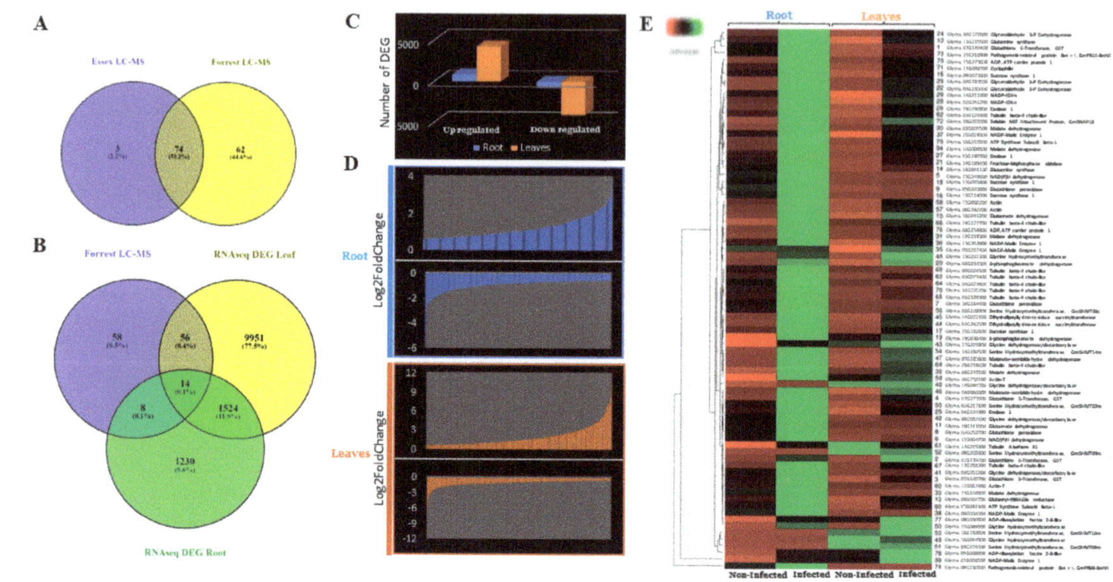

Figure 2. (**A**) Venn diagram showing relationships among different candidate genes obtained via mass spectrometry from Forrest and Essex. Sixty-two genes were identified in Forrest only, and 3 genes were identified in Essex only. The other 74 genes were common between Essex and Forrest. (**B**) Of 136 genes identified in Forrest MS, 78 were differentially expressed under SCN infection in the resistant Forrest roots and leaves. Of these 78 genes, 26 were differentially expressed in Forrest leaves, 8 were differentially expressed in Forrest roots, and 14 were differentially expressed in both Forrest roots and leaves. (**C**,**D**) Significantly up- and down-regulated genes by plant tissue. DEGs considered significant if p value < 0.05, log2foldchange no less than ±0.6. (**E**) Heatmap of all DEGs by log2FC (fold change) of induced response to SCN treatment of roots and leaves in Forrest. The expressions profiling was visualized through heatmap using Heatmapper [22].

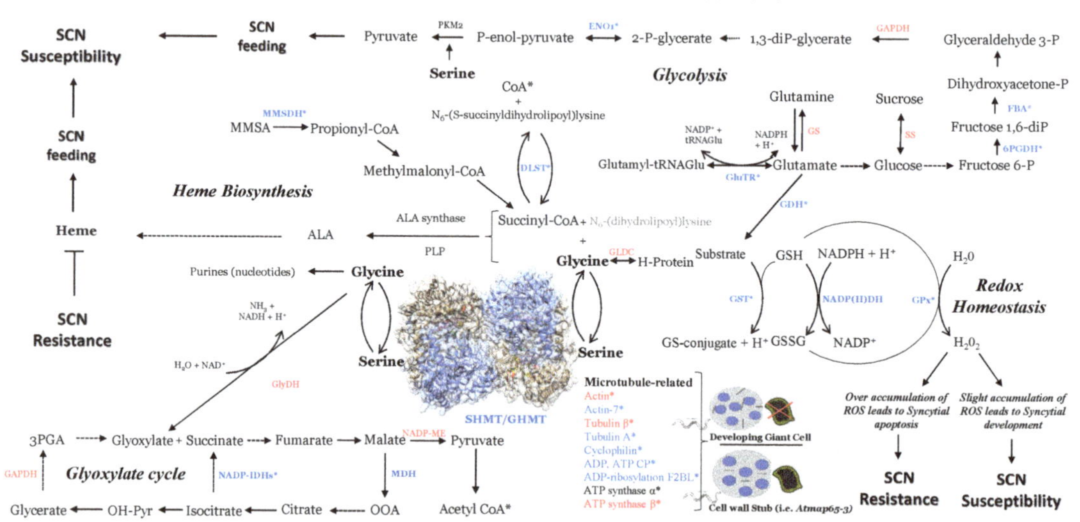

Figure 3. Summary of all proteins identified by mass spectrometry, revealing key components related to redox hemeostasis, serine/glycine conversion, glyoxylate cycle, glycolysis, succinyl-CoA, and

heme-biosynthesis-related enzymes. All identified proteins belong mainly to processes and cycles that use glycine/serine as a substrate/precursor and are therefore directly related to the interconversion reaction carried out by the serine hydroxymethyltransferase protein. Enolase (ENO1), glyceraldehyde 3-P dehydrogenase (GAPDH), fructose-bisphosphate aldolase (FBA), 6-phosphogluconate dehydrogenase (6PGDH), sucrose synthase (SS), glutamine synthase (GS), glutamyl-tRNAGlu reductase (GluTR), glutamate dehydrogenase (GDH), glycine decarboxylase (GLDC), dihydrolipoyllysine-residue succinyltransferase (DLST), methylmalonate-semialdehyde dehydrogenase (MMSDH), glutathione S-transferase (GST), NADP(H) dehydrogenase (NADP(H)DH), glutathione peroxidase (GPx), glycine dehydrogenase (GlyDH), serine/glycine hydroxymethyltransferase (SHMT), NADP-malic enzyme (NADP-ME), NADPH-specific isocitrate dehydrogenase (NADP-IDHs), malate dehydrogenase (MDH), carrier protein (CP), factor 2-B-like (F2BL). Blue are enzymes/genes that were identified in Forrest only. Red genes are present in both Essex and Forrest. * Genes identified within or very close to QTL for SCN resistance.

2.3. Correlation between the Identified Genes and the Previously Reported QTLs for SCN Resistance

Most of the identified genes were mapped to QTLs for resistance to SCN using different mapping populations. In fact, 34 genes were located within reported SCN QTLs; 21, 4, and 2 genes were located ~3Mbp, ~4–6 Mbp, and ~11Mbp away from reported SCN QTLs, respectively (Supplemental Table S2). The most reported gene that mapped to QTLs for resistance to SCN is the glutamine synthase (Glyma.18G041100) gene after the *GmSHMT08* at the *Rhg4* locus. The sucrose synthase 1 (Glyma.15G182600) was reported frequently in SCN QTL mapping analysis, followed by the 6-phosphogluconate dehydrogenase (Glyma.08G254500) and the mitochondrial GmSHMT08m (Glyma.08G274400), showing the contribution of the glycolysis cycle to SCN resistance (Supplemental Table S2).

Within the SCN QTLs, we were able to identify high frequency genes that belong to glycolysis (30 SCN QTLs), followed by cytoskeleton-related genes (18 SCN QTLs), glyoxylate cycle (13 SCN QTLs), redox hemeostasis (9 SCN QTLs), and ATP-mitochondrial-related genes (8 SCN QTLs) (Supplemental Table S2).

2.4. Identification of Genes Related to Redox Homeostasis

The current study revealed many candidate genes related to redox hemostasis (Supplemental Table S2, Figure 3). The annotation of the set containing these genes showed the presence of four genes that encode glutathione S-transferase (GST), three genes encoding glutathione peroxidases, two genes encoding NAD(P)H dehydrogenase, and two genes encoding glutamate dehydrogenases (Supplemental Table S2). The obtained mass spectrometry data are coherent with previous studies showing that modulation of the SHMT serine/glycine interconversion impact important maintenance of redox homeostasis occurs via both glutathione synthase and glutathione peroxidases [23]. RNAseq analysis showed that transcripts from the previous 11 genes were significantly induced under SCN infection in both root and leaves (Figure 2E, Supplemental Figure S1A).

2.5. Glycolysis Cycle in Response to SCN Infection

Glycolysis cycle provides several products that support nematodes growth. This study showed several candidate genes related to the glycolysis cycle including four sucrose synthase 1-related genes, three glyceraldehyde 3-phosphate dehydrogenase, three enolases, two glutamine synthetases, two 6-phosphogluconate dehydrogenases, 1 fructose-bisphosphate aldolase, and one glutamyl-tRNAGlu reductase (Supplemental Table S2). Transcripts from all these 16 genes were induced under SCN infection in both root and leaves (Figure 2E, Supplemental Figure S1B).

2.6. Identification of Gene-Related Glyoxylate Cycle

Glyoxylate products support early nematode development. Mass spectrometry analysis revealed the presence of many candidate genes related to the glyoxylate cycle. The annotation of the set containing these genes showed the presence of five malate dehydrogenases (MDH1), five NADP-dependent malic enzymes, four glycine dehydrogenases (GLDC, gcvP), and two NADPH-specific isocitrate dehydrogenases (Supplemental Table S2). Transcripts from all these 16 genes were induced under SCN infection in both root and leaves (Figure 2E, Supplemental Figure S1C).

2.7. Identification of Succinyl-CoA, Serine/Glycine, and Heme-Related Genes

Several genes related to the serine and glycine synthesis were identified by mass spectrometry including six serine hydroxymethyltransferases (GmSHMT08c, GmSHMT02m, GmSHMT08m, GmSHMT09m, GmSHMT14m, and GmSHMT18m), three glycine hydroxymethyltransferases, and one glycine decarboxylase. Additionally, mass spectrometry showed the presence of two dihydrolipoyllysine-residue succinyltransferases and two methylmalonate-semialdehyde dehydrogenases (Supplemental Table S2). Dihydrolipoyllysine-residue succinyltransferase and Methylmalonate-semialdehyde dehydrogenase are key enzyme to synthesize Succinyl-CoA that—together with glycine, the SHMT product—produce the ALA, an important component of and precursor to the production of heme. It is well known that SCN requires heme source for its survival. RNAseq data showed that transcripts from all previous 12 enzymes were induced under SCN infections (Figure 2E, Supplemental Figure S1D).

2.8. Identification of Cytoskeleton-Related and ATP-Mitochondrial-Related Genes

Several components of cytoskeleton-related genes were found, including two actin, two actin-7, one tubulin-A, and nine tubulin beta-4 in addition to several ATP- and ADP-mitochondrial-related genes, such as two ADP/ATP carrier protein 1, an ATP synthase subunit beta-1, an ATP synthase subunit alpha, and two ADP-ribosylation factor 2-B-like genes that regulate the interaction of tubulin-folding cofactor D with native tubulin (Supplemental Table S2). Interestingly, transcripts from all the 21 genes were induced under SCN infection in both root and leaves (Figure 2E, Supplemental Figure S1E,F).

2.9. In silico Analysis of the GmSHMT08 THF Cofactor Binding Sites and PLP Cofactor Binding and Catalysis Sites

Mass spectrometry analysis pointed to the importance of the GmSHMT08 in the interconversion of serine and glycine, two substrates that are essential for redox hemeostasis, serine/glycine conversion, glyoxylate cycle, glycolysis, succinyl-CoA, and heme biosynthesis (Figure 3). This interconversion relies on two essential sites at the GmSHMT08 enzyme, the PLP and the THF cofactor sites. To carry out the GmSHMT08 protein homology modeling, an available SHMT crystal structure from of serine hydroxymethyltransferase from glycine max cultivar Essex already complexed with PLP-glycine and 5-formyltetrahydrofolate residues was used as the template (Figure 4). Next, all four THF cofactor and six PLP cofactor sites at the GmSHMT08 were mapped against the model (Figure 4). To visualize the effect of the site directed mutagenesis on the THF and PLP residues, rotamers tools have been used to mutate the four THF and 6 PLP residues on the GmSHMT08 protein model (Figure 5). The PLP molecule binds to different residues in the GmSHMT08 PLP binding pocket, as shown in Figure 4. Lys-244 forms a covalent Schiff base linkage (internal aldimine) with PLP (Figure 4). Nearby residues from both chains of the obligate dimer (Tyr-59′, Glu-61′, Ser-107, Asp-215, Thr-241, Arg-250 (prime indicates chain B) (Figure 4)) assure conserved interactions with the phosphate, N1, and O3 hydroxyl of PLP, whereas the pyridine ring of PLP stacks against His-134. The SHMT8-PLP-Gly complex represents an intermediate step of the THF-dependent catalytic mechanism, in which L-Ser attacks the Schiff base linkage between Lys-244 (Figure 4) and PLP to form a PLP-Ser external aldimine. Formaldehyde is next liberated when the active site general base deprotonates the hydroxyl side chain of L-Ser. Once synthesized, formaldehyde is next attacked by THF N5, transferring the side chain of L-Ser to THF, resulting in an external aldimine/quinonoid product called PLP-Gly.

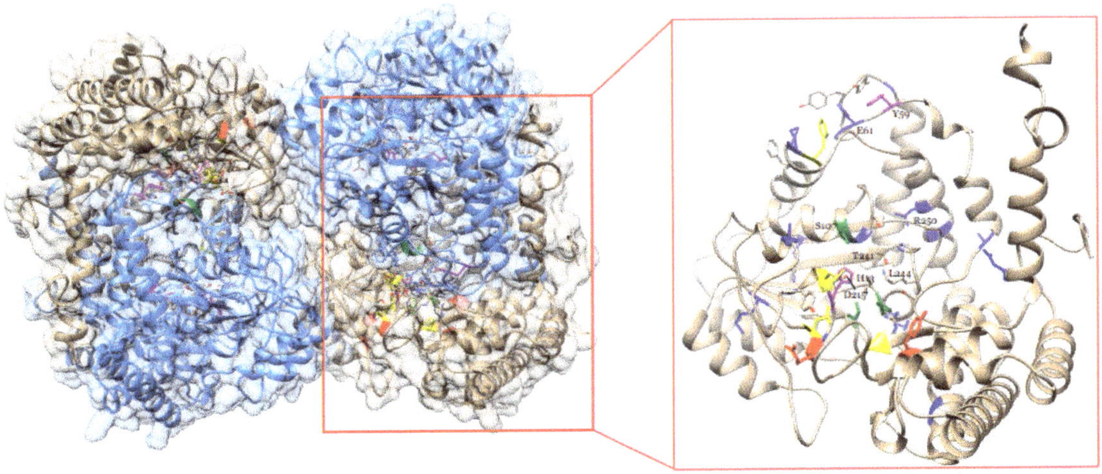

Figure 4. Homology modeling of an available SHMT crystal tetramer structure of a serine hydroxymethyltransferase from glycine max cultivar Essex complexed with PLP-glycine and 5-formyltetrahydrofolate residues used as template. On the right, on SHMT08 homomer showing all four mapped THF binding sites (yellow), four PLP binding sites (Green), two PLP catalysis sites (purple), two polymorphisms (red), and *Gmshmt08* EMS mutants (Blue).

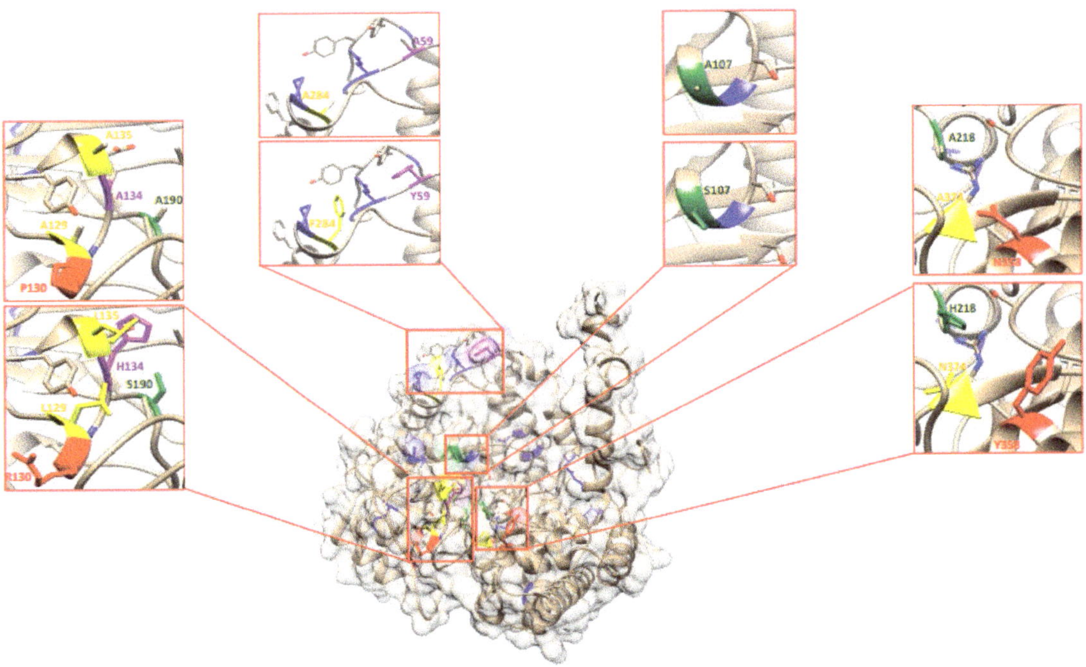

Figure 5. Homology modeling of one GmSHMT08 homomer showing mutated PLP and THF residues. Rotamers tools have been used to mutate the four THF and six PLP residues on the GmSHMT08 protein model to visualize the effect of the site directed mutagenesis on the THF and PLP residues. Four THF binding sites (yellow), four PLP binding sites (Green), two PLP catalysis sites (purple), two polymorphisms (red), and *Gmshmt08* EMS mutants (Blue).

In silico analysis revealed that the mutated GmSHMT08$^{\Delta L129A}$, GmSHMT08$^{\Delta L135A}$, GmSHMT08$^{\Delta F284A}$, GmSHMT08$^{\Delta N374A}$, GmSHMT08$^{\Delta Y59A}$, GmSHMT08$^{\Delta G106A}$, GmSHMT08$_{\Delta G107A}$, GmSHMT08$_{\Delta H134A}$, GmSHMT08$_{\Delta S190A}$, and GmSHMT08$_{\Delta H218A}$ alleles and the two polymorphisms between the SCN resistant cultivar Forrest and the SCN susceptible Essex are predicted to impact negatively their conserved interactions with the phosphate, N1, and O3 hydroxyl of PLP near the PLP binding pocket (Figure 5). These mutations are expected to affect the GmSHMT08's ability to bind PLP substrate and the interconversion of serine/glycine and tetrahydrofolate (THF)/5,10-methylene THF, which may impact resistance to SCN.

2.10. Re-Analysis of the EMS-Induced GmSHMT08 Mutations Reveal Their Potential Impact on PLP/THF Cofactor Binding and Catalysis

To gain more insight into the impact of the eighteen EMS *Gmshmt08* mutants identified earlier [2,6,7] on the PLP/THF cofactor binding and catalysis, all previous mutants were mapped and mutated using the rotamers tool. E61K and G71D are found close to the Tyr59 residue, that is required for PLP cofactor catalysis. E61K, G62S, and P285S are located very close to the Phe284 that is required for THF binding. Disruption of this site profoundly altered substrate binding and catalytic activity in *E. coli* [24]. Therefore, *Gmshmt08*$_{E61K}$, *Gmshmt08*$_{G62S}$, and *Gmshmt08*$_{p285S}$ mutations are likely to have the same conformational deficiency by impacting the THF cofactor binding and PLP cofactor catalysis at the GmSHMT08.

The Forrest polymorphism R130P and *Gmshmt08*$_{G132D}$ mutant were located close to the Leu129 residue that is required for THF binding. *Gmshmt08*$_{G132D}$ is also close to the two essential and conserved histidine residues: His134 and His137. Since proline has a conformational rigidity due to its direct incorporation of the α-carbon into its side chain, this may cause drastic conformational changes, interfering with this catalysis (Figure 6).

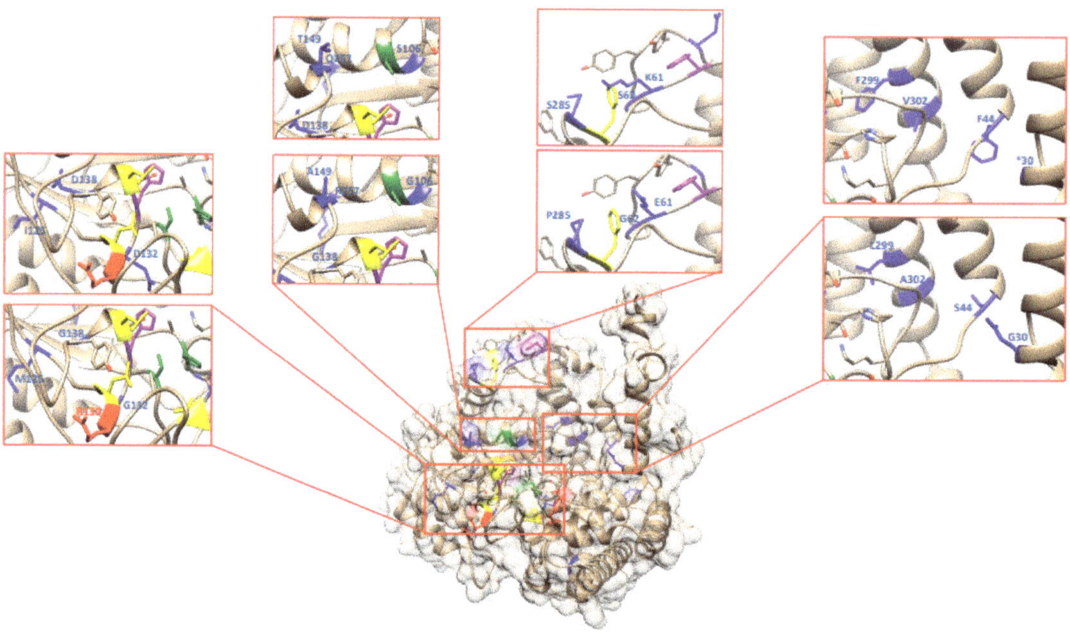

Figure 6. Homology modeling of one GmSHMT08 homomer, mapping EMS mutations at the GmSHMT08 protein. Rotamers tools have been used to mutate the 18 GmSHMT08 mutations. THF and PLP residues are shown. Four THF binding sites (yellow), four PLP binding sites (Green), two PLP catalysis sites (purple), two polymorphisms (red), and *Gmshmt08* EMS mutants (Blue).

Both polymorphic substitutions between Essex and Forrest (P130R and N358Y) were shown to impact the mobility of a loop near the entrance of the THF binding site at the GmSHMT08 protein, resulting in reduced affinity for folate substrate, subsequently impairing the enzymatic activity of GmSHMT08 [20]. $Gmshmt08_{G357R}$ mutation is located one residue away from the Forrest polymorphic substitution N358Y and therefore is predicted to impact the THF site's binding to folate. Another mutation, $Gmshmt08_{Q226*}$, which resulted in a loss of SCN resistance in Forrest, was mapped close to His218 residues involved in PLP cofactor binding.

$Gmshmt08_{G106S}$ is located at the PLP cofactor binding site Gly106, in addition to being located very close to the other PLP cofactor binding site Ser107. $Gmshmt08_{G106S}$, $Gmshmt08_{A302V}$, and $Gmshmt08_{L299F}$, were closely located to the Thr-241 and Arg-250 where the pyridine ring of PLP stacks against His-134 (PLP catalysis). In addition, the previous mutations were mapped close to the Lys-244 that forms a covalent Schiff base linkage (internal aldimine) with PLP.

2.11. Functional Validation of the GmSHMT08 THF Cofactor Binding Sites and Their Role in SCN Resistance

The four THF cofactor binding sites (L129, L135, F284, N374) at the GmSHMT08 protein contribute to the interconversion of tetrahydrofolate (THF) and 5,10-methylene THF [7]. To test the effect of each THF binding sites on SCN resistance, we introduced independent mutations in each one of the four THF-related residues at the *GmSHMT08* coding sequence from the resistant Forrest allele, then overexpressed it in the *ExF12* RIL, carrying the SCN-resistant *GmSNAP18*$^{+}$ from Forrest and the SCN-susceptible *GmSHMT08*$^{-}$ allele from Essex. To conduct the GmSHMT08$^{\Delta L129A}$, GmSHMT08$^{\Delta L135A}$, GmSHMT08$^{\Delta F284A}$, GmSHMT08$^{\Delta N374A}$, and GmSHMT08 overexpression analysis, the 1416-bp nucleotide coding sequence of the different *GmSHMT08* alleles were overexpressed under the control of a soybean ubiquitin promoter using a transgenic hairy root system. Interestingly, unlike the *GmSHMT08* wild-type allele from Forrest that reduced the cyst numbers of the *ExF12* RILs by 91% in the susceptible *ExF12* background, reductions of the cyst numbers at the induced mutations, including GmSHMT08$^{\Delta L129A}$, GmSHMT08$^{\Delta L135A}$, GmSHMT08$^{\Delta N374A}$, and GmSHMT08$^{\Delta F284A}$, were limited to 50%, 58%, 66%, and 78%, respectively (Figure 7). Thus, induced mutations at GmSHMT08$^{\Delta L129A}$, GmSHMT08$^{\Delta L135A}$, GmSHMT08$^{\Delta N374A}$, and GmSHMT08$^{\Delta F284A}$ affected the GmSHMT08's ability to reduce the number of cysts by 40%, 32%, 24%, and 12%, respectively. Statistical analysis showed that of the four THF sites, site directed mutagenesis of the GmSHMT08$^{\Delta L129A}$, GmSHMT08$^{\Delta L135A}$, and GmSHMT08$^{\Delta F284A}$ were significantly different ($p < 0.0001$) from the GmSHMT08 wild-type allele. The mutation at the GmSHMT08$^{\Delta N374A}$ THF residue presented the lowest reduction of cyst numbers.

2.12. Functional Validation of the GmSHMT08 PLP Cofactor Binding and Catalysis Sites Points to Their Involvement in SCN Resistance

Four PLP cofactor binding sites (G106, G107, S190A, H218) and two PLP cofactor catalysis sites (Y59 and H134) at the GmSHMT08 protein are involved in the interconversion of serine and glycine [7]. To test the real effect of each PLP cofactor binding and catalysis sites on SCN resistance, we introduced independent mutations in each one of the six PLP related residues at the *GmSHMT08* coding sequence from the resistant Forrest allele, then overexpressed it in the *ExF12* RIL. To conduct the GmSHMT08$^{\Delta Y59A}$, GmSHMT08$^{\Delta G106A,107A}$, GmSHMT08$^{\Delta H134A}$, GmSHMT08$^{\Delta S190A}$, GmSHMT08$^{\Delta H218A}$ and GmSHMT08 overexpression analysis, the 1416-bp nucleotide coding sequences of the different *GmSHMT08* alleles were overexpressed under the control of a soybean ubiquitin promoter using a transgenic hairy root system. Surprisingly, unlike the *GmSHMT08* wild-type allele from Forrest that reduced the cyst numbers of the *ExF12* RILs by 91% in the SCN-susceptible *ExF12* background, reductions of the cyst number at the induced mutations, including GmSHMT08$^{\Delta S190A}$, GmSHMT08$^{\Delta Y59A}$, GmSHMT08$^{\Delta G106A,107A}$, GmSHMT08$^{\Delta H218A}$, and GmSHMT08$^{\Delta H134A}$ were limited to 6%, 42%, 61%, 61%, and 63%, respectively (Figure 7).

Thus, induced mutations at GmSHMT08$^{\Delta S190A}$, GmSHMT08$^{\Delta Y59A}$, GmSHMT08$^{\Delta G106A,107A}$, GmSHMT08$^{\Delta H218A}$, and GmSHMT08$^{\Delta H134A}$ affected the GmSHMT08's ability to reduce the number of cysts by more than 84%, 48%, 29%, 29%, and 27%, respectively. Most importantly, the PLP cofactor binding site (S190A) and the PLP cofactor catalysis site (Y59) presented higher impact on SCN resistance when compared to the four THF binding sites that were tested previously.

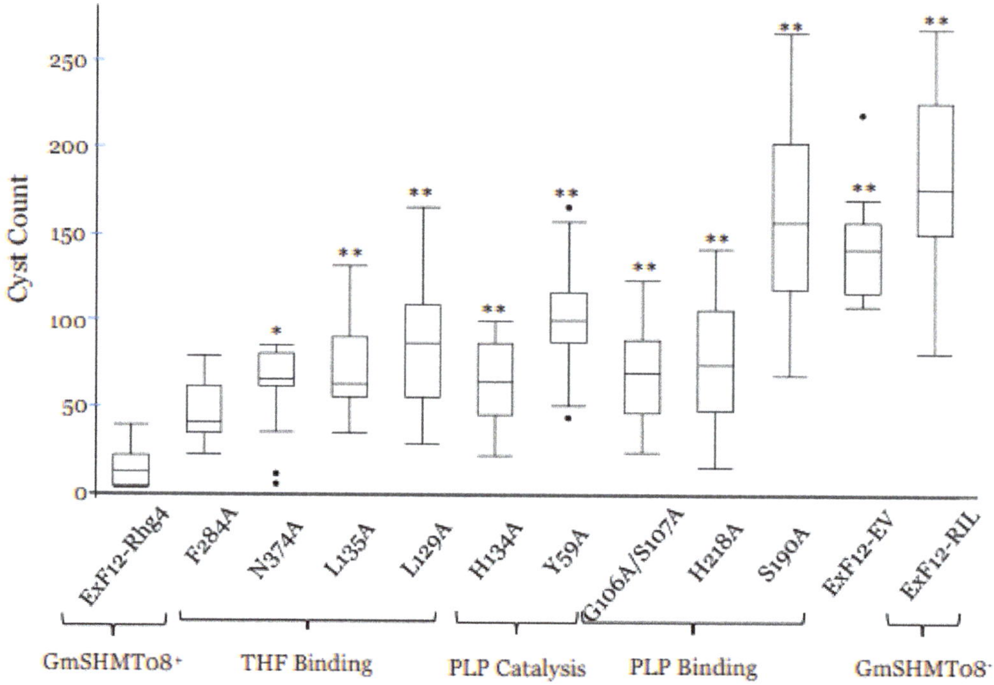

Figure 7. Composite hairy root transformation, overexpression analysis of the GmSHMT08$^{\Delta PLP}$ and GmSHMT08$^{\Delta THF}$. Overexpression analysis in transgenic composite roots (ExF12 RIL GmSNAP18^{+}/GmSHMT08^{-} transformed by pG2RNAi2::GmSHMT08$^{\Delta L129A}$, pG2RNAi2::GmSHMT08$^{\Delta L135A}$, pG2RNAi2::GmSHMT08$^{\Delta F284A}$, pG2RNAi2::GmSHMT08$^{\Delta N374A}$, pG2RNAi2::GmSHMT08$^{\Delta Y59A}$, pG2RNAi2::GmSHMT08$^{\Delta G106A}$, pG2RNAi2::GmSHMT08$^{\Delta G107A}$, pG2RNAi2::GmSHMT08$^{\Delta H134A}$, pG2RNAi2::GmSHMT08$^{\Delta S190A}$, and pG2RNAi2::GmSHMT08$^{\Delta H218A}$. The experiments were repeated three times, and similar results were obtained. The data shown represent the averages and SD from all three biological repeats ($n > 15$). Asterisks and connecting letters indicate significant differences between the tested lines and the pG2RNAi2::GmSHMT08 (used as positive control) as determined by ANOVA (** $p < 0.0001$, * $p < 0.01$).

3. Discussion

Soybean cyst nematode is the most destructive pathogen to soybeans [3]. Most of the efforts to understand the SCN resistance mechanism were focused on deciphering the genes for resistance to SCN within two types of SCN resistance: the PI88788, which uses the *rhg-1b*; and Peking-type resistance, which uses a combination of *rhg1-a* and *Rhg4* loci [2,6]. Although the gene that confers resistance to SCN at the *Rhg4* locus has been cloned and identified a decade ago [2]; the involvement of the four THF cofactor binding sites, four PLP cofactor binding sites, two PLP cofactor catalysis sites at the GmSHMT08 in resistance to SCN has not been revealed yet.

The current study revealed the presence of enriched cycles that mainly use glycine as a substrate, including the glyoxylate cycle, redox homeostasis, and heme biosynthesis. Several key enzymes involved in the glycolysis cycle were also identified, which is coherent with QTL analysis studies that were reported earlier (Supplemental Table S2). Although glucose and glutamine are the main sources that are used to maintain the glycolysis pathway, serine plays an essential role in the glycolysis pathway through de novo serine biosynthesis. Serine derived from a branch of glycolysis can be reintegrated into the glycolysis pathway to synthesize pyruvate but can also be converted to glycine, which provides carbon units for one carbon metabolism. One carbon metabolism is essential for the synthesis of proteins, lipids, nucleic acids, and other precursors through a complex metabolite network based on the chemical reactions of folate compounds. Thus, the main products of SHMT, serine, and glycine provide precursors for the biosynthesis of proteins, nucleic acids, and lipids, which are essential for both host and pathogen growth. The war between nematodes and soybean for metabolites and how plants can fight underground attacks is complex and still requires investigation [25].

Proteomic and transcriptomic assays of SCN-infected soybeans identified seven enzymes that play essential roles in the glycolysis and four enzymes in the glyoxylate cycle. Previous studies reported the ability of nematodes to steal nutrients from host plants [26]. Soluble sugars, such as fructose, glucose, and sucrose, were previously found to increase significantly in tomato leaves and roots during early infection by root-knot nematodes (RKNs) [27], which is coherent with the 7 genes identified at the glycolysis cycle in the current study. Another plant-parasitic nematode, *Heterodera schachtii*, has been shown to stimulate plant root cells to form syncytial feeding structures which synthesize all nutrients required for successful nematode development [27,28]. During glycolysis, a series of enzymatic reactions will convert sugars, typically sucrose to glucose, fructose, and then to pyruvate [29]. Products derived from glycolysis and glyoxylate cycle support early nematode development [30]. Nematodes will metabolize energy through the standard metabolic pathways, which is reflected by high metabolic activity, elevated sucrose levels, and the formation of starch [31,32]. The root cells affected by nematode attack show altered metabolisms—especially increased allocation of soluble sugars. Sugar importation into syncytia follows the symplasmic path during later stages of development [33–35].

On the other hand, it is known that glycine powers the biosynthesis of heme. Since SHMT catalyzes the conversion of serine to glycine, any disruption of the PLP/THF cofactor binding/catalysis sites, as shown in the EMS *GmSHMT08* mutants and by site-directed mutagenesis, may negatively impact the production of glycine and therefore the biosynthesis of heme. Heme is considered a major nutrient for nematodes from the plant host. Nematodes such as *Rhabditis maupasi*, *Caenorhabditis elegans*, and *Heterodera glycines* require heme source or any related iron porphyrin for feeding and survival [27,36,37]. This may explain the presence of four heme-related genes that were obtained via LC-MS and their differential expression during SCN infection. Two out of the four identified genes (Glyma.08G066600 and Glyma.07G183600, belonging to the malonate-semialdehyde dehydrogenase gene family), were mapped at two SCN QTLs in previous studies (Supplemental Table S2).

Several genes belonging to the redox hemostasis pathway that were identified by LC-MS in the current study, such as glutathione peroxidases, NADP(H) dehydrogenase, glutathione S-transferases (GSTs), and glycine dehydrogenases, were differentially expressed under SCN infection. GSTs catalyze the conjugation of glutathione (GSH) to xenobiotic substrates for detoxification [38–40]. GST activity is dependent upon GSH supply from the glutathione synthetase enzyme and the activity of some transporters to remove GSH conjugates from the cell [41,42]. Most of the identified ROS proteins from LC-MS were differentially expressed under SCN infection. This is coherent with previous transcriptomic analysis, in which both glutathione peroxidase and glutathione transferase transcripts, among other ROS-scavenging enzymes, were shown to be significantly modulated under SCN infection (in syncytia) [43].

Recently, mitochondrial *OsSHMT* and *NbSHMT* have been demonstrated to play a role in broad-spectrum resistance via the ROS pathway [44]. Cytoskeletons (i.e., microtubules) play an important role during the intracellular transport of mitochondria [45,46]. The interaction of some mitochondrial components with certain cytoskeletal proteins was found to be involved in the coordination of mitochondrial function [47,48]. In fact, interaction between the microtubule-associated C4HC3-type E3 Ligase (MEL) and the mitochondrial SHMT1 leads to SHMT1-dependent mitochondrial ROS generation, activation of MAPK cascades, and reprogramming of defense-related transcripts, ultimately leading to attenuated pathogen invasion. The interacting MEL-SHMT1 complex mediates regulation of plant immunity involving microtubules and mitochondria. Infections by multiple pathogens induce MEL transcription. This is followed by the formation of MEL homodimers, which activate MEL E3 ligase activity, subsequently triggering SHMT1 degradation by the 26S [44]. Cytoskeleton-including actin filaments are dynamic structures that can grow and shrink rapidly via the addition or removal of tubulin proteins. During cellular homeostasis responses, mitochondria organelles are considered the major source for the generation of intracellular ROS by supplying ATP and biosynthetic intermediates for redox, cell death, and energy metabolism [49–52].

The current study found several potential substrates/components of cytoskeleton including Actins, Tubulin A, Tubulin beta-4, and several ATP and ADP mitochondrial related genes including ADP/ATP carrier protein 1, ATP synthase subunit beta-1, ATP synthase subunit alpha, and ADP-ribosylation factor 2-B-like that regulates the interaction of tubulin-folding cofactor D with native tubulin. Tubulin and actin cytoskeletons have been continuously reported to be implicated in plant defense against pathogenic fungi, oomycetes, and bacteria [52–55]. We also found the presence of cyclophilin, which are known to be modulated by microtubules [44]. The role of cyclophilin in plant pathogenesis has been reported earlier [56]. The cyclophilin GmCYP1 (Glyma.11G098700) has been suggested to play a role in soybean defense via its interaction with the isoflavonoid regulator GmMYB176 [57], which is known to play major roles in resistance to cyst nematodes in *Arabidopsis* [58] and in SCN [59]. Microtubule disruption of hematopoietic cells cause a dramatic subcellular redistribution of cyclophilin-A and pin1 from the nucleus to the cytosol and plasma membrane [60]. Another microtubule, MAP65-3 microtubule-associated protein, has been shown to be essential for cytokinesis in somatic cells and also play an important role during nematode-induced giant cell ontogenesis in *Arabidopsis* [61]. In fact, MAP65-3 is associated with mini cell plates that are required for the formation of a functional nematode feeding cell. In giant cell *map65-3* mutants, a defect in mini cell plate formation prevents the development of functional feeding cells, which resulted in the death of the nematode [61]. The identification of several cytoskeleton components from the current study reinforces their involvement in resistance to SCN, which is coherent with QTL SCN analysis where the identified 22 cytoskeleton-related and ATP mitochondrial-related genes were mapped to 26 reported SCN QTLs.

4. Material and Methods

4.1. Protein Extractions and Immunoprecipitation Using GmSHMT08 Antibodies

Forrest and Essex soybean cultivars were infected using SCN (HG0), as described earlier [62]. Root and leaf samples from three biological replicates containing five SCN (HG0)-infected and five non-SCN-infected soybeans were washed and frozen in liquid nitrogen three days after infection. Total root proteins from SCN-infected and non-infected soybean "Forrest" and "Essex" cultivars were extracted in a lysis buffer containing 5mM DTT, 1% (*v*/*v*) NP40, 1mM sodium molybdate, 1 mM NaF, 1 mM PMSF, 1.5 mM Na3VO4, 100 mM NaCl, 2 mM EDTA, 50 mM Tris–HCl at pH 7.5, 10% (*v*/*v*) glycerol, and one tablet from the plant protease and phosphatase inhibitors at 1:100 mL (Thermo Scientific), as previously shown [21]. Coomassie Bradford Protein Assay Kit was used to quantify protein concentrations (Thermo Fisher Scientific, Waltham, MA, USA). For in planta immunoprecipitation analysis, anti-GmSHMT08 polyclonal antibodies [21] were immobilized in a column (Pierce Co-Immunoprecipitation Kit). Then, immunoblot analysis of root protein fraction samples from soybean Forrest and Essex was incubated overnight with the immobilized antibodies. After three washes, the associated proteins were eluted as described by the Pierce Co-Immunoprecipitation Kit. The eluted fraction was then used for mass spectrometry analysis.

4.2. Mass Spectrometry Analysis

Peptide digestion, microsequencing analyses, and protein characterization of the SHMT-associated proteins from non-infected and SCN-infected Forrest and Essex roots 3 DAI were carried out in the Charles W Gehrke Proteomics Center at the University of Missouri-Columbia, as previously shown [21]. The eluted fractions obtained from the immunoprecipitation experiment using anti-GmSHMT08 polyclonal antibodies were briefly subjected to lyophilization. Then, all proteins were subsequently digested with trypsin, resulting in one main fraction representing the three biological replicates. Furthermore, samples were acidified, lyophilized, and re-suspended in 21 µL of a 5% acetonitrile, 0.1% formic acid solution, and peptides were analyzed via LC-MS (18 µL injection), as previously described [63]. Liquid chromatography gradient conditions were carried out as previously shown [63]. The Proxeon Easy nLC HPLC system was attached to an LTQ Orbitrap XL mass spectrometer. BSA was used for quality control on the column. Searches of Swiss-Prot-all species and NCBI-Gmax were conducted using Sorcerer-Sequest.

4.3. RNA-seq Library Preparation and Analysis

Four plant soybean tissues were used for RNA-seq, including SCN-infected (3 DAI) soybean root, non-SCN-infected soybean root, SCN-infected (3 DAI) soybean leaves, and non-SCN-infected soybean leaves. Three biological replicates that correspond to three independent experiments where each experiment contained five SCN (HG0) infected and five non-SCN-infected soybean plants were washed and frozen in liquid nitrogen three days after infection. Total RNA for each sample was extracted from 100 mg of frozen grounded samples using RNeasy QIAGEN KIT (Cat. No./ID: 74004, Germantown, Maryland). Total RNA was treated with DNase I (Invitrogen, Carlsbad, CA, USA). RNA-seq libraries preparation and sequencing were performed at Novogene INC. (Cambridge, UK) using Illumina NovaSeq 6000. The four libraries were multiplexed and sequenced in two different lanes generating 20 million raw pair end reads per sample (150 bp). Quality assessment of sequenced reads was performed using fastqc version 0.11.9 [64]. After removing the low-quality reads and adapters with trimmomatic version V0.39 [64], the remaining high-quality reads were mapped to the soybean reference genome Wm82.a2.v1 using STAR, version v2.7.9 [65,66]. Uniquely mapped reads were counted using Python package HTseq v0.13.5 [67]. Read count normalization and differential gene expression analysis were conducted using the Deseq2 package v1.30.1 [68] integrated in the OmicsBox platform from BioBam (Valencia, Spain). DEGs were considered significant if p value < 0.05, Log_2FoldChange no less than ± 0.6. Expression profiling was visualized through a heatmap using Heatmapper [22].

4.4. Cloning the Forrest GmSHMT08 WT and Site Directed Mutagenesis

The *GmSHMT08* coding sequence from the Forrest WT (*Rhg4*) was amplified from soybean Forrest root cDNA via RT-PCR using the GmSHMT08c-AscI-Fw primers (ggcgcgc-cATGGATCCAGTAAGCGTGTGGGTA) and the GmSHMT08c-AvrII-Rv primers (ggatcc-CTAATCCTTGTACTTCATTTCAGATACC) and cloned into the *pG2RNAi2* vector under the control of the soybean ubiquitin (GmUbi) promoter [21,23]. Cloning was carried out between *AscI* and *AvrII* cloning sites at the *pG2RNAi2* vector to generate *pG2RNAi2::GmSHMT08*, which was used as a positive control (Supplemental Figure S2). All mutations described in this study were introduced using site directed mutagenesis. The following mutated residues, GmSHMT08$^{\Delta L129A}$, GmSHMT08$^{\Delta L135A}$, GmSHMT08$^{\Delta F284A}$, GmSHMT08$^{\Delta N374A}$, GmSHMT08$^{\Delta Y59A}$, GmSHMT08$^{\Delta G106A}$, GmSHMT08$^{\Delta G107A}$, GmSHMT08$^{\Delta H134A}$, GmSHMT08$^{\Delta S190A}$, and GmSHMT08$^{\Delta H218A}$, were cloned into the pG2RNAi2 vector to generate *pG2RNAi2::GmSHMT08*$^{\Delta L129A}$, *pG2RNAi2::GmSHMT08*$^{\Delta L135A}$, *pG2RNAi2::GmSHMT08*$^{\Delta F284A}$, *pG2RNAi2::GmSHMT08*$^{\Delta N374A}$, *pG2RNAi2::GmSHMT08*$^{\Delta Y59A}$, *pG2RNAi2::GmSHMT08*$^{\Delta G106A}$, *pG2RNAi2::GmSHMT08*$^{\Delta G107A}$, *pG2RNAi2::GmSHMT08*$^{\Delta H134A}$, *pG2RNAi2::GmSHMT08*$^{\Delta S190A}$, and *pG2RNAi2::GmSHMT08*$^{\Delta H218A}$ constructs, respectively (Supplemental Figures S3–S11). All clones were target-sequenced to confirm that the genes and associated mutations were inserted correctly (Supplemental Figures S2–S11).

4.5. Genotyping of ExF RIL Population

The ExF12 RIL used for composite hairy root soybean transformation carrying the resistant *GmSNAP18*$^{+}$ allele from Forrest but the susceptible *GmSHMT08*$^{-}$ allele from Essex was developed and genotyped as described by [62].

4.6. Transgenic Soybean Composite Hairy Root

The functional characterization of the four THF binding sites (L129, L135, F284, N374), four PLP binding sites (G106, G107, S190A, H218), and the two PLP catalysis sites (Y59 and H134) at the GmSHMT08 protein has been validated using the transgenic hairy root system experiment. Williams 82 composite hairy roots transformed with pG2RNAi2:: empty vectors were used as a negative control. The pG2RNAi2 vector has a sGFP-selectable marker in planta [21,23]. Transgenic Williams 82 composite hairy roots transformed with *pG2RNAi2::GmSHMT08* and the ten different mutated GmSHMT08 PLP and THF cofactor binding/catalysis sites (*pG2RNAi2::GmSHMT*$^{\Delta PLP}$ and *pG2RNAi2::GmSHMT*$^{\Delta THF}$) were produced by injecting agrobacterium bacterial suspensions three times into the hypocotyl directly below soybean cotyledons using a 3 mL needle (BD#309578) as shown earlier [21]. After injection, composite hairy roots from at least 50 independent soybean transgenic plants per construct were grown and propagated in medium vermiculite. Transgenic soybeans were covered with plastic humidity domes sprayed consistently with water, maintained in a growth chamber for 1–2 weeks, and fertilized once per week with NPK 20-20-20 fertilizer. GFP-positive composite hairy roots at ~2–3 inches long were transferred into a steam-pasteurized sandy soil and packed into plastic containers as mentioned earlier [62]. Each container held 25 tubes and was suspended over water baths maintained at 27 °C. At least 15 plants from the control lines (WI82 and ExF12) were arranged in a randomized complete block design. Two days after transplanting, each plant was inoculated with ~2000 SCN (HG0) eggs. After 30 days, cysts were counted under a stereomicroscope. The experiment was independently conducted three times to obtain a minimum of 15 to 20 independent composite hairy root lines per construct per experiment. The results were plotted and analyzed for statistical significance by using analysis of variance (ANOVA) using the JMP Pro V12 software as described earlier.

4.7. GmSHMT08 TILLING Mutants

The availability of new crystal structure of the GmSHMT08 in soybeans [20] enhanced our knowledge of how the previously identified 18 EMS *Gmshmt08* TILLING mutants can affect the PLP/THF cofactor binding and catalysis sites. Thus, we performed in silico mutational analysis of the thirteen *Gmshmt08* EMS mutants that were identified using forward genetic screening [6], the three *Gmshmt08* EMS mutants identified using forward genetics [7], and the two *Gmshmt08* EMS mutants that were identified using Gel-TILLING [2].

4.8. Modeling of GmSHMT08 Protein, PLP and THF Cofactor Sites

Homology modeling of a putative GmSHMT08 protein structure was conducted using Deepview and Swiss-Model Workspace software, as previously shown [21,23]. Briefly, protein sequences from Forrest and the available SHMT crystal structure from soybean Glycine max cultivar Essex (PDB accession 6uxj.1) were used as templates. Residues 2–470 were modelled against their corresponding template with a sequence identity of

99.57% (according to the Protein Data Bank database). The structure of serine hydroxymethyltransferase from Glycine max cultivar Essex was complexed with PLP-glycine and 5-formyltetrahydrofolate residues [20]. Visualization of the THF cofactor binding sites (L129, L135, F284, N374), PLP cofactor binding and catalysis residues (Y59, G106, G107, H134, S190A, H218)—in addition to the two polymorphisms (P130R and N358Y)—and the 18 EMS-induced GmSHMT08 mutations was performed using the UCSF Chimera package [69]. To study the impact on the THF/PLP cofactor binding/catalysis and EMS mutations that were located close to the PLP/THF cofactor sites, the mapped induced mutations at the PLP cofactor sites, THF cofactor sites, and EMS-induced mutations were mutated using the structural editing tool from the UCSF Chimera package. Then, the rotamers tool that is incorporated within the Chimera package software was used to mutate the corresponding residues [24]. The rotamers tool allows amino acid sidechain rotamers to be viewed, evaluated, and incorporated into structures in which a given residue can be changed into different amino acids to predict the impact and effect of the mutations on the adjacent residues surrounding the mutated residue.

5. Conclusions

Our data are coherent with previous studies showing that glutathione peroxidase transcription, among other ROS scavenging enzymes, was significantly modulated under SCN infection in syncytia [51]. The *Arabidopsis thaliana Atshmt1-1* mutant showed a greater accumulation of H_2O_2, which is known to induce salicylic acid biosynthesis [70,71]. The implications of phytohormones, such as SA and CK, have been previously shown to be involved in a crosstalk between SCN-resistant genes (*GmSHMT08* and *GmSNAP18*) and SCN defense genes (*GmPR08-Bet VI*) [21]. Maintenance of a certain level of ROS homeostasis at low levels is required for parasitic nematodes to cause and maintain pathogenic disease [72,73]. However, disruption of this homeostasis (overaccumulation of ROS) can cause termination of syncytial formation or syncytial apoptosis [72–74]. Taken together, modulation of the SHMT serine/glycine interconversion may impact important maintenance of redox homeostasis that occurs via ROS. Maintenance of balanced SHMT expression appears to be highly important in plants. The current study uncovered for the first time the involvement of the interconversion reaction carried out by the serine hydroxymethyltransferase protein involving the two cofactors at the GmSHMT08c protein, the four THF cofactor biding sites, and the six PLP cofactor binding/catalysis sites in resistance to SCN.

Supplementary Materials: The following supporting information can be downloaded at: https://www.mdpi.com/article/10.3390/ijms231911278/s1. References [75–105] are cited in Supplementary Materials.

Author Contributions: N.L. conceived and wrote the manuscript, performed data analysis and interpretation, molecular cloning, co-immunoprecipitation experiments, composite hairy root transformation, SCN phenotyping, mass spectrometry, protein homology modelling, RNAseq analysis, and mutational analysis. D.K. performed SCN QTL mapping. A.E.B. assisted on soybean transformation. D.K., A.E.B., A.L. and J.M. performed data analysis and edited the manuscript. K.M. supervised and conceived the work, performed data interpretation, designed the experiment, and edited the manuscript. All authors reviewed and commented on the manuscript. All authors have read and agreed to the published version of the manuscript.

Funding: This research was supported in part by the United States Department of Agriculture (USDA-NIFA) project #2018-08232 and by the United Soybean Board (USB) Project #2120-172-0153 to KM and NL.

Institutional Review Board Statement: Not applicable.

Informed Consent Statement: Not applicable.

Data Availability Statement: Data about GmSHMT08 EMS mutations were deposited at the NCBI.

Acknowledgments: We thank all the student workers at Southern Illinois University Carbondale who assisted in planting the soybean lines and in greenhouse and field work.

Conflicts of Interest: The authors declare no conflict of interest.

References

1. Shahbandeh, M. *Production Value of Soybeans in the U.S. 2000–2020*; Statista: Hamburg, Germany, 2021.
2. Liu, S.; Kandoth, P.K.; Warren, S.D.; Yeckel, G.; Heinz, R.; Alden, J.; Yang, C.; Jamai, A.; El-Mellouki, T.; Juvale, P.S.; et al. A soybean cyst nematode resistance gene points to a new mechanism of plant resistance to pathogens. *Nature* **2012**, *492*, 256–260. [CrossRef]
3. Mitchum, M.G. Soybean Resistance to the Soybean Cyst Nematode Heterodera glycines: An Update. *Phytopathology* **2016**, *106*, 1444–1450. [CrossRef]
4. Howland, A.; Monnig, N.; Mathesius, J.; Nathan, M.; Mitchum, M.G. Survey of Heterodera glycines Population Densities and Virulence Phenotypes During 2015–2016 in Missouri. *Plant Dis.* **2018**, *102*, 2407–2410. [CrossRef] [PubMed]
5. Meksem, K.; Pantazopoulos, P.; Njiti, V.N.; Hyten, L.D.; Arelli, P.R.; Lightfoot, D.A. 'Forrest' resistance to the soybean cyst nematode is bigenic: Saturation mapping of the Rhg1and Rhg4 loci. *Theor. Appl. Genet.* **2001**, *103*, 710–717. [CrossRef]
6. Kandoth, P.K.; Liu, S.; Prenger, E.; Ludwig, A.; Lakhssassi, N.; Heinz, R.; Zhou, Z.; Howland, A.; Gunther, J.; Eidson, S.; et al. Systematic Mutagenesis of Serine Hydroxymethyltransferase Reveals an Essential Role in Nematode Resistance. *Plant Physiol.* **2017**, *175*, 1370–1380. [CrossRef] [PubMed]
7. Lakhssassi, N.; Patil, G.; Piya, S.; Zhou, Z.; Baharlouei, A.; Kassem, M.A.; Lightfoot, D.A.; Hewezi, T.; Barakat, A.; Nguyen, H.T.; et al. Genome reorganization of the GmSHMT gene family in soybean showed a lack of functional redundancy in resistance to soybean cyst nematode. *Sci. Rep.* **2019**, *9*, 1506. [CrossRef] [PubMed]
8. Patil, G.B.; Lakhssassi, N.; Wan, J.; Song, L.; Zhou, Z.; Klepadlo, M.; Vuong, T.D.; Stec, A.O.; Kahil, S.S.; Colantonio, V.; et al. Whole-genome re-sequencing reveals the impact of the interaction of copy number variants of the rhg1 and Rhg4 genes on broad-based resistance to soybean cyst nematode. *Plant Biotechnol. J.* **2019**, *17*, 1595–1611. [CrossRef]
9. Stover, P.; Schirch, V. Serine hydroxymethyltransferase catalyzes the hydrolysis of 5,10-methenyltetrahydrofolate to 5-formyltetrahydrofolate. *J. Biol. Chem.* **1990**, *265*, 14227–14233. [CrossRef]
10. Hanson, A.D.; Gage, D.A.; Shachar-Hill, Y. Plant one-carbon metabolism and its engineering. *Trends Plant Sci.* **2000**, *5*, 206–213. [CrossRef]
11. Appaji Rao, N.; Ambili, M.; Jala, V.R.; Subramanya, H.S.; Savithri, H.S. Structure-function relationship in serine hydroxymethyltransferase. *Biochim. Biophys. Acta* **2003**, *1647*, 24–29. [CrossRef]
12. Schirch, L. Serine hydroxymethyltransferase. *Adv. Enzym. Relat. Areas Mol. Biol.* **1982**, *53*, 83–112. [CrossRef]
13. Shuvalov, O.; Petukhov, A.; Daks, A.; Fedorova, O.; Vasileva, E.; Barlev, N.A. One-carbon metabolism and nucleotide biosynthesis as attractive targets for anticancer therapy. *Oncotarget* **2017**, *8*, 23955–23977. [CrossRef]
14. Parra, M.; Stahl, S.; Hellmann, H. Vitamin B_6 and Its Role in Cell Metabolism and Physiology. *Cells* **2018**, *7*, 84. [CrossRef] [PubMed]
15. Narkewicz, M.R.; SAULS, S.D.; TJOA, S.S.; TENG, C.; FENNESSEY, P.V. Evidence for intracellular partitioning of serine and glycine metabolism in Chinese hamster ovary cells. *Biochem. J.* **1996**, *313*, 991–996. [CrossRef]
16. Voll, L.M.; Jamai, A.; Renneé, P.; Voll, H.; McClung, C.R.; Weber, A.P.M. The Photorespiratory Arabidopsis shm1 Mutant Is Deficient in SHM1. *Plant Physiol.* **2006**, *140*, 59–66. [CrossRef] [PubMed]
17. Skibola, C.F.; Smith, M.T.; Hubbard, A.; Shane, B.; Roberts, A.C.; Law, G.R.; Rollinson, S.; Roman, E.; Cartwright, R.A.; Morgan, G.J. Polymorphisms in the thymidylate synthase and serine hydroxymethyltransferase genes and risk of adult acute lymphocytic leukemia. *Blood* **2002**, *99*, 3786–3791. [CrossRef]
18. Lim, U.; Peng, K.; Shane, B.; Stover, P.J.; Litonjua, A.A.; Weiss, S.T.; Gaziano, J.M.; Strawderman, R.L.; Raiszadeh, F.; Selhub, J.; et al. Polymorphisms in Cytoplasmic Serine Hydroxymethyltransferase and Methylenetetrahydrofolate Reductase Affect the Risk of Cardiovascular Disease in Men. *J. Nutr.* **2005**, *135*, 1989–1994. [CrossRef]
19. Amelio, I.; Cutruzzolá, F.; Antonov, A.; Agostini, M.; Melino, G. Serine and glycine metabolism in cancer. *Trends Biochem. Sci* **2014**, *39*, 191–198. [CrossRef]
20. Korasick, D.A.; Kandoth, P.K.; Tanner, J.J.; Mitchum, M.G.; Beamer, L.J. Impaired folate binding of serine hydroxymethyltransferase 8 from soybean underlies resistance to the soybean cyst nematode. *J. Biol. Chem.* **2020**, *295*, 3708–3718. [CrossRef]
21. Lakhssassi, N.; Piya, S.; Bekal, S.; Liu, S.; Zhou, Z.; Bergounioux, C.; Miao, L.; Meksem, J.; Lakhssassi, A.; Jones, K.; et al. A pathogenesis-related protein GmPR08-Bet VI promotes a molecular interaction between the GmSHMT08 and GmSNAP18 in resistance to Heterodera glycines. *Plant Biotechnol. J.* **2020**, *18*, 1810–1829. [CrossRef]
22. Babicki, S.; Arndt, D.; Marcu, A.; Liang, Y.; Grant, J.R.; Maciejewski, A.; Wishart, D.S. Heatmapper: Web-enabled heat mapping for all. *Nucleic Acids Res.* **2016**, *44*, W147–W153. [CrossRef]
23. Lakhssassi, N.; Piya, S.; Knizia, D.; El Baze, A.; Cullen, M.A.; Meksem, J.; Lakhssassi, A.; Hewezi, T.; Meksem, K. Mutations at the Serine Hydroxymethyltransferase Impact its Interaction with a Soluble NSF Attachment Protein and a Pathogenesis-Related Protein in Soybean. *Vaccines* **2020**, *8*, 349. [CrossRef] [PubMed]
24. Yang, Z.; Lasker, K.; Schneidman-Duhovny, D.; Webb, B.; Huang, C.C.; Pettersen, E.F.; Goddard, T.D.; Meng, E.C.; Sali, A.; Ferrin, T.E. UCSF Chimera, MODELLER, and IMP: An integrated modeling system. *J. Struct. Biol.* **2012**, *179*, 269–278. [CrossRef] [PubMed]

25. Kandoth, P.K.; Mitchum, M.G. War of the worms: How plants fight underground attacks. *Curr. Opin. Plant Biol.* **2013**, *16*, 457–463. [CrossRef] [PubMed]
26. Zhao, D.; You, Y.; Fan, H.; Zhu, X.; Wang, Y.; Duan, Y.; Xuan, Y.; Chen, L. The Role of Sugar Transporter Genes during Early Infection by Root-Knot Nematodes. *Int. J. Mol. Sci.* **2018**, *19*, 302. [CrossRef]
27. Hofmann, J.; Hess, P.H.; Szakasits, D.; Blöchl, A.; Wieczorek, K.; Daxböck-Horvath, S.; Bohlmann, H.; van Bel, A.J.E.; Grundler, F.M.W. Diversity and activity of sugar transporters in nematode-induced root syncytia. *J. Exp. Bot.* **2009**, *60*, 3085–3095. [CrossRef]
28. Sijmons, P.C.; Grundler, F.M.; von Mende, N.; Burrows, P.R.; Wyss, U. Arabidopsis thaliana as a new model host for plant-parasitic nematodes. *Plant J.* **1991**, *1*, 245–254. [CrossRef]
29. Braeckman, B.; Houthoofd, K.; Vanfleteren, J. Intermediary metabolism. In *WormBook: The Online Review of C. elegans Biology*; WormBook: Pasadena, CA, USA, 2009.
30. Zečić, A.; Dhondt, I.; Braeckman, B.P. The nutritional requirements of Caenorhabditis elegans. *Genes Nutr.* **2019**, *14*, 15. [CrossRef]
31. Hofmann, J.; Wieczorek, K.; Blöchl, A.; Grundler, F.M. Sucrose supply to nematode-induced syncytia depends on the apoplasmic and symplasmic pathways. *J. Exp. Bot.* **2007**, *58*, 1591–1601. [CrossRef]
32. Hofmann, J.; Szakasits, D.; Blöchl, A.; Sobczak, M.; Daxböck-Horvath, S.; Golinowski, W.; Bohlmann, H.; Grundler, F.M.W. Starch serves as carbohydrate storage in nematode-induced syncytia. *Plant Physiol.* **2008**, *146*, 228–235. [CrossRef]
33. Hofmann, J.; Grundler, F.M. Females and males of root-parasitic cyst nematodes induce different symplasmic connections between their syncytial feeding cells and the phloem in Arabidopsis thaliana. *Plant Physiol. Biochem.* **2006**, *44*, 430–433. [CrossRef]
34. Hoth, S.; Schneidereit, A.; Lauterbach, C.; Scholz-Starke, J.; Sauer, N. Nematode infection triggers the de novo formation of unloading phloem that allows macromolecular trafficking of green fluorescent protein into syncytia. *Plant Physiol.* **2005**, *138*, 383–392. [CrossRef]
35. Hoth, S.; Stadler, R.; Sauer, N.; Hammes, U.Z. Differential vascularization of nematode-induced feeding sites. *Proc. Natl. Acad. Sci. USA* **2008**, *105*, 12617–12622. [CrossRef]
36. Ko, M.P.; Huang, P.Y.; Huang, J.S.; Barker, K.R. The occurrence of phytoferritin and its relationship to effectiveness of soybean nodules. *Plant Physiol.* **1987**, *83*, 299–305. [CrossRef] [PubMed]
37. Goheen, S.; Campbell, J.; Donal, P. Nutritional Requirements of Soybean Cyst Nematodes. In *Soybean—Pest Resistance*; El-Shemy, H., Ed.; IntechOpen: London, UK, 2013.
38. Sheehan, D.; Meade, G.; Foley, V.M.; Dowd, C.A. Structure, function and evolution of glutathione transferases: Implications for classification of non-mammalian members of an ancient enzyme superfamily. *Biochem. J.* **2001**, *360*, 1–16. [CrossRef] [PubMed]
39. Udomsinprasert, R.; Pongjaroenkit, S.; Wongsantichon, J.; Oakley, A.J.; Prapanthadara, L.-a.; Wilce, M.C.J.; Ketterman, A.J. Identification, characterization and structure of a new Delta class glutathione transferase isoenzyme. *Biochem. J.* **2005**, *388*, 763–771. [CrossRef]
40. Allocati, N.; Federici, L.; Masulli, M.; Di Ilio, C. Glutathione transferases in bacteria. *FEBS J.* **2009**, *276*, 58–75. [CrossRef]
41. Hayes, J.D.; Flanagan, J.U.; Jowsey, I.R. Glutathione transferases. *Annu. Rev. Pharmacol. Toxicol.* **2005**, *45*, 51–88. [CrossRef] [PubMed]
42. Josephy, P.D. Genetic variations in human glutathione transferase enzymes: Significance for pharmacology and toxicology. *Hum. Genom. Proteom.* **2010**, *2010*, 876940. [CrossRef]
43. Kandoth, P.K.; Ithal, N.; Recknor, J.; Maier, T.; Nettleton, D.; Baum, T.J.; Mitchum, M.G. The Soybean Rhg1 locus for resistance to the soybean cyst nematode Heterodera glycines regulates the expression of a large number of stress- and defense-related genes in degenerating feeding cells. *Plant Physiol.* **2011**, *155*, 1960–1975. [CrossRef] [PubMed]
44. Vivoli, M.; Angelucci, F.; Ilari, A.; Morea, V.; Angelaccio, S.; di Salvo, M.L.; Contestabile, R. Role of a conserved active site cation-pi interaction in Escherichia coli serine hydroxymethyltransferase. *Biochemistry* **2009**, *48*, 12034–12046. [CrossRef] [PubMed]
45. De Rossi, M.C.; Levi, V.; Bruno, L. Retraction of rod-like mitochondria during microtubule-dependent transport. *Biosci. Rep.* **2018**, *38*, BSR20180208. [CrossRef] [PubMed]
46. Melkov, A.; Abdu, U. Regulation of long-distance transport of mitochondria along microtubules. *Cell Mol. Life Sci.* **2018**, *75*, 163–176. [CrossRef]
47. Rostovtseva, T.K.; Gurnev, P.A.; Chen, M.Y.; Bezrukov, S.M. Membrane lipid composition regulates tubulin interaction with mitochondrial voltage-dependent anion channel. *J. Biol. Chem.* **2012**, *287*, 29589–29598. [CrossRef]
48. Kuznetsov, A.V.; Javadov, S.; Grimm, M.; Margreiter, R.; Ausserlechner, M.J.; Hagenbuchner, J. Crosstalk between Mitochondria and Cytoskeleton in Cardiac Cells. *Cells* **2020**, *9*, 222. [CrossRef]
49. West, A.P.; Shadel, G.S.; Ghosh, S. Mitochondria in innate immune responses. *Nat. Rev. Immunol.* **2011**, *11*, 389–402. [CrossRef]
50. Lisowski, P.; Kannan, P.; Mlody, B.; Prigione, A. Mitochondria and the dynamic control of stem cell homeostasis. *EMBO Rep.* **2018**, *19*, e45432. [CrossRef] [PubMed]
51. Dan Dunn, J.; Alvarez, L.A.; Zhang, X.; Soldati, T. Reactive oxygen species and mitochondria: A nexus of cellular homeostasis. *Redox Biol.* **2015**, *6*, 472–485. [CrossRef]
52. Li, Z.; Ding, B.; Zhou, X.; Wang, G.L. The Rice Dynamin-Related Protein OsDRP1E Negatively Regulates Programmed Cell Death by Controlling the Release of Cytochrome c from Mitochondria. *PLoS Pathog.* **2017**, *13*, e1006157. [CrossRef]
53. Tang, C.; Deng, L.; Chang, D.; Chen, S.; Wang, X.; Kang, Z. TaADF3, an Actin-Depolymerizing Factor, Negatively Modulates Wheat Resistance Against Puccinia striiformis. *Front. Plant Sci.* **2016**, *6*, 1214. [CrossRef] [PubMed]

54. Day, B.; Henty, J.L.; Porter, K.J.; Staiger, C.J. The Pathogen-Actin Connection: A Platform for Defense Signaling in Plants. *Annu. Rev. Phytopathol.* **2011**, *49*, 483–506. [CrossRef]
55. Miklis, M.; Consonni, C.; Bhat, R.A.; Lipka, V.; Schulze-Lefert, P.; Panstruga, R. Barley MLO Modulates Actin-Dependent and Actin-Independent Antifungal Defense Pathways at the Cell Periphery. *Plant Physiol.* **2007**, *144*, 1132–1143. [CrossRef] [PubMed]
56. Singh, H.; Kaur, K.; Singh, M.; Kaur, G.; Singh, P. Plant Cyclophilins: Multifaceted Proteins With Versatile Roles. *Front. Plant Sci.* **2020**, *11*, 585212. [CrossRef] [PubMed]
57. Mainali, H.R.; Vadivel, A.K.A.; Li, X.; Gijzen, M.; Dhaubhadel, S. Soybean cyclophilin GmCYP1 interacts with an isoflavonoid regulator GmMYB176. *Sci. Rep.* **2017**, *7*, 39550. [CrossRef] [PubMed]
58. Piya, S.; Kihm, C.; Rice, J.H.; Baum, T.J.; Hewezi, T. Cooperative Regulatory Functions of miR858 and MYB83 during Cyst Nematode Parasitism. *Plant Physiol.* **2017**, *174*, 1897–1912. [CrossRef] [PubMed]
59. Lei, P.; Qi, N.; Zhou, Y.; Wang, Y.; Zhu, X.; Xuan, Y.; Liu, X.; Fan, H.; Chen, L.; Duan, Y. Soybean miR159-GmMYB33 Regulatory Network Involved in Gibberellin-Modulated Resistance to Heterodera glycines. *Int. J. Mol. Sci.* **2021**, *22*, 13172. [CrossRef]
60. Bane, F.T.; Bannon, J.H.; Pennington, S.R.; Campiani, G.; Williams, D.C.; Zisterer, D.M.; Mc Gee, M.M. The Microtubule-Targeting Agents, PBOX-6 [Pyrrolobenzoxazepine 7-[(dimethylcarbamoyl)oxy]-6-(2-naphthyl)pyrrolo-[2,1-d] (1,5)-benzoxazepine] and Paclitaxel, Induce Nucleocytoplasmic Redistribution of the Peptidyl-Prolyl Isomerases, Cyclophilin A and Pin1, in Malignant Hematopoietic Cells. *J. Pharmacol. Exp. Ther.* **2009**, *329*, 38. [CrossRef] [PubMed]
61. Caillaud, M.-C.; Lecomte, P.; Jammes, F.; Quentin, M.; Pagnotta, S.; Andrio, E.; de Almeida Engler, J.; Marfaing, N.; Gounon, P.; Abad, P.; et al. MAP65-3 microtubule-associated protein is essential for nematode-induced giant cell ontogenesis in Arabidopsis. *Plant Cell* **2008**, *20*, 423–437. [CrossRef]
62. Lakhssassi, N.; Liu, S.; Bekal, S.; Zhou, Z.; Colantonio, V.; Lambert, K.; Barakat, A.; Meksem, K. Characterization of the Soluble NSF Attachment Protein gene family identifies two members involved in additive resistance to a plant pathogen. *Sci. Rep.* **2017**, *7*, 45226. [CrossRef]
63. Moraes, J.G.N.; Behura, S.K.; Bishop, J.V.; Hansen, T.R.; Geary, T.W.; Spencer, T.E. Analysis of the uterine lumen in fertility-classified heifers: II. *Proteins and metabolites*†. *Biol. Reprod.* **2020**, *102*, 571–587. [CrossRef]
64. Bolger, A.M.; Lohse, M.; Usadel, B. Trimmomatic: A flexible trimmer for Illumina sequence data. *Bioinformatics* **2014**, *30*, 2114–2120. [CrossRef] [PubMed]
65. Dobin, A.; Gingeras, T.R. Mapping RNA-seq Reads with STAR. *Curr. Protoc. Bioinform.* **2015**, *51*, 11.14.11–11.14.19. [CrossRef]
66. Trapnell, C.; Pachter, L.; Salzberg, S.L. TopHat: Discovering splice junctions with RNA-Seq. *Bioinformatics* **2009**, *25*, 1105–1111. [CrossRef]
67. Anders, S.; Pyl, P.T.; Huber, W. HTSeq–a Python framework to work with high-throughput sequencing data. *Bioinformatics* **2015**, *31*, 166–169. [CrossRef] [PubMed]
68. Love, M.I.; Huber, W.; Anders, S. Moderated estimation of fold change and dispersion for RNA-seq data with DESeq2. *Genome Biol.* **2014**, *15*, 550. [CrossRef]
69. Pettersen, E.F.; Goddard, T.D.; Huang, C.C.; Couch, G.S.; Greenblatt, D.M.; Meng, E.C.; Ferrin, T.E. UCSF Chimera–a visualization system for exploratory research and analysis. *J. Comput. Chem.* **2004**, *25*, 1605–1612. [CrossRef]
70. Leon, J.; Lawton, M.A.; Raskin, I. Hydrogen Peroxide Stimulates Salicylic Acid Biosynthesis in Tobacco. *Plant Physiol.* **1995**, *108*, 1673–1678. [CrossRef] [PubMed]
71. Moreno, J.I.; Martín, R.; Castresana, C. Arabidopsis SHMT1, a serine hydroxymethyltransferase that functions in the photorespiratory pathway influences resistance to biotic and abiotic stress. *Plant J.* **2005**, *41*, 451–463. [CrossRef]
72. Melillo, M.T.; Leonetti, P.; Bongiovanni, M.; Castagnone-Sereno, P.; Bleve-Zacheo, T. Modulation of reactive oxygen species activities and H2O2 accumulation during compatible and incompatible tomato-root-knot nematode interactions. *New Phytol.* **2006**, *170*, 501–512. [CrossRef]
73. Siddique, S.; Matera, C.; Radakovic, Z.S.; Hasan, M.S.; Gutbrod, P.; Rozanska, E.; Sobczak, M.; Torres, M.A.; Grundler, F.M. Parasitic worms stimulate host NADPH oxidases to produce reactive oxygen species that limit plant cell death and promote infection. *Sci. Signal.* **2014**, *7*, ra33. [CrossRef] [PubMed]
74. Hewezi, T. Cellular Signaling Pathways and Posttranslational Modifications Mediated by Nematode Effector Proteins. *Plant Physiol.* **2015**, *169*, 1018–1026. [CrossRef]
75. Webb, D.M.; Baltazar, B.M.; Rao-Arelli, A.P.; Schupp, J.; Clayton, K.; Keim, P.; Beavis, W.D. Genetic mapping of soybean cyst nematode race-3 resistance loci in the soybean PI 437.654. *Theoret. Appl. Genet.* **1995**, *91*, 574–581. [CrossRef]
76. Ferreira, M.; Cervigni, G.; Ferreira, A.; Schuster, I.; Santana, F.; Pereira, W.; Barros, E.; Moreira, M. QTL for resistance to soybean cyst nematode races 3, 9, and 14 in cultivar Hartwig. *Pesq. Agropec. Bras.* **2011**, *46*, 420–428. [CrossRef]
77. Qiu, B.X.; Arelli, P.R.; Sleper, D.A. RFLP markers associated with soybean cyst nematode resistance and seed composition in a 'Peking' × 'Essex' population. *Theor. Appl. Genet.* **1999**, *98*, 356–364. [CrossRef]
78. Chang, W.; Dong, L.; Wang, Z.; Hu, H.; Han, Y.; Teng, W.; Zhang, H.; Guo, M.; Li, W. QTL underlying resistance to two HG types of Heterodera glycines found in soybean cultivar 'L-10'. *BMC Genom.* **2011**, *12*, 233. [CrossRef]
79. Yue, P.; Arelli, P.; Sleper, D. Molecular characterization of resistance to Heterodera glycines in soybean PI 438489B. *Theor. Appl. Genet.* **2001**, *102*, 921–928. [CrossRef]

80. Abdelmajid, K.; Ramos, L.; Hyten, D.; Bond, J.; Bendahmane, A.; Arelli, P.; Njiti, V.; Cianzio, S.; Kantartzi, S.; Meksem, K. Quantitative Trait Loci (QTL) that Underlie SCN Resistance in Soybean [Glycine max (L.) Merr.] PI438489B by 'Hamilton' Recombinant Inbred Line (RIL) Population. *Atlas J. Plant Biol.* **2014**, *1*, 29–38. [CrossRef]
81. Concibido, V.C.; Denny, R.L.; Boutin, S.R.; Hautea, R.; Orf, J.H.; Young, N.D. DNA marker analysis of loci underlying resistance to soybean cyst nematode (Heterodera glycines Ichinohe). *Crop Sci.* **1994**, *34*, 240–246. [CrossRef]
82. Guo, B.; Sleper, D.; Nguyen, H.; Arelli, P.; Shannon, J. Quantitative Trait Loci Underlying Resistance to Three Soybean Cyst Nematode Populations in Soybean PI 404198A. *Crop Sci.* **2006**, *46*, 224–233. [CrossRef]
83. Wu, X.; Blake, S.; Sleper, D.; Shannon, J.; Cregan, P.; Nguyen, H. QTL, additive and epistatic effects for SCN resistance in PI 437654. *Theor. Appl. Genet.* **2009**, *118*, 1093–1105. [CrossRef]
84. Chang, S.J.C.; Doubler, T.W.; Kilo, V.Y.; Abu-Thredeih, J.; Prabhu, R.; Freire, V.; Suttner, R.; Klein, J.; Schmidt, M.E.; Gibson, P.T.; et al. Association of loci underlying field-resistance to soybean sudden-death syndrome (SDS) and cyst-nematode (SCN) RACE-3. *Crop Sci.* **1997**, *37*, 966–971. [CrossRef]
85. Vuong, T.; Sleper, D.; Shannon, J.; Nguyen, H. Novel quantitative trait loci for broad-based resistance to soybean cyst nematode (Heterodera glycines Ichinohe) in soybean PI 567516C. *Theor. Appl. Genet.* **2010**, *121*, 1253–1266. [CrossRef]
86. Arriagada, O.; Mora, F.; Dellarosa, J.; Ferreira, M.; Cervigni, G.; Schuster, I. Bayesian mapping of quantitative trait loci (QTL) controlling soybean cyst nematode resistant. *Euphytica* **2012**, *186*, 907–917. [CrossRef]
87. Jiao, Y.; Vuong, T.; Liu, Y.; Meinhardt, C.; Liu, Y.; Joshi, T.; Cregan, P.; Xu, D.; Shannon, J.; Nguyen, H. Identification and evaluation of quantitative trait loci underlying resistance to multiple HG types of soybean cyst nematode. *Theor. Appl. Genet.* **2015**, *128*, 15–23. [CrossRef]
88. Silva, M.; Schuster, I.; Silva, J.; Ferreira, A.; Barros, E.; Moreira, M. Validation of microsatellite markers for assisted selection of soybean resistance to cyst nematode races 3 and 14. *Pesq. Agropec. Bras.* **2007**, *42*, 1143–1150. [CrossRef]
89. Vaghchhipawala, Z.; Bassuner, R.; Clayton, K.; Lewers, K.S.; Shoemaker, R.C.; Mackenzie, S. Modulations in gene expression and mapping of genes associated with cyst nematode infection of soybean. *Mol. Plant Microbe Interac.* **2001**, *14*, 42–54. [CrossRef]
90. Valdes-Lopez, O.; Thibivilliers, S.; Qiu, J.; Xu, W.; Nguyen, T.; Libault, M.; Le, B.; Goldberg, R.; Hill, C.; Hartman, G.; et al. Identification of Quantitative Trait Loci Controlling Gene Expression during the Innate Immunity Response of Soybean. *Plant Physiol.* **2011**, *157*, 1975–1986. [CrossRef] [PubMed]
91. Carlson, S.R.; Diers, B.W.; Kabelka, E.A. Localization of Two Loci that Confer Resistance to Soybean Cyst Nematode from Glycine soja PI 468916. *Crop Sci.* **2005**, *45*, 2473–2481.
92. Yue, P.; Sleper, D.A.; Arelli, P.R. Mapping Resistance to Multiple Races of Heterodera glycines in Soybean PI 89772. *Crop Sci.* **2001**, *41*, 1589–1595. [CrossRef]
93. Guo, B.; Sleper, D.; Arelli, P.; Shannon, J.; Nguyen, H. Identification of QTLs associated with resistance to soybean cyst nematode races 2, 3 and 5 in soybean PI 90763. *Theor. Appl. Genet.* **2005**, *111*, 965–971. [CrossRef]
94. Swaminathan, S.; Abeysekara, N.; Knight, J.; Liu, M.; Dong, J.; Hudson, M.; Bhattacharyya, M.; Cianzio, S. Mapping of new quantitative trait loci for sudden death syndrome and soybean cyst nematode resistance in two soybean populations. *Theor. Appl. Genet.* **2018**, *131*, 1047–1062. [CrossRef]
95. Vuong, T.; Sleper, D.; Shannon, J.; Wu, X.; Nguyen, H. Confirmation of quantitative trait loci for resistance to multiple-HG types of soybean cyst nematode (Heterodera glycines Ichinohe). *Euphytica* **2011**, *181*, 101–113. [CrossRef]
96. Mahalingam, R.; Skorupska, H.T. DNA markers for resistance to Heterodera glycines I. race 3 in soybean cultivar Peking. *Breed. Sci.* **1995**, *45*, 435–443.
97. Wang, D.; Diers, B.W.; Arelli, P.R.; Shoemaker, R.C. Loci underlying resistance to Race 3 of soybean cyst nematode in Glycine soja plant introduction 468916. *Theor. Appl. Genet.* **2001**, *103*, 561–566. [CrossRef]
98. Kazi, S.; Shultz, J.; Afzal, J.; Hashmi, R.; Jasim, M.; Bond, J.; Arelli, P.; Lightfoot, D. Iso-lines and inbred-lines confirmed loci that underlie resistance from cultivar 'Hartwig' to three soybean cyst nematode populations. *Theor. Appl. Genet.* **2010**, *120*, 633–644. [CrossRef]
99. Schuster, I.; Abdelnoor, R.V.; Marin, S.R.R.; Carvalho, V.P.; Kiihl, R.A.S.; Silva, J.F.V.; Sediyama, C.S.; Barros, E.G.; Moreira, M.A. Identification of a new major QTL associated with resistance to soybean cyst nematode (Heterodera glycines). *Theor. Appl. Genet.* **2001**, *102*, 91–96. [CrossRef]
100. Winter, S.; Shelp, B.; Anderson, T.; Welacky, T.; Rajcan, I. QTL associated with horizontal resistance to soybean cyst nematode in Glycine soja PI464925B. *Theor. Appl. Genet.* **2007**, *114*, 461–472. [CrossRef]
101. Prabhu, R.R.; Njiti, V.N.; Bell-Johnson, B.; Johnson, J.E.; Schmidt, M.E.; Klein, J.H.; Lightfoot, D.A. Selecting soybean cultivars for dual resistance to soybean cyst nematode and sudden death syndrome using two DNA markers. *Crop Sci.* **1999**, *39*, 982–987. [CrossRef]
102. Concibido, V.C.; Lange, D.A.; Denny, R.L.; Orf, J.H.; Young, N.D. Genome mapping of soybean cyst nematode resistance genes in Peking, PI90763, and PI88788 using DNA markers. *Crop Sci.* **1997**, *37*, 258–264. [CrossRef]
103. Vierling, R.A.; Faghihi, J.; Ferris, V.R.; Ferris, J.M. Association of RFLP markers with loci conferring broad-based resistance to the soybean cyst nematode (Heterodera glycines). *Theor. Appl. Genet.* **1996**, *92*, 83–86. [CrossRef] [PubMed]

104. Yuan, J.; Njiti, V.N.; Meksem, K.; Iqbal, M.J.; Triwitayakorn, K.; Kassem, M.A.; Davis, G.T.; Schmidt, M.E.; Lightfoot, D.A. Quantitative trait loci in Two Soybean Recombinant Inbred Line Populations Segregating for Yield and Disease Resistance. *Crop Sci.* **2002**, *42*, 271–277. [CrossRef] [PubMed]
105. Ferdous, S.; Watanabe, S.; Suzuki-Orihara, C.; Tanaka, Y.; Kamiya, M.; Yamanaka, N.; Harada, K. QTL Analysis of Resistance to Soybean Cyst Nematode Race 3 in Soybean Cultivar Toyomusume. *Breed. Sci.* **2006**, *56*, 155–163. [CrossRef]

International Journal of
Molecular Sciences

MDPI

Article

NAC Transcription Factor *GmNAC12* Improved Drought Stress Tolerance in Soybean

Chengfeng Yang [1,†], Yanzhong Huang [1,2,†], Peiyun Lv [1], Augustine Antwi-Boasiako [1,3], Naheeda Begum [1], Tuanjie Zhao [1,*] and Jinming Zhao [1,*]

1 National Center for Soybean Improvement, Key Laboratory of Biology and Genetics and Breeding for Soybean, Ministry of Agriculture, State Key Laboratory of Crop Genetics and Germplasm Enhancement, Nanjing Agricultural University, Nanjing 210095, China
2 National Forage Breeding Innovation Base (JAAS), Key Laboratory for Saline-Alkali Soil Improvement and Utilization (Coastal Saline-Alkali Lands), Ministry of Agriculture and Rural Affairs, Institute of Animal Science, Jiangsu Academy of Agricultural Sciences, Nanjing 210014, China
3 Crops Research Institute, Council for Scientific and Industrial Research, Kumasi AK420, Ghana
* Correspondence: tjzhao@njau.edu.cn (T.Z.); jmz3000@njau.edu.cn (J.Z.)
† These authors contributed equally to this work.

Abstract: NAC transcription factors (TFs) could regulate drought stresses in plants; however, the function of NAC TFs in soybeans remains unclear. To unravel NAC TF function, we established that *GmNAC12*, a NAC TF from soybean (*Glycine max*), was involved in the manipulation of stress tolerance. The expression of *GmNAC12* was significantly upregulated more than 10-fold under drought stress and more than threefold under abscisic acid (ABA) and ethylene (ETH) treatment. In order to determine the function of *GmNAC12* under drought stress conditions, we generated *GmNAC12* overexpression and knockout lines. The present findings showed that under drought stress, the survival rate of *GmNAC12* overexpression lines increased by more than 57% compared with wild-type plants, while the survival rate of *GmNAC12* knockout lines decreased by at least 46%. Furthermore, a subcellular localisation analysis showed that the *GmNAC12* protein is concentrated in the nucleus of the tobacco cell. In addition, we used a yeast two-hybrid assay to identify 185 proteins that interact with *GmNAC12*. Gene ontology (GO) and KEGG analysis showed that *GmNAC12* interaction proteins are related to chitin, chlorophyll, ubiquitin–protein transferase, and peroxidase activity. Hence, we have inferred that *GmNAC12*, as a key gene, could positively regulate soybean tolerance to drought stress.

Keywords: *Glycine max*; NAC transcription factors; drought tolerance; CRISPR/Cas9; protein interaction

1. Introduction

As the world's population continues to rise, there is a growing requirement for increased crop yield in order to ensure agricultural production. Consequently, crop growers have been under much stress due to biotic and abiotic pressures [1]. Abiotic stresses, described as unfavourable non-living environmental factors, such as drought, salinity, flooding, and extreme temperatures, have a negative impact on crop plant survival and productivity [2]. Drought is considered to become the most serious hazard of the abiotic stressors, posing difficult obstacles to yield performance and productivity around the world [3].

Soybean (*Glycine max* [L.] Merr.) is one of the world's most economically significant crops. Soybeans are used for human food and livestock feed and are highly valued for their high protein and oil contents and their numerous uses in industrial products [4–6]; however, soybean production and quality are threatened by multiple abiotic stressors including drought, salinity, and extreme temperatures. Drought is becoming a major constraint for crop production throughout the world. In soybean, drought stress mainly occurs during the growing season, leading to considerable yield loss and quality deterioration, mainly in arid and semi-arid zones. To adapt to the fast-changing climatic conditions and the prevailing threats caused by drought, many plants have established a variety of defence techniques ranging from altering the architecture of the roots and leaves to changing the metabolite composition and controlling the expression of genes involved in resistance to drought [7,8]. Recently, Huang et al. [9] reported that advancements in genomics and molecular biological approaches are useful for discovering stress-related genes, improving the soybean's ability to adapt to environmental changes, and providing critical genetic resources in soybeans.

Several recent studies have shown that a complex network consisting of diverse proteins and metabolites is deployed in plant defence [10]. Transcription factors (TFs) are known for their crucial roles among the established participants in promoting plant responses to environmental stresses [11]. Some of these TF family (NAC, AP2/EREBP, WRKY, bZIP, bHLH, etc.) members are involved in response to adaptive stress, regulating stress-related genes by influencing the plant capacity in responding to stressful conditions, while other members engage in coordinating the growth and development of plants [12–14]. NAC proteins are composed of an extremely preserved NAC domain that consists of approximately 150 amino acids at the N-terminus and a highly variable transcription region on the C-terminus [15,16]. The NAC domain is classified into five subdomains ranging from A to E. Although C and D are highly conserved, positively charged subdomains associated with DNA binding [17,18], the function of subdomain A may be involved in the formation of functional dimers and subdomains B and E may play a role in the diversified function of the protein [19–21].

Several plant species, especially the terrestrials, are known to have NAC TF families that are extensively dispersed [22]. Since the initial discovery of NAC TFs in Arabidopsis thaliana [23], findings of NAC TFs have progressed rapidly, and their presence has been detected in rice soybeans, as well as in other plants [24,25]. Moreover, previous studies have demonstrated that the NAC TF family participates in responses to stress caused by biotic and abiotic factors, hormone signal transduction pathway, and cell apoptosis [26–28], and it plays important regulatory roles in various processes, including plant cell secondary wall formation, plant senescence, and lateral root development [29,30]. For instance, some researchers enhanced plant tolerance of drought stress by overexpressing Arabidopsis NAC TF family genes [31,32]. In rice (*Oryza sativa*), overexpression of *OsNAC5*, *OsNAC6*, *OsNAC10*, and *SNAC1* genes improved drought tolerance [33–36]. Furthermore, the overexpression of *GmSNAC49* in Arabidopsis improved drought tolerance by upregulating the genes related to drought and the ABA signal pathway [37]. In addition, *MusaNAC042* positively regulates drought and salinity tolerance in bananas [38], and *MusaSNAC1* improves drought tolerance by modifying stomatal closure and H_2O_2 content [27]. Recently, Ju et al. [26] found that overexpression of *VvNAC17* in transgenic Arabidopsis increases resistance to drought, salinity, and freezing, and upregulates the expression of ABA- and stress-related genes.

Recently, there has been growing interest in the characterisation of the NAC TFs in soybean, emphasising stress responses. The overexpression of *GmNAC8* in soybean plants improves tolerance to drought stress by interacting with a drought-induced protein (GmDi19-3) [39]. In addition, the overexpression of *GmNAC109* in Arabidopsis improves tolerance to drought and salt stress by upregulating the expression of stress-related genes [40]. Another study reported that soybean-dehydration-induced *GmNAC085* plays a positive role in regulating plant drought tolerance [24]. In addition, the transient expression of *GmNAC065* and *GmNAC085* induce the appearance of leaf senescence features, which include loss of chlorophyll, yellowing of leaves, peroxidation of liquids, and the accumulation of H_2O_2 [41]. This evidence suggests that NAC TFs play significant and crucial roles in improving the growth of soybean plants and increasing their tolerance to biotic and abiotic stress. Therefore, to explore excellent germplasm resources, it is necessary to understand the function of NAC and its mechanism in the soybean.

Clustered regularly interspaced short palindromic repeats/CRISPR-associated protein 9 (CRISPR/Cas9) system-mediated genome editing technology, which has been applied to various plants, has developed rapidly over the last few years and was successfully carried out in the soybean for the first time in 2015 [42]. Thus far, the CRISPR/Cas9 system has been used in soybeans for genetic transformation and hairy root genetic transformation to validate the functions of various genes related to diverse traits of interest in soybeans [43]. Successful application of the CRISPR/Cas9 system results in induced targeted mutagenesis of *GmFT2a*. Although *GmFT2a* regulates flowering in soybeans, the homozygous *GmFT2a* mutant delayed flowering [44]. These studies confirm the efficacy of applying the CRISPR/Cas9 system. The system is a highly specific and simple genome editing tool conducive to the functional verification of soybean genes and provides excellent potential for creating new soybean germplasm resources.

Previously, *GmNAC12* (*Glyma.16G043200*) in soybean plants has been reported with induced upregulated expression from stress caused by drought [25,45]. Our results showed that *GmNAC12* could positively regulate the tolerance of soybean to drought stress. The functional analysis of the *GmNAC12* gene could provide a basis for the response mechanism of soybean under drought stress, as well as provide a theoretical basis and germplasm resources for crop resistance genetic engineering breeding. In addition, using the yeast two-hybrid (Y2H) assay, we found that *GmNAC12* interacts with many functional proteins related to chitin, ubiquitin-protein transferase, peroxidase activity, etc. These results provided evidence support for the study of the function of soybean *GmNAC12* protein and the regulatory pathways involved and were conducive to the analysis of the regulatory network of *GmNAC12* in soybean stress tolerance. Therefore, our findings indicate that *GmNAC12* is a key regulator in the soybean plant.

2. Results

2.1. *GmNAC12* Responded to Drought Stress as Well as ABA and ETH Hormone Treatments

To investigate the expression of *GmNAC12* in the leaves, roots, land cotyledons, and stems of soybean plants, we implemented the qRT-PCR assay. The results showed that *GmNAC12* was expressed in all the tissues, but the expression level of *GmNAC12* was highest and lowest in the cotyledons and the stems, respectively (Figure 1A). To explore the consequences of drought stress on the expression of *GmNAC12*, the soybean seedlings were exposed to drought conditions, and the gene's expression was estimated by qRT-PCR. The findings indicated that the expression of *GmNAC12* in roots, stems, and leaves changed significantly and was significantly upregulated after 3, 5, and 7 days of drought stress, respectively (Figure 1B). These results indicate that *GmNAC12* may be involved in regulating drought stress.

Further, to investigate whether *GmNAC12* is involved in phytohormone regulation signals, qRT-PCR was used to detect the transcription level of *GmNAC12* in soybean seedlings after treatments with ABA and ETH. The results revealed that after the ABA treatment, the expression level of *GmNAC12* was significantly upregulated at 3 h and 12 h (Figure 1C), while under ETH treatment, the expression of *GmNAC12* was significantly upregulated at 3 h (Figure 1D). The results indicate that *GmNAC12* is also involved in the ABA and ETH regulation pathways.

2.2. Targeted Mutagenesis of *GmNAC12* Produced Using the CRISPR/Cas9 System

The analysis of the gene structure demonstrated that the *GmNAC12* gene is made up of three exons and two introns, totalling 2348 bp. At the second exon, a site was designed to target mutagenesis (Figure 2A), and its matching CRISPR/Cas9 vector was converted into the soybean cultivar Tianlong 1 soybean cultivar for *GmNAC12* gene knockout using *Agrobacterium*-mediated soybean genetic transformation. The transformation resulted in 42 T_0 generations of *GmNAC12* transgenic lines with a transformation efficiency of 5.6%. Among them, 26 lines were *GmNAC12* gene knockout lines with an edited rate of 61.9%.

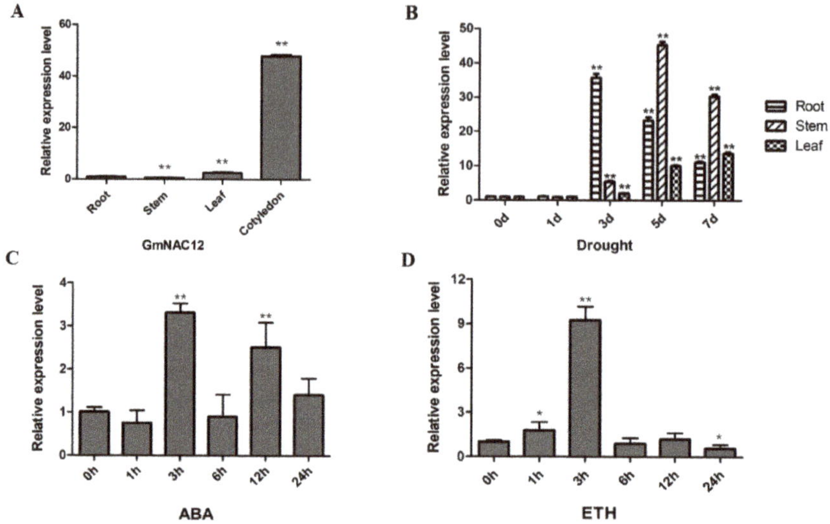

Figure 1. Expression pattern of *GmNAC12* under drought stress and ABA and ETH hormone treatment by qRT-PCR. (**A**) Expression of the *GmNAC12* in cotyledon, roots, stems, and leaves in normal condition by qRT-PCR. (**B**) Expression of the *GmNAC12* in roots, stems, and leaves under drought stress by qRT-PCR. (**C**,**D**) Expression of *GmNAC12* in roots under ABA (150 μM) and ETH (2 mM) treatments determined by qRT-PCR. The data represent the means ± SEs, $n = 3$. * $p < 0.05$ (Student's t-test). ** $p < 0.01$ (Student's t-test).

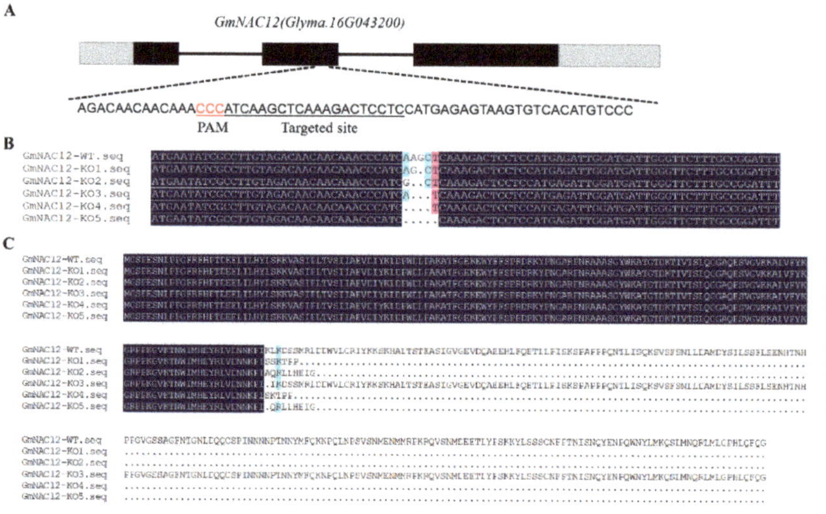

Figure 2. Targeted mutagenesis of *GmNAC12* induced by CRISPR/Cas9. (**A**) Gene structures of *GmNAC12* with a CRISPR/Cas9 target site designed in the second exon. Black stripe, black line, and grey stripe represent exon, intron, and UTR (untranslated regions), respectively. The underlined nucleotides indicate the target site. Nucleotides in red represent PAM sequences. PAM, protospacer adjacent motif. (**B**,**C**) DNA sequences and amino acid sequences of wild-type and representative mutation types, *GmNAC12-KO1*, *GmNAC12-KO2*, *GmNAC12-KO3*, *GmNAC12-KO4*, and *GmNAC12-KO5*, induced at the target site. Underline, insertions. Dashes, deletions.

After analysing 26 *GmNAC12* gene knockout lines, five gene editing types were identified (Figure 2B): *GmNAC12-KO1* (deletion 1 bp), *GmNAC12-KO2* (deletion 2 bp), *GmNAC12-KO3* (3 bp deletion), *GmNAC12-KO4* (4 bp deletion), and *GmNAC12-KO5* (5 bp deletion). In comparing these with the amino acid sequence of wild-type soybeans, it was noted that frameshift mutations occurred in *GmNAC12-KO1*, *GmNAC12-KO2*, *GmNAC12-KO4*, and *GmNAC12-KO5*, which lead to early termination of translation. *GmNAC12-KO3* lacks 3 bp in the amino acid sequence, resulting in one amino acid change and one amino acid deletion (Figure 2C).

2.3. *GmNAC12* Enhanced Drought Tolerance in Soybean Plants

We generated *GmNAC12* overexpression (OE1 and OE4) and *GmNAC12* knockout (KO1 and KO2) lines to investigate the reaction of *GmNAC12* when exposed to drought conditions. *GmNAC12* overexpression lines (OE1 and OE4), gene knockout lines (KO1 and KO2), and wild-type plants were exposed to drought treatment (stress) after 14 days of standard growth. After 10 days of drought conditions, the leaves of the *GmNAC12* gene knockout lines (KO1 and KO2) were severely wilted compared to the leaves of the wild-type plants; however, the *GmNAC12* overexpression lines (OE1 and OE4) only showed minor wilting symptoms (Figure 3A). In addition, after rewatering, only 59% of the wild-type plants recovered, while 92% and 94% of the OE1 and OE4 *GmNAC12* overexpression lines recovered, respectively, and they exhibited a significantly higher survival rate compared to the wild type. Furthermore, 29% and 33% of the *GmNAC12* gene knockout lines KO1 and KO2 recovered, respectively, and the survival rate was significantly lower than the wild type (Figure 3B). The above results indicate that *GmNAC12* plays a positive role in regulating soybean drought tolerance.

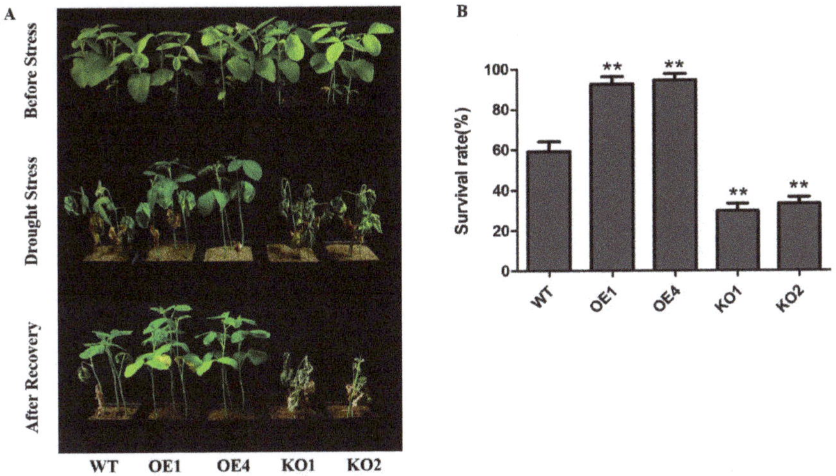

Figure 3. *GmNAC12* positively regulated drought tolerance in soybean. (**A**) Performance of wild-type (WT) plants, *GmNAC12* overexpression lines (OE1 and OE4), and *GmNAC12* knockout lines (KO1 and KO2) under drought stress and after recovery. (**B**) Survival rate of wild-type plants, *GmNAC12* overexpression lines, and *GmNAC12* knockout lines after recovery ($n = 3$). Over 30 plants in each line were used for survival rate analysis. The data represent the means ± SEs. ** $p < 0.01$ (Student's *t*-test).

2.4. *GmNAC12* Encoded a Nuclear Localisation Protein

In establishing the subcellular localisation, the domain and nuclear localisation signal of the *GmNAC12* protein were predicted online. An amino acid sequence analysis showed that *GmNAC12* encoded 353 amino acids, with a NAM domain at 9–138 amino acids (Figure S1A) and two NLSs, RPKRQVSNMDEETLYPSKKYLSS and RPKRQVSNMDEET-

LYPSKKYLSSS (Figure S1B), at 291–314 amino acids. It was predicted that *GmNAC12* is a nuclear localisation protein.

To verify the prediction results of the subcellular localisation vector of *GmNAC12*, namely, *pBinGPF4-GmNAC12*, it was constructed, as shown in Figure S1C. The Agrobacterium-mediated transformation of tobacco transferred the expression vector into tobacco leaves for transient expression. The green fluorescent protein signal was observed through laser confocal microscopy. The results showed that a green fluorescent protein signal of an empty vector was detected on the cytoplasm, cell membrane, and nucleus (Figure 4E–H). In contrast, the fusion protein expressed by *pBinGPF4-GmNAC12* was only detected in the nucleus (Figure 4A–D). These results confirm that the *GmNAC12* protein is a nuclear-localised protein.

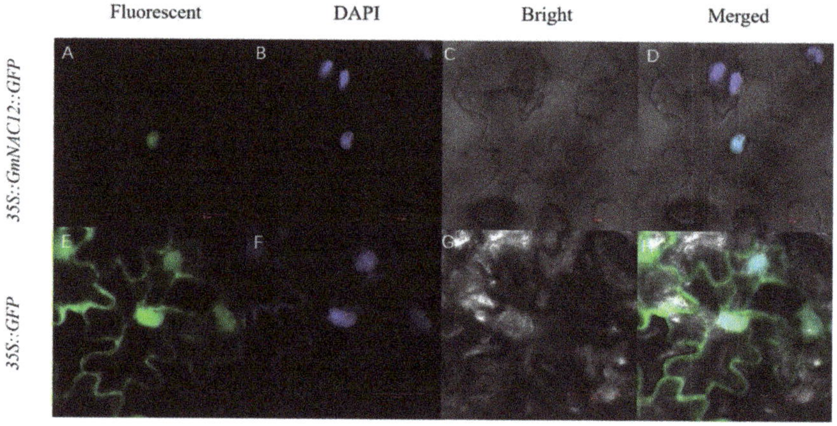

Figure 4. Subcellular localisation of *GmNAC12* in a tobacco cell. **(A–D)** *35S::GmNAC12::GFP* fluorescence images in a tobacco cell. **(E–H)** *35S::GFP* fluorescence images in a tobacco cell. *Nicotiana benthamiana* leaves were transiently infiltrated with A. tumefaciens EHA105-containing vector expressing *35S::GFP* or *35S::GmNAC12::GFP*. All images were collected using the Zeiss confocal microscope after agroinfiltration for 48 h. DAPI images indicate nuclear staining. Scale bars are 50 μm.

2.5. Candidate Interaction Protein Analysis of GmNAC12

The study of interaction proteins analyses the action mechanism of the targeted protein and helps explore new functions of the protein [46]. To further understand the action mechanism of *GmNAC12*, we used yeast two-hybrid technology to screen the interaction proteins of *GmNAC12*. After screening and identification, 185 candidate proteins were determined to interact with *GmNAC12*.

The 185 candidate interaction proteins of *GmNAC12* were annotated with GO functions (Figure 5A, Table S2). The identified interaction proteins related to one or more of the following three GO categories: biological processes (BP), cellular components (CC), and molecular function (MF). An analysis of the biological processes indicated that the interaction proteins were mainly involved in translation (GO: 0006412), fatty acid beta-oxidation (GO: 0006635), plant-type secondary cell wall biogenesis (GO: 0009834), hydrogen peroxide catabolic process (GO: 0042744), defensive responses to the bacterium (GO: 0042742), defensive responses to the virus (GO: 0051607), and defensive responses to fungus (GO: 0050832) and other processes. An analysis of cell composition indicated that the interaction proteins were mainly composed of the cytosolic small ribosomal subunit (GO: 0022627), ribosome (GO: 0005840), photosystem I (GO: 0009522), and photosystem II (GO: 0009523). Furthermore, a molecular biological function analysis determined that the interaction proteins mainly included structural components of the ribosome (GO: 0003735), chitin-binding activity (GO: 0008061), chlorophyll-binding activity (GO: 0016168), ubiquitin-protein transferase activity (GO: 0004842), and peroxidase activity (GO: 0004601), among others.

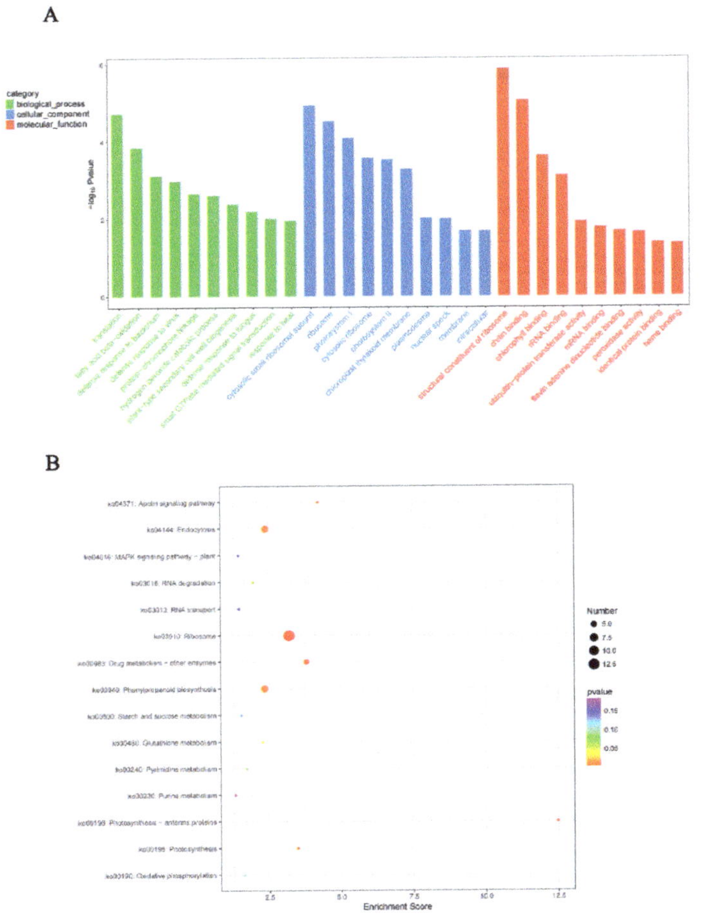

Figure 5. GO and KEGG analysis of candidate interaction proteins of *GmNAC12*. (**A**) Diagram showing GO enrichment analysis of candidate interaction proteins of *GmNAC12*. BP, biological processes; CC, cellular components; MF, molecular function. (**B**) KEGG analysis of candidate interaction proteins of *GmNAC12*.

On the basis of the results of previous studies, we know that secondary metabolites and signals mediated by hormones perform crucial roles in improving stress tolerance. Considering this knowledge, we performed a KEGG enrichment analysis to broaden our understanding of the metabolic pathways of genes in plant cells. The KEGG enrichment analysis of the 185 candidate interaction proteins of *GmNAC12* indicated that the proteins mainly involved 15 pathways (Figure 5B, Table S3), including ribosomes (ko03010), photosynthesis-antenna proteins (ko00196), drugs metabolic-other enzymes (ko00983), photosynthesis (ko00195), endocytosis (ko04144), phenylpropanoid biosynthesis (ko00940), and other regulatory pathways.

2.6. Expression Analysis of *GmNAC12* under Biotic-Stress-Related Hormones

On the basis of the GO and KEGG analyses, we speculated that *GmNAC12* might also be involved in biotic stresses. Previous studies have shown that SA and MeJA are the key phytohormone signals for plants to respond to biotic stress. The expression levels of *GmNAC12* in diverse tissues of soybean seedlings under SA and MeJA phytohormone treatments were detected by qRT-PCR. Under the SA treatment, the expression of *GmNAC12* was significantly upregulated at 3 h (Figure 6A). Under the MeJA treatment, the expression of *GmNAC12* was significantly downregulated at 1 h, then significantly upregulated at 12 h (Figure 6B). In summary, *GmNAC12* might help regulate soybean response to biotic stress in a way that depends on plant hormones.

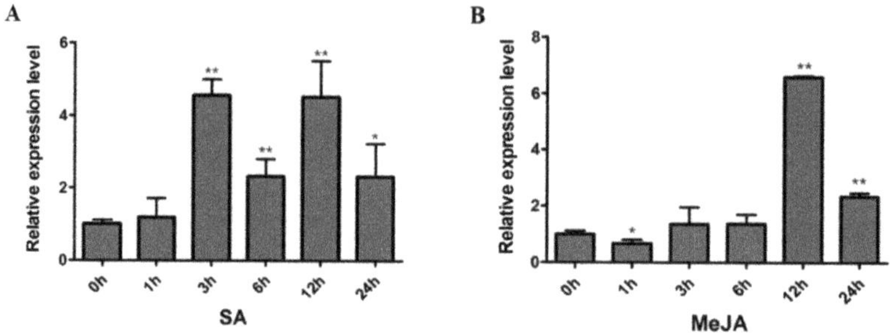

Figure 6. Expression of *GmNAC12* under SA and MeJA hormone treatments. (**A**,**B**) Expression of *GmNAC12* in roots under SA (2 mM) and MeJA (100 μM) treatments determined by qRT-PCR ($n = 3$). The data represent the means ± SEs. * $p < 0.05$ (Student's *t*-test). ** $p < 0.01$ (Student's *t*-test).

3. Discussion

In the context of rapidly changing climate, drought is considered one of the major environmental factors limiting plant growth and development, which adversely affects a plant's yield. However, many strategies have been devised and implemented to prevent drought's damaging effects on plant growth, thus improving productivity. As upstream genes for gene expression regulation, the NAC TF family can respond quickly when plants are subjected to unfavourable environmental pressures [20]. Hence, the functional identification of NAC TFs in soybean plants is important for studying how the soybean adapts to drought stress. From this perspective, we analyzed the expression of *GmNAC12* in the tissues (roots, stems, and leaves) of soybean seedlings after drought stress by qRT-PCR and noted substantial upregulation of *GmNAC12* in the various tissues (Figure 1B). The present results support the findings of previous researchers [25,45], which suggest that *GmNAC12* may regulate drought tolerance in soybeans.

The ABA and ETH regulation pathways are often involved in abiotic stress in plants [47,48]. Under drought stress conditions, the concentration of ABA in the plant increases, which promotes stomata closure and reduces transpiration, ultimately enhancing the drought resistance of plants [49]. In this study, *GmNAC12* was significantly upregulated under the induction of ABA and ETH (Figure 1C,D). The results are similar to the findings reported by Yang et al. [39] in that *GmNAC8*, a *GmNAC12* homologous gene, is induced by drought, ABA, and ETH. Therefore, we inferred that *GmNAC12* may regulate drought stress through plant hormones' signal transduction pathway.

Since the advent of CRISPR/Cas9, genome editing has enabled quick and easy transformations in plants to characterise gene functions and improve traits, mainly through double-strand breaks generated by CRISPR/Cas9 to induce mutations [50]. Due to the advantages of using CRISPR/Cas9 technology to knockout, insert, or replace important genes in plants, precise improvements of plant traits or varieties can be achieved to cultivate new varieties suitable for various geographical and environmental conditions, which can help to alleviate the current soybean dilemma. The realisation of soybean gene editing depends on soybean genetic transformation technology; therefore, establishing a stable and efficient soybean genetic transformation system is essential. Applying gene editing technology to soybeans provides an essential tool for analysing soybean gene function and molecular mechanisms [51].

In this study, to explore the specific function of *GmNAC12* under drought stress, five different *GmNAC12* gene knockouts were created by the CRISPR/Cas9 system. Of the five gene knockouts, *GmNAC12-KO1*, *GmNAC12-KO2*, *GmNAC12-KO4*, and *GmNAC12-KO5* had frameshift mutations due to base deletion (Figure 2B), which led to early termination of translation and resulted in the almost complete deletion of the transcriptional regulatory domain at the C-terminal of *GmNAC12* (Figure 2C), as well as loss of function [20]. The remaining knockout gene, *GmNAC12-KO3*, lacked three bases, which resulted in an amino acid deletion and an amino acid mutation; however, its function in the transcriptional regulatory activity of *GmNAC12* remains unknown.

Through the Agrobacterium-mediated genetic transformation of the soybean, three stable *GmNAC12* overexpression lines were obtained after screening and identifying. Drought treatment was performed on the *GmNAC12* gene of overexpression, gene knockout, and wild-type plants (Figure 3A). The treatment findings demonstrated that the survival rate of the *GmNAC12* overexpressed lines was greater than that of the wild-type plants, and the plants showed more substantial drought tolerance. Conversely, the survival rate of the *GmNAC12* gene knockout plants was lower compared to the survival rate of the wild-type plants, indicating limited drought tolerance (Figure 3B). These results indicate that *GmNAC12* positively regulates drought tolerance in soybeans, which is consistent with recent findings indicating that GmNAC8, the homologous gene of *GmNAC12*, positively regulates the drought stress tolerance of soybeans [39].

We used *GmNAC12* as the bait protein and screened 185 candidate interacting proteins through yeast two hybrid (Y2H). Interestingly, although *GmNAC12* was a transcription factor expressed in the nucleus (Figure 4), its interacting proteins were expressed not only in the nucleus but also in the cytoplasm and cell membrane. It was shown that the rose transcription factor PTM could change its expression position in response to drought stress [52], which might suggest that *GmNAC12* was not exclusively expressed in the soybean nucleus. In addition, GmHsp90A2 could change its protein expression region by interacting with GmHsp90A1 in soybean [9]. Therefore, *GmNAC12* might change its expression position by interacting with proteins that function in different regions of the cell.

Furthermore, we performed a GO annotation of 185 interaction proteins of *GmNAC12* and found that some interaction proteins related to *GmNAC12* functions involve peroxidase activity (Figure 5A). The regulation of drought stress in plants involves a complex signal transduction network, and reactive oxygen species (ROS) have a signal interface function in plants adapting to drought stress [53]. Previous studies on *GmNAC8* showed that it could improve the soybean's tolerance of drought by increasing the activity of intracellular peroxidase to remove excess ROS [39]. During this process, peroxidase can partially eliminate the damage plants under drought stress experience due to ROS accumulation [15,53], thereby improving the drought tolerance of *GmNAC12* in overexpressed plants. Therefore, the action mechanism of *GmNAC12* that enhances the drought stress tolerance of soybeans by removing excessive ROS may be the same as that of *GmNAC8*.

NAC transcription factors play important roles in the various signalling pathways that govern how a plant responds to biotic and abiotic stress and developmental activities [20,54]. Many studies have reported that SA and MeJA are the main signals used in the system to obtain resistance signal pathways [55]. The current work exhibited that *GmNAC12* expression was considerably upregulated by the induction of the plant hormones such as SA and MeJA (Figure 6A,B). As a result of binding protein interactions, during which *GmNAC12* interacts with proteins involved in responses to bacteria, fungi, and viruses, we speculate *GmNAC12* could participate in regulating biotic stresses by using a soybean's plant hormone pathway. Additionally, *GmNAC12* interacts with proteins involved in plant secondary wall synthesis and photosynthetic systems, indicating that *GmNAC12* may also be a key regulator for improving plant growth and development [56]. We are certain that the *GmNAC12* gene in soybeans controls functions related to abiotic stresses such as drought and biotic stresses caused by pathogens such as bacteria, fungi, and viruses.

4. Materials and Methods

4.1. Plant Materials, Growth Conditions, and Stress Treatments

The soybean cultivar, Tianlong 1, was used as the plant material in various experiments. Soybean seeds were cultivated in the soil in a controlled growth chamber at a temperature of 25 °C under long day (16/8 h light/dark) conditions and a relative humidity of 70%. Additionally, various stress treatments were imposed on the plants when they reached the first trifoliate growth stage. We initiated the drought treatment of soybean seedlings by depriving the plants of water for 0, 1, 3, 5, and 7 days, and we harvested the roots, stems, and leaves at different time intervals. For various hormone treatments, soybean seedlings were transferred in 1/2 Hoagland nutrient solution with 100 mM MeJA (methyl jasmonate), 2 μM ETH (ethylene), 2 mM SA (salicylic acid), or 150 μM ABA (abscisic acid). The roots were harvested at 0, 1, 3, 6, 12, and 24 h time intervals and were immediately preserved in liquid nitrogen for further investigation.

4.2. Quantitative Real-Time PCR Assay

Total RNA was obtained from the samples using an RNAprep Pure Plant Kit (Tiangen, Beijing, China). The total RNA's purity and concentration were established using a Nanodrop UV spectrophotometer (Thermo Fisher Scientific, Waltham, USA) and an RNA Nanochip on an Agilent Bioanalyzer 2100 (Agilent Technologies, Palo Alto, CA, USA). The Prime Script^{TM} RT Reagent Kit (TaKaRa, Kyoto, Japan) was used for cDNA synthesis. A quantitative real-time PCR (qRT-PCR) assay was performed for each cDNA template following the standard protocol using AceQ qPCR SYBR Green Master Mix (Vazyme, Nanjing, China). In addition, *GmActin11* (*Glyma.18g290800*) was used as an internal control by following the Ct-method to normalise expression levels [57]. NCBI Primer-BLAST designed all the primers used, and all primers are listed in Table S1. Three biological replications were used for qRT-PCR assays, and three measurements were performed on each replicate.

4.3. Subcellular Localisation of GmNAC12 Protein

The nuclear localisation signal (NLS) of *GmNAC12* was predicted through the online software cNLS Mapper (http://nls-mapper.iab.keio.ac.jp/cgi-bin/NLS_Mapper_form.cgi) using its amino acid sequence, which accessed on 1 June 2019. To verify the localisation of the *GmNAC12* protein in plant cells, we constructed a fusion vector using the *pBIN-GFP4* vector. Following the manufacturer's protocol, we cloned the CDS of the *GmNAC12* gene into the pBIN-GFP4 vector, minus the terminator (Figure S1C), then injected the constructed and empty vectors into tobacco leaves. The fluorescence of GFP was detected at 488 nm and 405 nm by an upright confocal microscope (Carl Zeiss, Thornwood, NY, USA) at 48–72 h post-inoculation.

4.4. GmNAC12 Knockout in Soybean Plants Using the CRISPR/Cas9 System

We used the CRISPR/Cas9 system to knockout *GmNAC12*. The CRISPR-P web tool targeted one site and was selected in the second exon of *GmNAC12* (Figure 3A). For the targeted knockout of *GmNAC12*, a 20 bp exon sequence from the target site of the *GmNAC12* gene was duplicated into CRISPR/Cas9 vector *p0645*. The soybean cultivar Tianlong 1 was used to transform the CRISPR/Cas9 knockout vector using the protocol established by Yang et al. [39]. Using the modified CTAB method, we removed the genomic DNA from the young leaves of soybean plants from the T_0 generation. Subsequently, PCR amplified an area extending across the target site, and the PCR products were sequenced (GenScript, Nanjing, China). Table S1 lists the primers used in the study. For heterozygous mutations, overlapping peaks were found at the target site, while the wild-type and homozygous mutations did not record overlapping peaks at the target site. The homozygous mutations at the target site were discovered by using DNAMAN software to execute a sequence alignment with the wild-type sequence. For the detection of targeted mutations, the same protocol was used at T_1 and T_2 generations.

4.5. GmNAC12 Overexpression in Soybean Plants

The full-length CDS of the *GmNAC12* gene was cloned into the *pTF101.1* vector under the control of the *CaMV 35S* promoter. After the vector sequence was confirmed by sequencing, the recombinant *pTF101.1–GmNAC12* plasmid vector was transformed into soybean cultivar Tianlong 1 using *Agrobacterium tumefaciens* strain EHA101. The selection marker gene (*bar*) and a *35S* promoter were amplified by PCR to verify the positive transgenic plants. The homozygous transgenic lines' phenotypic evaluation was conducted at the T_3 generation.

4.6. Yeast Two-Hybrid Assay

The whole coding sequence for the *GmNAC12* was cloned into vector *pGBKT7* to produce the recombinant *pGBKT7-GmNAC12* plasmid vector as bait while using the soybean yeast library plasmids as prey. Following the manufacturer's protocol (Takara, Kyoto, Japan), the *pGBKT7-GmNAC12* construct and library plasmids were co-transformed into the Y2H Gold yeast strain. Subsequently, these yeast cells were streaked onto media containing SD/-Trp/-Leu plates, SD/-Trp/-Leu/-Ade/-His+AbA+10 mM 3-AT plates, and SD/-Trp/-Leu/-Ade/-His+AbA+10 mM 3-AT + X-α-Gal plates.

4.7. GO Annotation and KEGG Enrichment Analysis

The GO annotation comes from the GO (Gene Ontolog) database. The software Goatools (https://github.com/tanghaibao/GOatools), which accessed on 21 September 2021, was used for enrichment analysis, and the method was Fisher's exact test. In controlling the calculated false-positive rate, the *p*-value was corrected using four multiple-test approaches: Bonferroni, Holm, Sidak, and the false discovery rate. Usually, when the *p*-value is lower than 0.05, the GO function is considered significant enrichment. The KEGG (Kyoto Encyclopaedia of Genes and Genomes) database assists with the systematic examination of gene roles, contact genomics, and functional information. Using the KEGG database, genes can be categorised by their pathways or by the functions they are involved in. The analysis of the KEGG uses KOBAS (http://kobas.cbi.pku.edu.cn/home.do), which accessed on 21 September 2021, to determine interacting proteins pathway enrichment analysis. To control the false positive rate calculation, the Benjamini–Hochberg (BH) method was used for many tests. The BH method sets the *p*-value at 0.05 as the threshold. When the KEGG pathway meets the set conditions, this is defined as significant enrichment.

The primers and sequences used in this study were from Genscript (Nanjing, China).

5. Conclusions

This study explored the potential functions of *GmNAC12*, a member of the soybean NAC family, in response to abiotic stress. The present results revealed that *GmNAC12* overexpressed lines demonstrated improved tolerance to drought compared to wild-type plants, whereas *GmNAC12* knockout lines were sensitive under drought conditions. Furthermore, the findings demonstrated that *GmNAC12* positively regulates drought stress in soybeans. In addition, the 185 candidate interaction proteins of *GmNAC12* identified by yeast two-hybrid assay were performed on GO analysis and KEGG enrichment, showing that some interaction proteins of *GmNAC12* are related to peroxidase activity. Hence, we inferred that *GmNAC12*, as a key gene, could positively regulate soybean tolerance to drought stress.

Supplementary Materials: The following supporting information can be downloaded at https://www.mdpi.com/article/10.3390/ijms231912029/s1.

Author Contributions: C.Y., Y.H., J.Z. and T.Z. designed and conceived the research. C.Y. constructed the CRISPR/Cas9 vectors. C.Y., Y.H. and P.L. performed the experiments. C.Y. and Y.H. wrote the draft paper. A.A.-B., N.B., J.Z. and T.Z. revised the paper. All authors have read and agreed to the published version of the manuscript.

Funding: This work was supported by National Key R and D Program of China (2021YFD1201603), the National Natural Science Foundation of China (grant nos. 31571691 and 31871646), the Science and Technology Innovation Project for Hebei Province Modern Seed Industry (21326313D-1), the China Agriculture Research System of MOF and MARA (CARS-04), the Fundamental Research Funds for the Central Universities (KYZZ2022003), the Jiangsu Collaborative Innovation Centre for Modern Crop Production (JCIC-MCP), and the Program and Cyrus Tang Innovation Center for Seed Industry (CTIC-SI201902).

Conflicts of Interest: The authors declare no conflict of interest. The funders had no role in the design of the study; in the collection, analyses, or interpretation of data; in the writing of the manuscript; or in the decision to publish the results.

References

1. Shao, H.; Wang, H.; Tang, X. NAC transcription factors in plant multiple abiotic stress responses: Progress and prospects. *Front. Plant Sci.* **2015**, *6*, 902. [CrossRef] [PubMed]
2. Hasanuzzaman, M.; Bhuyan, M.H.M.B.; Zulfiqar, F.; Raza, A.; Mohsin, M.S.; Mahmud, A.J.; Fujita, M.; Fotopoulos, V. Reactive oxygen species and antioxidant defense in plants under abiotic stress: Revisiting the crucial role of a universal defense regulator. *Antioxidants* **2020**, *9*, 681. [CrossRef] [PubMed]
3. Bhat, M.A.; Mir, R.A.; Kumar, V.; Shah, A.A.; Zargar, M.A.; Rahman, S.; Jan, T.A. Mechanistic insights of CRISPR/Cas-mediated genome editing towards enhancing abiotic stress tolerance in plants. *Physiol. Plant* **2021**, *172*, 1255–1268. [CrossRef] [PubMed]
4. Wang, J.; Chen, P.; Wang, D.; Shannon, G.; Zeng, A.; Orazaly, M.; Wu, C. Identification and mapping of stable QTL for protein content in soybean seeds. *Mol. Breed.* **2015**, *35*, 92. [CrossRef]
5. Karikari, B.; Li, S.; Bhat, J.A.; Cao, Y.; Kong, J.; Yang, J.; Gai, J.; Zhao, T. Genome-wide detection of major and epistatic effect QTLs for seed protein and oil content in soybean under multiple environments using high-density bin map. *Int. J. Mol. Sci.* **2019**, *20*, 979. [CrossRef]
6. Zhang, W.; Liao, X.; Cui, Y.; Ma, W.; Zhang, X.; Du, H.; Ma, Y.; Ning, L.; Wang, H.; Huang, F.; et al. A cation diffusion facilitator, GmCDF1, negatively regulates salt tolerance in soybean. *PLoS Genet.* **2019**, *15*, e1007798. [CrossRef] [PubMed]
7. Chimungu, J.G.; Brown, K.M.; Lynch, J.P. Reduced root cortical cell file number improves drought tolerance in maize. *Plant Physiol.* **2014**, *166*, 1943–1955. [CrossRef] [PubMed]
8. Per, T.S.; Khan, N.A.; Reddy, P.S.; Masood, A.; Hasanuzzaman, M.; Khan, M.I.R.; Anjum, N.A. Approaches in modulating proline metabolism in plants for salt and drought stress tolerance: Phytohormones, mineral nutrients and transgenics. *Plant Physiol. Biochem.* **2017**, *115*, 126–140. [CrossRef] [PubMed]
9. Huang, Y.; Xuan, H.; Yang, C.; Guo, N.; Wang, H.; Zhao, J.; Xing, H. GmHsp90A2 is involved in soybean heat stress as a positive regulator. *Plant Sci.* **2019**, *285*, 26–33. [CrossRef]
10. He, M.; He, C.Q.; Ding, N.Z. Abiotic stresses: General defenses of land plants and chances for engineering multistress tolerance. *Front. Plant Sci.* **2018**, *9*, 1771. [CrossRef] [PubMed]
11. Zhu, J. Abiotic stress signaling and responses in plants. *Cell* **2016**, *167*, 313–324. [CrossRef] [PubMed]
12. Morran, S.; Eini, O.; Pyvovarenko, T.; Parent, B.; Singh, R.; Ismagul, A.; Eliby, S.; Langride, P.; Lopato, S. Improvement of stress tolerance of wheat and barley by modulation of expression of DREB/CBF factors. *Plant Biotechnol. J.* **2011**, *9*, 230–249. [CrossRef] [PubMed]
13. Zhong, H.; Guo, Q.; Chen, L.; Ren, F.; Wang, Q.; Zheng, Y.; Li, X. Two Brassica napus genes encoding NAC transcription factors are involved in response to high-salinity stress. *Plant Cell Rep.* **2012**, *31*, 1991–2003. [CrossRef]
14. Zhang, H.; Ma, F.; Wang, X.; Liu, S.; Saeed, U.H.; Hou, X.; Zhang, Y.; Luo, D.; Meng, Y.; Zhang, W.; et al. Molecular and functional characterization of CaNAC035, an NAC transcription factor from pepper (Capsicum annuum L.). *Front. Plant Sci.* **2020**, *11*, 14. [CrossRef] [PubMed]
15. Sharma, P.; Jha, A.B.; Dubey, R.S.; Pessarakli, M. Reactive oxygen species, oxidative damage, and antioxidative defense mechanism in plants under stressful conditions. *J. Bot.* **2012**, *2012*, 217037. [CrossRef]
16. Rauf, M.; Arif, M.; Fisahn, J.; Xue, G.P.; Balazadeh, S.; Mueller-Roeber, B. NAC transcription factor speedy hyponastic growth regulates flooding-induced leaf movement in Arabidopsis. *Plant Cell* **2013**, *25*, 4941–4955. [CrossRef] [PubMed]
17. Jensen, M.K.; Rung, J.H.; Gregersen, P.L.; Gjetting, T.; Fuglsang, A.T.; Hansen, M.; Joehnk, N.; Lynkjaer, M.F.; Collinge, D.B. The Hv NAC6 transcription factor: A positive regulator of penetration resistance in barley and Arabidopsis. *Plant Mol. Biol.* **2007**, *65*, 37–150. [CrossRef] [PubMed]
18. Ernst, H.A.; Olsen, N.A.; Skriver, K.; Larsen, S.; Leggio, L.L. Structure of the conserved domain of ANAC, a member of the NAC family of transcription factors. *EMBO Rep.* **2004**, *5*, 297–303. [CrossRef]
19. Ooka, H.; Satoh, K.; Doi, K.; Nagata, T.; Otomo, Y.; Murakami, K.; Osato, N.; Kawai, J.; Carninci, P.; Hayashizaki, Y.; et al. Comprehensive analysis of NAC family genes in Oryza sativa and Arabidopsis thaliana. *DNA Res.* **2003**, *10*, 239–247. [CrossRef] [PubMed]
20. Puranik, S.; Sahu, P.P.; Srivastava, P.S.; Prasad, M. NAC proteins: Regulation and role in stress tolerance. *Trends Plant Sci.* **2012**, *17*, 369–381. [CrossRef]

21. Na, C.; Shuanghua, W.; Jinglong, F.; Bihao, C.; Jianjun, L.; Changming, C.; Jin, J. Overexpression of the eggplant (Solanum melongena) NAC family transcription factor S mNAC suppresses resistance to bacterial wilt. *Sci. Rep.* **2016**, *6*, 31568. [CrossRef] [PubMed]
22. Shen, H.; Yin, Y.; Chen, F.; Xu, Y.; Dixon, R.A. A bioinformatic analysis of NAC genes for plant cell wall development in relation to lignocellulosic bioenergy production. *Bioenergy Res.* **2009**, *2*, 217–232. [CrossRef]
23. Aida, M.; Ishida, T.; Fukaki, H.; Fujisawa, H.; Tasaka, M. Genes involved in organ separation in Arabidopsis: An analysis of the cup-shaped cotyledon mutant. *Plant Cell* **1997**, *9*, 841–857. [CrossRef] [PubMed]
24. Nguyen, H.K.; Mostofa, G.M.; Li, W.; Ha, V.C.; Watanabe, Y.; Le, D.T.; Thao, N.P.; Tran, L.S.P. The soybean transcription factor GmNAC085 enhances drought tolerance in Arabidopsis. *Environ. Exp. Bot.* **2018**, *151*, 12–20. [CrossRef]
25. Le, D.T.; Nishiyama, R.; Watanabe, Y.; Mochida, K.; Yamaguchi-Shinozaki, K.; Shinozaki, K.; Tran, L.S.P. Genome-wide survey and expression analysis of the plant-specific NAC transcription factor family in soybean during development and dehydration stress. *DNA Res.* **2011**, *18*, 263–276. [CrossRef] [PubMed]
26. Ju, Y.; Yue, X.; Min, Z.; Wang, X.; Fang, Y.; Zhang, J. VvNAC17, a novel stress-responsive grapevine (Vitis vinifera L.) NAC transcription factor, increases sensitivity to abscisic acid and enhances salinity, freezing, and drought tolerance in transgenic Arabidopsis. *Plant Physiol. Biochem.* **2020**, *146*, 98–111. [CrossRef] [PubMed]
27. Negi, S.; Tak, H.; Ganapathi, T.R. A banana NAC transcription factor (MusaSNAC1) impart drought tolerance by modulating stomatal closure and H2O2 content. *Plant Mol. Biol.* **2018**, *96*, 57–471. [CrossRef] [PubMed]
28. Hao, Y.; Wei, W.; Song, Q.; Chen, H.; Zhang, Y.; Wang, F.; Zou, H.; Lei, G.; Tian, A.; Zhang, W.; et al. Soybean NAC transcription factors promote abiotic stress tolerance and lateral root formation in transgenic plants. *Plant J.* **2011**, *68*, 302–313. [CrossRef] [PubMed]
29. Kou, X.; Liu, C.; Han, L.; Wang, S.; Xue, Z. NAC transcription factors play an important role in ethylene biosynthesis, reception and signaling of tomato fruit ripening. *Mol. Genet. Genom.* **2016**, *291*, 1205–1217. [CrossRef] [PubMed]
30. Pimenta, M.R.; Silva, P.A.; Mendes, G.C.; Alves, J.R.; Caetano, H.D.N.; Machado, J.P.B.; Brustolini, O.J.B.; Carpinetti, P.A.; Melo, B.P.; Silva, J.C.F.; et al. The stress-induced soybean NAC transcription factor GmNAC81 plays a positive role in developmentally programmed leaf senescence. *Plant Cell Physiol.* **2016**, *57*, 1098–1114. [CrossRef] [PubMed]
31. Wu, Y.; Deng, Z.; Lai, J.; Zhang, Y.; Yang, C.; Yin, B.; Zhao, Q.; Zhang, L.; Li, Y.; Yang, C.; et al. Dual function of Arabidopsis ATAF1 in abiotic and biotic stress responses. *Cell Res.* **2009**, *19*, 1279–1290. [CrossRef] [PubMed]
32. Tran, L.S.P.; Nakashima, K.; Sakuma, Y.; Simpson, S.D.; Fujita, Y.; Maruyama, K.; Fujita, M.; Seki, M.; Shinozaki, K.; Yamaguchi-Shinozaki, K. Isolation and functional analysis of Arabidopsis stress-inducible NAC transcription factors that bind to a drought-responsive cis-element in the early responsive to dehydration stress 1 promoter. *Plant Cell* **2004**, *16*, 2481–2498. [CrossRef] [PubMed]
33. Nakashima, K.; Tran, L.S.P.; Nguyen, V.D.; Fujita, M.; Maruyama, K.; Todaka, D.; Ito, Y.; Hayashi, N.; Shinozaki, K.; Yamaguchi-Shinozaki, K. Functional analysis of a NAC-type transcription factor OsNAC6 involved in abiotic and biotic stress-responsive gene expression in rice. *Plant J.* **2007**, *51*, 617–630. [CrossRef] [PubMed]
34. Hu, H.; Dai, M.; Yao, J.; Xiao, B.; Li, X.; Zhang, Q.; Xiong, L. Overexpressing a NAM, ATAF, and CUC (NAC) transcription factor enhances drought resistance and salt tolerance in rice. *Proc. Natl. Acad. Sci. USA* **2006**, *103*, 2987–12992. [CrossRef] [PubMed]
35. Jeong, J.S.; Kim, Y.S.; Baek, K.H.; Jung, H.; Ha, S.H.; Choi, Y.D.; Kim, M.; Reuzeau, C.; Kim, J.K. Root-specific expression of OsNAC10 improves drought tolerance and grain yield in rice under field drought conditions. *Plant Physiol.* **2010**, *153*, 185–197. [CrossRef] [PubMed]
36. Takasaki, H.; Maruyama, K.; Kidokoro, S.; Ito, Y.; Fujita, Y.; Shinozaki, K.; Yamaguchi-Shinozaki, K.; Nakashima, K. The abiotic stress-responsive NAC-type transcription factor OsNAC5 regulates stress-inducible genes and stress tolerance in rice. *Mol. Genet. Genomics* **2010**, *284*, 173–183. [CrossRef]
37. So, H.A.; Lee, J.H. NAC transcription factors from soybean (Glycine max L.) differentially regulated by abiotic stress. *J. Plant Biol.* **2019**, *62*, 147–160. [CrossRef]
38. Tak, H.; Negi, S.; Ganapathi, T.R. Banana NAC transcription factor MusaNAC042 is positively associated with drought and salinity tolerance. *Protoplasma* **2017**, *254*, 803–816. [CrossRef]
39. Yang, C.; Huang, Y.; Lv, W.; Zhang, Y.; Bhat, J.A.; Kong, J.; Xing, H.; Zhao, J.; Zhao, T. GmNAC8 acts as a positive regulator in soybean drought stress. *Plant Sci.* **2020**, *293*, 110442. [CrossRef]
40. Yang, X.; Kim, M.Y.; Ha, J.; Lee, S.H. Overexpression of the soybean NAC gene GmNAC109 increases lateral root formation and abiotic stress tolerance in transgenic Arabidopsis plants. *Front. Plant Sci.* **2019**, *10*, 1036. [CrossRef]
41. Melo, B.P.; Fraga, O.T.; Silva, J.C.F.; Ferreira, D.O.; Brustolini, O.J.; Carpinetti, P.A.; Machado, J.P.B.; Reis, P.A.B.; Fontes, E.P.B. Revisiting the soybean GmNAC superfamily. *Front. Plant Sci.* **2018**, *9*, 1864. [CrossRef] [PubMed]
42. Cai, Y.; Chen, L.; Liu, X.; Sun, S.; Wu, C.; Jiang, B.; Han, T.; Hou, W. CRISPR/Cas9-mediated genome editing in soybean hairy roots. *PLoS ONE* **2015**, *10*, e0136064. [CrossRef] [PubMed]
43. Bao, A.; Chen, H.; Chen, L.; Chen, S.; Hao, Q.; Guo, W.; Qiu, D.; Shan, Z.; Yang, Z.; Yuan, S.; et al. CRISPR/Cas9-mediated targeted mutagenesis of GmSPL9 genes alters plant architecture in soybean. *BMC Plant Biol.* **2019**, *19*, 131. [CrossRef] [PubMed]
44. Cai, Y.; Chen, L.; Liu, X.; Guo, C.; Sun, S.; Wu, C.; Jiang, B.; Han, T.; Hou, W. CRISPR/Cas9-mediated targeted mutagenesis of GmFT2a delays flowering time in soya bean. *Plant Biotechnol. J.* **2018**, *16*, 176–185. [CrossRef] [PubMed]

45. Hussain, R.M.; Ali, M.; Feng, X.; Li, X. The essence of NAC gene family to the cultivation of drought-resistant soybean (Glycine max L. Merr.) cultivars. *BMC Plant Biol.* **2017**, *17*, 55. [CrossRef]
46. Rao, V.S.; Srinivas, K.; Sujini, G.N.; Kumar, G.N.S. Protein-protein interaction detection: Methods and analysis. *J. Proteom.* **2014**, *2014*, 147648. [CrossRef] [PubMed]
47. Wani, S.H.; Kumar, V.; Shriram, V.; Sah, S.K. Phytohormones and their metabolic engineering for abiotic stress tolerance in crop plants. *Crop J* **2016**, *4*, 162–176. [CrossRef]
48. Verma, V.; Ravindran, P.; Kumar, P.P. Plant hormone-mediated regulation of stress responses. *BMC Plant Biol.* **2016**, *16*, 86. [CrossRef] [PubMed]
49. Qian, D.; Zhang, Z.; He, J.; Zhang, P.; Ou, X.; Li, T.; Niu, L.; Nan, Q.; Niu, Y.; He, W.; et al. Arabidopsis ADF5 promotes stomatal closure by regulating actin cytoskeleton remodeling in response to ABA and drought stress. *J. Exp. Bot.* **2019**, *70*, 435–446. [CrossRef]
50. Jiang, W.; Zhou, H.; Bi, H.; Fromm, M.; Yang, B.; Weeks, D.P. Demonstration of CRISPR/Cas9/sgRNA-mediated targeted gene modification in Arabidopsis, tobacco, sorghum and rice. *Nucleic Acids Res.* **2013**, *41*, e188. [CrossRef]
51. Mushtaq, M.; Bhat, J.A.; Mir, Z.A.; Sakina, A.; Ali, S.; Singh, A.K.; Tyagi, A.; Salgotra, R.K.; Dar, A.A.; Bhat, R. CRISPR/Cas approach: A new way of looking at plant-abiotic interactions. *J. Plant Physiol.* **2018**, *224*, 156–162. [CrossRef] [PubMed]
52. Zhang, S.; Feng, M.; Chen, W.; Zhou, X.; Lu, J.; Wang, Y.; Li, Y.; Jiang, C.; Gan, S.; Ma, N.; et al. In rose, transcription factor PTM balances growth and drought survival via PIP2;1 aquaporin. *Nat. Plants* **2019**, *5*, 290–299. [CrossRef] [PubMed]
53. Baxter, A.; Mittler, R.; Suzuki, N. ROS as key players in plant stress signalling. *J. Exp. Bot.* **2014**, *65*, 1229–1240. [CrossRef] [PubMed]
54. Li, Y.; Chen, Q.; Nan, H.; Li, X.; Lu, S.; Zhao, X.; Liu, B.; Guo, C.; Kong, F.; Cao, D. Overexpression of GmFDL19 enhances tolerance to drought and salt stresses in soybean. *PLoS ONE* **2017**, *12*, e0179554. [CrossRef] [PubMed]
55. Clarke, J.D.; Volko, S.M.; Ledford, H.; Ausubel, F.M.; Dong, X. Roles of salicylic acid, jasmonic acid, and ethylene in cpr-induced resistance in Arabidopsis. *Plant Cell* **2000**, *12*, 2175–2190. [CrossRef] [PubMed]
56. Zhang, J.; Huang, G.; Zou, D.; Yan, J.; Li, Y.; Hu, S.; Li, X. The cotton (Gossypium hirsutum) NAC transcription factor (FSN1) as a positive regulator participates in controlling secondary cell wall biosynthesis and modification of fibers. *New Phytol.* **2018**, *217*, 625–640. [CrossRef] [PubMed]
57. Livak, J.K.; Schmittgen, D.T. Analysis of Relative Gene Expression Data Using Real-Time Quantitative PCR and the $2^{-\Delta\Delta CT}$ method. *Methods* **2001**, *25*, 402–408. [CrossRef]

International Journal of
Molecular Sciences

MDPI

Article

Systematic Characterization of the *OSCA* Family Members in Soybean and Validation of Their Functions in Osmotic Stress

Congge Liu [1,2,†], Hong Wang [3,†], Yu Zhang [2], Haijing Cheng [2], Zhangli Hu [2], Zhen-Ming Pei [4,*] and Qing Li [2,3,*]

1 Longping Branch, College of Biology, Hunan University, Changsha 410125, China
2 College of Life Sciences and Oceanography, Shenzhen University, Shenzhen 518060, China
3 State Key Laboratory of Rice Biology, China National Rice Research Institute, Hangzhou 311401, China
4 Department of Biology, Duke University, Durham, NC 27708, USA
* Correspondence: zpei@duke.edu (Z.-M.P.); liqing1986102@163.com or liqing02@caas.cn (Q.L.)
† These authors contributed equally to this work.

Abstract: Since we discovered OSCA1, a hyperosmolarity-gated calcium-permeable channel that acted as an osmosensor in *Arabidopsis*, the *OSCA* family has been identified genome-wide in several crops, but only a few *OSCA* members' functions have been experimentally demonstrated. Osmotic stress seriously restricts the yield and quality of soybean. Therefore, it is essential to decipher the molecular mechanism of how soybean responds to osmotic stress. Here, we first systematically studied and experimentally demonstrated the role of *OSCA* family members in the osmotic sensing of soybean. Phylogenetic relationships, gene structures, protein domains and structures analysis revealed that 20 *GmOSCA* members were divided into four clades, of which members in the same cluster may have more similar functions. In addition, *GmOSCA* members in clusters III and IV may be functionally redundant and diverged from those in clusters I and II. Based on the spatiotemporal expression patterns, *GmOSCA1.6*, *GmOSCA2.1*, *GmOSCA2.6*, and *GmOSCA4.1* were extremely low expressed or possible pseudogenes. The remaining 16 *GmOSCA* genes were heterologously overexpressed in an *Arabidopsis osca1* mutant, to explore their functions. Subcellular localization showed that most GmOSCA members could localize to the plasma membrane (PM). Among 16 *GmOSCA* genes, only overexpressing *GmOSCA1.1*, *GmOSCA1.2*, *GmOSCA1.3*, *GmOSCA1.4*, and *GmOSCA1.5* in cluster I could fully complement the reduced hyperosmolality-induced $[Ca^{2+}]_i$ increase (OICI) in *osca1*. The expression profiles of *GmOSCA* genes against osmotic stress demonstrated that most *GmOSCA* genes, especially *GmOSCA1.1*, *GmOSCA1.2*, *GmOSCA1.3*, *GmOSCA1.4*, *GmOSCA1.5*, *GmOSCA3.1*, and *GmOSCA3.2*, strongly responded to osmotic stress. Moreover, overexpression of *GmOSCA1.1*, *GmOSCA1.2*, *GmOSCA1.3*, *GmOSCA1.4*, *GmOSCA1.5*, *GmOSCA3.1*, and *GmOSCA3.2* rescued the drought-hypersensitive phenotype of *osca1*. Our findings provide important clues for further studies of *GmOSCA*-mediated calcium signaling in the osmotic sensing of soybean and contribute to improving soybean drought tolerance through genetic engineering and molecular breeding.

Keywords: osmotic stress; osmosensor; calcium; *OSCA*; soybean

1. Introduction

As sessile organisms, plants need to adapt to the changing environments around them, especially biotic and abiotic stresses, to sustain their growth and development. Common abiotic stresses include drought, salinity, waterlogging, high temperature, cold, acid rain, and heavy metal pollution. Among these, drought and salinity, which both induce osmotic stress in plant cells, are the two most prevalent environmental stresses that affect the geographical distribution of plants in nature and restrict crop growth and yield in agriculture [1]. It is estimated that global maize and wheat yields are reduced by about 40% and 21% due to drought stress, respectively [2]. In addition, more than one-third of the irrigated lands in the world are affected by salinization due to seawater intrusion and poor-quality agricultural management practices [3], which seriously threatens crop cultivation and food security. To reduce the damages of osmotic stress such as drought and salinity, plants have evolved different resistance mechanisms at the morphological, physiological, cellular, biochemical, and molecular levels [4].

Although the signal transduction pathway of plant response to osmotic stress is very complex, it can be divided into three stages in general, including signal perception, signal transduction, and adaptive responses to stress signals [5–7]. Firstly, plants perceive osmotic stress signals through their specific receptors. After the initial perception, the second messengers such as calcium (Ca^{2+}), reactive oxygen species (ROS), and inositol phosphates are generated during the early signaling responses, which further modulate the intracellular free Ca^{2+} concentrations ($[Ca^{2+}]_i$) and trigger $[Ca^{2+}]_I$ fluctuation. The $[Ca^{2+}]_i$ fluctuation is sensed by Ca^{2+} sensors, which then interact with their corresponding partners to activate a phosphorylation cascade. The signal cascade leads to altered expression of major stress-responsive genes and induced abscisic acid (ABA), an important plant stress signaling hormone, which accumulates during downstream signaling. Finally, the products of these stress genes result in plant adaptation to unfavorable conditions.

To date, the signal transduction process of osmotic stress in plants has been thoroughly studied, and the transcriptional regulation of osmotic-stress-responsive gene expression is mainly governed by ABA-dependent and ABA-independent pathways [8]. The cis-element of ABA-responsive element (ABRE), and a class of transcription factors, the ABRE-binding proteins/ABRE-binding factors (AREBs/ABFs), play critical roles in ABA-dependent gene expression. Under osmotic stress, AREBs/ABFs are phosphorylated and activated by SNF1-related kinase 2s (SnRK2s) in an ABA-dependent manner, then bind to ABRE in the promoter regions of target genes, and, thereby, induce the stress-responsive genes' expression [9–11]. In contrast, the cis-element of dehydration-responsive element/C-repeat (DRE/CRT) and DRE/CRT-binding protein 2 (DREB2) transcription factors have critical functions in ABA-independent gene expression. During osmotic stress, *DREB2* transcripts are highly induced, and DREB2 proteins are also stabilized, then bind to DRE, and, consequently, activate their target genes' expression [8,12,13]. In addition, there is a crosstalk between ABA-independent and ABA-dependent pathways [8]. Besides these master regulators, WRKY, MYB, MYC, and NF-Y transcription factors are also involved in osmotic response and tolerance [14].

In contrast to processes of signal transduction and adaptive responses to stress signals, the molecular mechanisms underlying how plants sense osmotic stress are not well understood. In *Arabidopsis*, OSCA1, a hyperosmolarity-gated calcium-permeable channel that acted as an osmosensor, was identified using a calcium-imaging-based forward-genetic screen in our previous study [15]. It is the first and only known class of plant osmosensors. OSCA1 was located in the plasma membrane (PM), and its dysfunctional *osca1* mutant exhibited reduced hyperosmolality-induced $[Ca^{2+}]_i$, increased (OICI), decreased root growth, and showed defective leaf transpiration under hyperosmotic stress. Phylogenetic analyses revealed that land plants had four ancient clades of *OSCA1* homologs and that *Arabidopsis* contains 15 *OSCAs* [15]. For this reason, *OSCA1* was named *AtOSCA1.1*, and the other members were called *AtOSCA1.2-1.8*, *AtOSCA2.1-2.5*, *AtOSCA3.1*, and *AtOSCA4.1*, according to their relative distance from *AtOSCA1.1*. *AtOSCA1.2* (also called *AtCSC1*), which shares the highest homology with *AtOSCA1.1*, also encodes a hyperosmolarity-gated calcium-permeable channel and responds to hyperosmotic stress [16]. In addition, AtOSCA1.3, located in the PM, is a BIK1-activated calcium-permeable channel and is specifically required for plant stomatal immunity [17]. However, the molecular functions of other *Arabidopsis OSCA* members are currently unknown.

As the first and only kind of osmosensors to be reported in plants, the *OSCA* gene family has been genome-wide identified and analyzed in several crops, such as rice [18], maize [19,20], and wheat [21]. However, only a few *OSCA* family members' functions have been experimentally demonstrated. In addition, there is no systematic study on the roles of *OSCA* family members in soybean. As an important crop to provide seed protein and edible oil, soybean is vulnerable to osmotic stress, leading to a severe decline in soybean yield and quality. Therefore, studying the molecular mechanism of osmotic stress sensing and response in soybean is of great significance. To explore whether the soybean *GmOSCA* family members are involved in osmotic stress perception or response, we performed a genome-wide identification and comprehensive characterization of *GmOSCA* members through phylogenetic relationships, gene structures, protein domains and structures, and spatiotemporal and stressed expression profiles in this study. More importantly, the functions of *GmOSCA* members were investigated by the heterologous expression of them in an *osca1* mutant. As a result, a total of 20 *GmOSCA* members were identified in the soybean genome, and they were classified into four classes based on phylogenetic analysis. Among them, only *GmOSCA1.1*, *GmOSCA1.2*, *GmOSCA1.3*, *GmOSCA1.4*, and *GmOSCA1.5* could functionally complement the reduced OICI phenotype in *osca1* under hyperosmotic stress. In addition, *GmOSCA1.1*, *GmOSCA1.2*, *GmOSCA1.3*, *GmOSCA1.4*, *GmOSCA1.5*, *GmOSCA3.1*, and *GmOSCA3.2* were significantly induced by osmotic stress, and their ectopic expression in *osca1* conferred enhanced drought tolerance on transgenic plants. Our findings will improve the understanding of the genetic and molecular basis for osmotic stress perception in soybean and provide valuable gene resources for improving soybean drought tolerance through genetic engineering and molecular breeding.

2. Results

2.1. Phylogenetic Tree, Gene Structures, Protein Domains and Structures of GmOSCA Genes

To identify all of the *OSCA* family members in soybean, the 15 AtOSCA protein sequences were used as baits to search the soybean genome (Wm82.a2.v1) in Phytozome (https://phytozome.jgi.doe.gov/pz/portal.html, accessed on 15 August 2022). Ultimately, 20 *GmOSCA* genes were screened after removing candidates with low protein homology, candidates without typical RSN1_7TM domain of OSCA family proteins, and candidates not localized on chromosomes. The 20 *GmOSCA* members were mapped onto the 15 chromosomes in the soybean genome (Supplementary Materials Table S1). Their proteins ranged from 500 to 803 amino acids in length, varied between 57,229.64 to 92,256.79 Da in relative molecular weight, and the predicted isoelectric points ranged from 6.35 to 9.38 (Supplementary Materials Table S1). These proteins varied widely, with 13.3% to 98.5% pairwise sequence identity (Supplementary Materials Figure S1), suggesting that they may have divergent functions.

To investigate the evolutionary relationships of *GmOSCA* members, a neighbor-joining phylogenetic tree using the OSCA protein sequences from *Arabidopsis* and soybean was constructed (Figure 1A). Similar to the classification of AtOSCA members, 20 GmOSCA proteins were divided into four clades, according to the bootstrap values and phylogenetic topology. Based on the similarity with the corresponding AtOSCA proteins, they were named *GmOSCA1.1* to *1.9* in cluster I, *GmOSCA2.1* to *2.7* in cluster II, *GmOSCA3.1* and *GmOSCA3.2* in cluster III, and *GmOSCA4.1* and *GmOSCA4.2* in cluster IV, respectively.

As a paleo-polyploid crop, soybean has undergone at least two whole-genome duplication (WGD) events, thereby generating a highly duplicated genome, with nearly 75% of the genes showing multi-copies [22]. The collinearity analysis revealed that all *GmOSCA* genes, except for *GmOSCA1.5* and *GmOSCA2.7*, had duplicated counterparts (Supplementary Materials Figure S2). The generation of duplicated genes could facilitate gene evolution through nonfunctionalization, neofunctionalization, and subfunctionalization.

The exon/intron structure divergence in duplicate genes played a crucial role during the evolution of some gene families [23]. Therefore, we compared the exon/intron structures of *OSCA* genes from *Arabidopsis* and soybean (Figure 1B). It showed that most *OSCA* genes contained multiple exons (>5), except for cluster IV. Moreover, the number of exons in the same cluster was almost identical. For instance, there were 9~11 exons in clusters I and II, whereas fewer exons were included in cluster III.

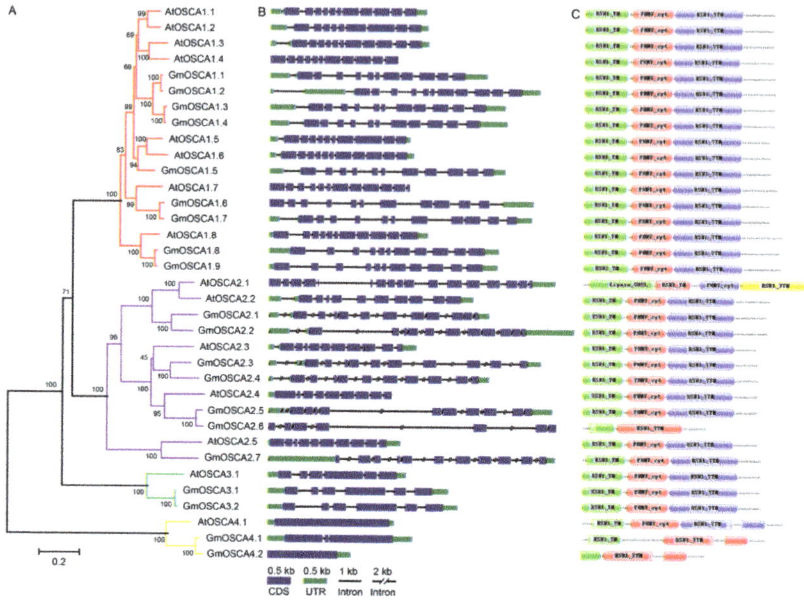

Figure 1. The phylogenetic relationships, gene structures, and conserved domains of *OSCA* genes from *Arabidopsis* and soybean. (**A**) A neighbor-joining tree was constructed using 35 OSCA protein sequences by MEGA 6.0 software with 1000 bootstrap replications. Clusters I, II, III, and IV were marked with red, blue, green, and yellow colors, respectively. (**B**) The exon-intron structures of *OSCA* genes. (**C**) The conserved domains in OSCA proteins.

As the primary executor of biological functions, proteins with similar motifs and structures in the same gene family are likely to have the same functions. Thus, the conserved domains and three-dimensional structures of OSCA proteins from *Arabidopsis* and soybean were predicted and compared. The results showed that these OSCA proteins not only contained the typical RSN1_7TM domain (calcium-dependent channel, 7TM region) at the carboxyl end but also had the RSN1_TM domain (late exocytosis, associated with Golgi transport) at the amino end and the PHM7_cyt domain (cytosolic domain of 10TM putative phosphate transporter) in the middle, except for GmOSCA2.6, GmOSCA4.1, and GmOSCA4.2 (Figure 1C). As expected, the three-dimensional structures of OSCAs proteins in clusters III and IV were significantly different from those in clusters I and II (Supplementary Materials Figure S3), implying that the OSCA members in clusters III and IV were likely to have functionally diverged from those in clusters I and II.

In sum, the comparative analysis of homologs between soybean and *Arabidopsis* contributed to exploring the *OSCA* functions in soybean. The *OSCA* members in the same cluster might have more similar functions. In addition, *GmOSCA* genes in clusters III and IV might be functionally redundant because they had more copies than those in *Arabidopsis*.

2.2. Spatiotemporal Expression Patterns of GmOSCA Genes

Gene expression profiles are helpful in predicting gene functions. The previously reported Illumina RNA-seq raw data [24] were used to mine the expression profiles of *GmOSCA* genes in 25 samples (Figure 2). It showed that the expression patterns of *GmOSCA* genes were very different, even those genes from the same cluster, indicating that *GmOSCA* genes might function in different tissues at different developmental periods. For instance, *GmOSCA3.1*, *GmOSCA3.2*, and *GmOSCA4.2* showed high gene expression levels in almost every sample, suggesting they might play important roles throughout soybean growth and development. In contrast, *GmOSCA1.6*, *GmOSCA2.6*, and *GmOSCA4.1* were not detected in all samples, indicating that they were potential pseudogenes. Additionally, *GmOSCA1.2*, *GmOSCA1.7*, *GmOSCA2.1*, *GmOSCA2.4*, and *GmOSCA2.7* exhibited tissue-specific expression patterns, which were only detected in some specific tissues, implicating that they might have served a unique function at a particular time and place. Moreover, *GmOSCA1.1*, *GmOSCA1.3*, *GmOSCA1.4*, *GmOSCA1.5*, *GmOSCA1.8*, *GmOSCA1.9*, *GmOSCA2.2*, *GmOSCA2.3*, and *GmOSCA2.5* were expressed in most tissues with differential expression patterns. Apart from the RNA-seq data by Shen et. al, the expression profiles of *GmOSCA* genes were also explored in nine tissues, using the online soybean gene expression database in Phytozome (https://phytozome.jgi.doe.gov/pz/portal.html, accessed on 15 August 2022) The results showed that the expression patterns of *GmOSCA* genes were similar in the two-transcriptome data, except for *GmOSCA1.2* and *GmOSCA4.1* (Figure 2 and Supplementary Materials Figure S4).

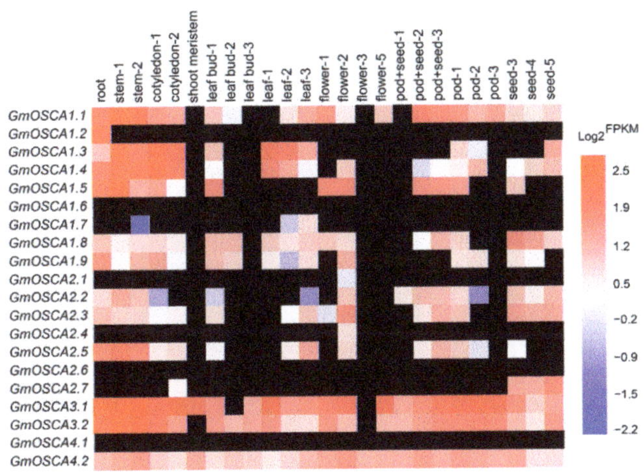

Figure 2. The spatiotemporal expression profiles of *GmOSCA* genes in soybean. The size number behind each sample indicated earlier to later developmental stages with the same sample. Gradient color blocks represented log2-transformed FPKM values. The black blocks indicated that no FPKM value was available.

2.3. *GmOSCA Proteins Are Mainly Located in Membrane Systems*

AtOSCA1.1 and AtOSCA1.2 have multiple transmembrane helices and are localized in the PM to form calcium-permeable channels [15,16]. Therefore, as Ca^{2+} channels, OSCA family members need to meet at least two traits: multiple transmembrane helices and localization on membrane structures. To determine whether GmOSCA proteins have these two characteristics, the transmembrane helices and subcellular localizations were predicted using MemPype (http://mu2py.biocomp.unibo.it/mempype/). It showed that all of the GmOSCA members have at least seven transmembrane helices and could be localized in the membrane system (Supplementary Materials Table S1), implicating their potential as Ca^{2+} channels. To further confirm the prediction accuracy of the subcellular localization, some GmOSCA members were selected to conduct subcellular localization experiments, using the protoplast transient transformation system. The results revealed that GmOSCA1.1 to GmOSCA1.4 and GmOSCA1.9 in cluster I, GmOSCA2.2 and GmOSCA2.3 in cluster II, and GmOSCA3.1 and GmOSCA3.2 in cluster III could localize to the PM, which was similar to the subcellular localization of AtOSCA1.1 (Figure 3). Notably, these OSCA proteins might be localized elsewhere besides PM, especially for GmOSCA2.3 (Figure 3).

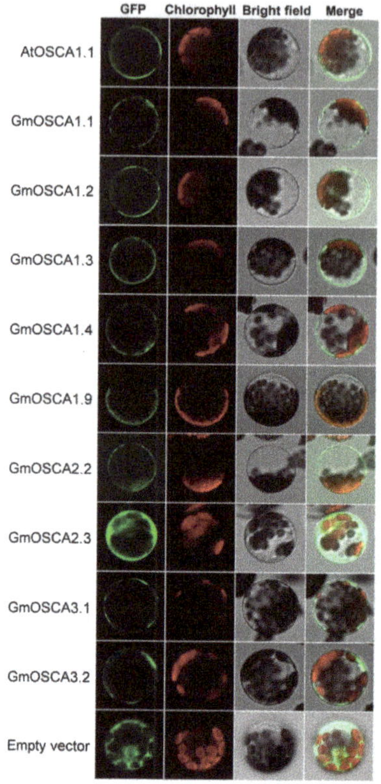

Figure 3. The subcellular localization of GmOSCA proteins.

2.4. *Overexpression of Some GmOSCA Members Rescues Decreased OICI in osca1 Mutant*

To further explore the biological functions of *GmOSCA* genes, except for *GmOSCA1.6*, *GmOSCA2.1*, *GmOSCA2.6*, and *GmOSCA4.1*, which were extremely low expressed or possible pseudogenes, the remaining 16 *GmOSCA* genes were cloned. Currently, no such aequorin-based calcium imaging detection system is established in soybean as in *Arabidopsis*. In addition, the efficiency of soybean genetic transformation is lower than that of *Arabidopsis*, so it is hard to transform all *GmOSCA* genes into soybean. Therefore, we planned to investigate the functions of *GmOSCA* genes by overexpression of *GmOSCAs* in the *osca1* mutant. In addition, the CDS of *AtOSCA1.1* and the empty vector pfgc5941 were transformed into the *osca1* mutant, to generate transgenic *Arabidopsis AtOSCA1.1* and *pfgc*, as the positive and negative controls, respectively. By observing the calcium imaging phenotype under 600 mM sorbitol treatment, a lower OICI was observed in *osca1* than in the wildtype (WT) (Figure 4A). Meanwhile, we found that the decreased OICI in *osca1* was fully complemented in transgenic *Arabidopsis* lines carrying *GmOSCA1.1*, *GmOSCA1.2*, *GmOSCA1.3*, *GmOSCA1.4*, and *GmOSCA1.5* (Figure 4A), which are the closest homologs with *AtOSCA1.1* in cluster I. However, the transgenic *Arabidopsis* lines with other *GmOSCA* genes in clusters I, II, III, and IV did not rescue the reduced OICI in *osca1*. To exclude the influence of the difference in total aequorin, the total amount of aequorin in each sample was measured using a discharge solution, and used as an internal reference for quantification [15]. It further supported the above conclusion (Figure 4B). Taken together, *GmOSCA1.1*, *GmOSCA1.2*, *GmOSCA1.3*, *GmOSCA1.4*, and *GmOSCA1.5* might function in the osmotic stress sensing of soybean, similar to *AtOSCA1.1* in *Arabidopsis*.

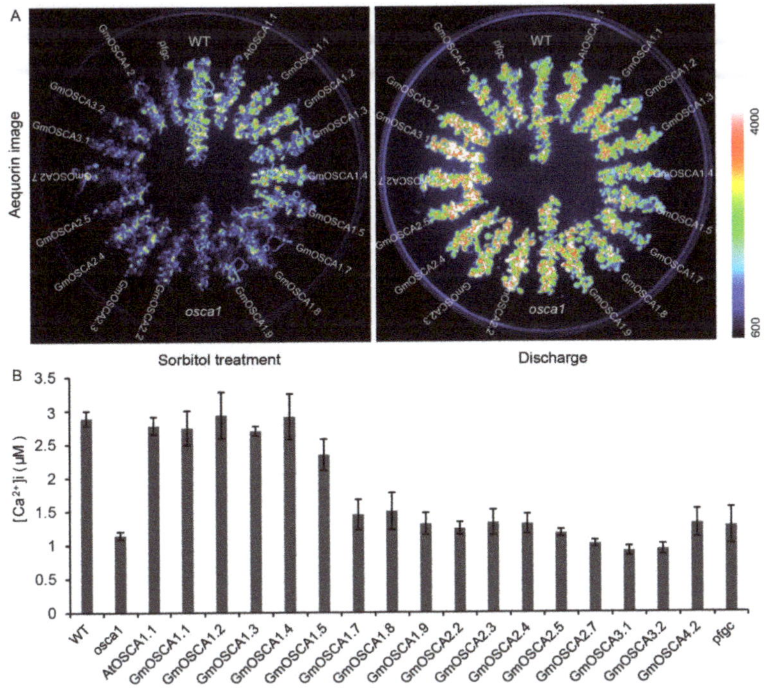

Figure 4. The calcium-imaging phenotype of *GmOSCA*s transgenic *Arabidopsis* seedlings. (**A**) Aequorin bioluminescence-based imaging of seedlings treated with 600 mM sorbitol (left) and discharge solution (right), respectively. (**B**) Quantification of increased $[Ca^{2+}]_i$ data from representative experiments (mean ± SD, n = 4 pools, 20–22 seedings per pool).

2.5. Expression Profiles of GmOSCA Genes against Osmotic Stress

As potential osmosensors, the expressions of *GmOSCA* genes are likely to respond to osmotic stress. To support this hypothesis, the *GmOSCA*s expression profiles against dehydration and salinity (NaCl) treatments in soybean roots were explored, using the previously released Illumina RNA-seq raw data [25,26]. The expression profiles displayed that *GmOSCA1.1*, *GmOSCA1.2*, *GmOSCA1.3*, *GmOSCA1.4*, *GmOSCA1.5*, *GmOSCA3.1*, and *GmOSCA3.2* were significantly induced in the early stage (<6 h) of dehydration and salinity treatments and then gradually returned to normal with the extension of treatment time (Figure 5A,B). In contrast, *GmOSCA1.7*, *GmOSCA1.8*, and *GmOSCA2.4* were down-regulated only in the early stage of dehydration and salinity treatments, but these results may not be reliable because of the superficial expression level of *GmOSCA1.7* and *GmOSCA2.4* (Figure 5A,B). Interestingly, *GmOSCA2.7* was down-regulated under dehydration but up-regulated under salt stress (Figure 5A,B). However, this result was not very convincing because of the shallow expression level of *GmOSCA2.7*. In addition, *GmOSCA1.9* was down-regulated, but *GmOSCA2.2* and *GmOSCA2.5* were up-regulated by salt stress (Figure 5B). And there was little change in the expression of *GmOSCA4.2* under drought and salt stresses (Figure 5A,B). To further verify the response of *GmOSCA* genes to osmotic stress, RT-qPCR experiments were performed. The results showed that the gene expression of *GmOSCA1.1*, *GmOSCA1.2*, *GmOSCA1.3*, *GmOSCA1.4*, *GmOSCA1.5*, *GmOSCA3.1*, and *GmOSCA3.2* were indeed induced by drought and salt stresses in soybean roots (Figure 5C–P). Similarly, these genes were also up-regulated by drought and salt stress in soybean leaves (Supplementary Materials Figure S5). Taken together, our data demonstrated that most *GmOSCA* genes, especially *GmOSCA1.1*, *GmOSCA1.2*,

GmOSCA1.3, *GmOSCA1.4*, *GmOSCA1.5*, *GmOSCA3.1*, and *GmOSCA3.2*, were osmotic-stress-responsive genes.

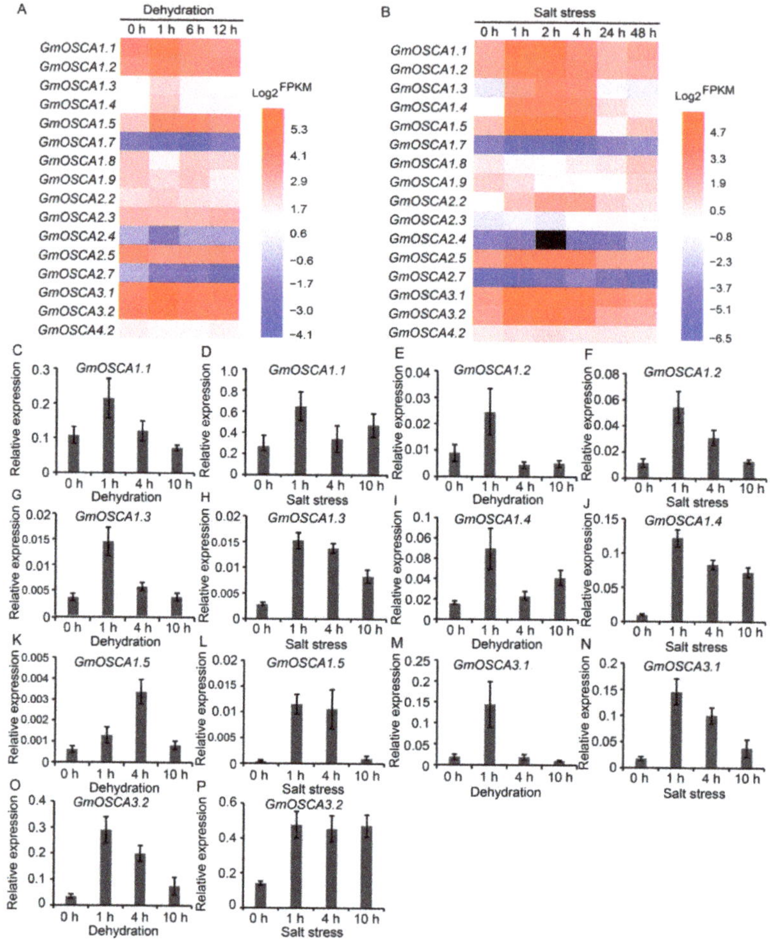

Figure 5. The expression patterns of *GmOSCA* genes in roots against dehydration and salt stress. (**A**) *GmOSCAs* expression profiles under dehydration using RNA-seq data. (**B**) *GmOSCAs* expression profiles under NaCl treatment using RNA-seq data. (**C–P**) *GmOSCAs* expression profiles in roots under PEG treatment and NaCl treatment using RT-qPCR data (mean ± SD, n = 3 or 4).

2.6. Overexpression of Some GmOSCA Members Complements the Drought-Hypersensitive Phenotype of *osca1* Mutant

Since *GmOSCA1.1*, *GmOSCA1.2*, *GmOSCA1.3*, *GmOSCA1.4*, and *GmOSCA1.5* were not only induced by osmotic stress but also could restore the low OICI phenotype of *osca1*, they were selected to further study the functions in response to osmotic stress at the whole-plant level. In addition, two other osmotic-stress-induced genes, *GmOSCA3.1* and *GmOSCA3.2*, also belonging to the early responsive to dehydration stress protein 4 (*ERD4*) family, were added for further functional studies too. We directly monitored the growth status of the WT, *osca1*, and transgenic *Arabidopsis* lines of *AtOSCA1.1*, *GmOSCA1.1*, *GmOSCA1.2*, *GmOSCA1.3*, *GmOSCA1.4*, *GmOSCA1.5*, *GmOSCA3.1*, *GmOSCA3.2*, and *pfgc* under drought stress in soil. During drought treatment, *osca1* and the empty *pfgc* transgenic plants exhibited earlier and more severe wilting than WT, whereas the *AtOSCA1.1*, *GmOSCA1.1*, *GmOSCA1.2*, *GmOSCA1.3*, *GmOSCA1.4*, *GmOSCA1.5*, *GmOSCA3.1*, and *GmOSCA3.2* transgenic lines could fully complement the drought-hypersensitive phenotype of *osca1* and showed the same or even stronger drought tolerance than WT (Figure 6A). After re-watering, the survival rates of *osca1* and the empty *pfgc* transgenic plants were much lower than that of WT, while the lower survival rate of *osca1* could be fully rescued by the overexpression of *GmOSCA1.1*, *GmOSCA1.2*, *GmOSCA1.3*, *GmOSCA1.4*, *GmOSCA1.5*, *GmOSCA3.1*, and *GmOSCA3.2*, just like *AtOSCA1.1* (Figure 6A,B). Despite the similar survival rates to WT, the restoring states of these transgenic *Arabidopsis* lines were different, among which some transgenic lines seemed to show better growth status than WT (Figure 6A).

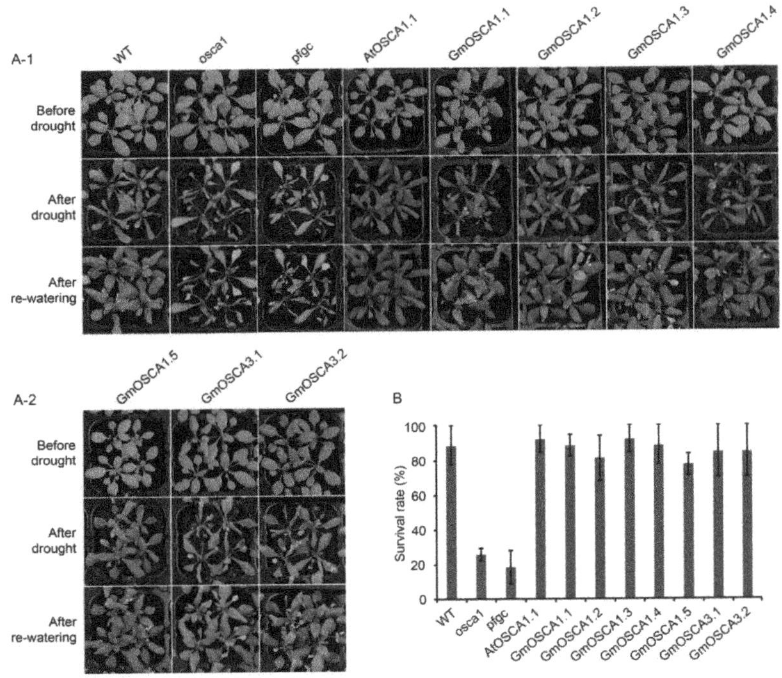

Figure 6. The drought-responsive phenotypes of *GmOSCAs* transgenic *Arabidopsis* plants. (**A**) Phenotypes of WT, *osca1*, and homozygous transgenic *Arabidopsis* lines. (**B**) The survival rates after re-watering were determined. Data were mean ± SE from three representative experiments.

3. Discussion

The signal transduction pathway of plant adaptation to osmotic stress, which seriously affects plant growth and crop yield, contains the processes of stress perception, signal transduction, and adaptive responses, such as stress-responsive gene expression and the production of metabolites. Compared with the well-studied processes of signal transduction and stress adaptation to osmotic stress, until we discover AtOSCA1.1, it is not clear who is the osmosensor responsible for osmotic sensing in plants. *AtOSCA1.1* belongs to the *OSCA* family, which contains four clades and 15 *OSCA* members in *Arabidopsis*. Among the 15 *AtOSCA* members, the two paralogs, AtOSCA1.1 and AtOSCA1.2, with the highest sequence identity are hyperosmolarity-gated calcium-permeable channels involved in osmotic stress signaling [15,16], suggesting that the *OSCA* family may be functionally conserved and redundant, especially for members with high similarity. Furthermore, with the resolution of the cryo-electron microscopy structures of AtOSCA1.1, AtOSCA1.2,

AtOSCA2.2, and AtOSCA3.1 [27–30], we know that the OSCA family belongs to a new group of mechanosensitive ion channels, which have quite a similar topological structure as the mammalian TMEM16 proteins. The conformation changes of OSCA channels lead to the opening of the ion channel pore under mechanical stimuli, such as gravity, touch, and osmotic pressure, which triggers the local tension and deformation of the membrane. These similar cryo-electron microscopies of the OSCA family further support that they may have conserved protein functions. However, electrophysiological examinations of different subclasses of *OSCA* family members expressed in HEK-P1KO cells revealed quite a distinct mechanically activated ion conductance, implicating the divergent channel properties in OSCA family members with different clades [31]. In this study, we demonstrated that GmOSCA1.1, GmOSCA1.2, GmOSCA1.3, GmOSCA1.4, and GmOSCA1.5 derived from cluster I have functions similar to AtOSCA1.1, at least at the protein level, by the heterologous expression of *GmOSCA* members in *osca1* (Figure 4), supporting the concept that the *OSCA* family members with high identity are likely functionally conserved and redundant. In contrast, the fact that the remaining *GmOSCA* genes derived from clusters I, II, III, and IV did not rescue the reduced OICI of *osca1* suggests the different functional properties between them with *AtOSCA1.1* (Figure 4). Similarly, *AtOSCA3.1* is an early inducible gene in response to drought stress, and its knockout mutant displayed wildtype OICI under osmotic stress, which also supports a different biological working mechanism between *AtOSCA3.1* and *AtOSCA1.1* [15]. Recently, AtOSCA1.3 was demonstrated as a Ca^{2+}-permeable channel required for stomatal immunity, and its activation depends on BIK1-mediated phosphorylation at its serine residue within a motif (Ser-X-X-Leu) [17]. Besides AtOSCA1.3, AtOSCA1.7 within OSCA cluster I also had a similar motif to AtOSCA1.3 at the same position and was activated by BIK1 activity. Furthermore, an *Atosca1.3/1.7* double mutant exhibited impaired stomatal closure and reduced flg22-induced Ca^{2+} increase upon treatment with flg22 [17]. Although it remains to be tested whether AtOSCA1.3 and AtOSCA1.7 are similarly mechanosensitive to other OSCAs, these results suggested that phosphorylation by BIK1 may represent an additional regulatory layer for this conserved family of Ca^{2+}-permeable channels in response to different stresses. Therefore, we used sequence alignment to explore whether GmOSCA members of cluster I have a similar motif (Ser-X-X-Leu) and found that only GmOSCA1.5 had this motif (Supplementary Materials Figure S6), indicating that, in addition to sensing osmotic stress, it may also regulate soybean immunity.

Besides the model plant *Arabidopsis*, the *OSCA* gene family has been identified genome-wide and described in several plant species, but only a very few *OSCA* genes have been functionally validated, except those in rice, of which 11 *OsOSCA* members were identified and divided into four clades, like *Arabidopsis* and soybean. The functions of 11 *OsOSCA* members, except for *OsOSCA4.1*, have been reported in several papers [32–34]. Of the 10 OsOSCA members, only OsOSCA1.4 was exclusively located in the PM, while the other 9 OsOSCAs were mainly localized in the endoplasmic reticulum (ER). In addition, *OsOSCA1.4* mediated both OICI and salt-stress-induced cytosolic [Ca^{2+}] increases ($SICI_{cyt}$) in HEK293 cells and *osca1* mutant, suggesting that OsOSCA1.4 may function as an osmosensor in rice [34]. Surprisingly, although different from AtOSCA1.1 localization, overexpressing of some of the remaining *OsOSCA* genes (*OsOSCA1.1*, *OsOSCA1.2*, *OsOSCA1.3*, *OsOSCA2.1*, and *OsOSCA2.2*) could restore the reduced OICI and $SICI_{cyt}$ in *osca1* [33]. Recently, the biological function of *OsOSCA1.1* in rice was reported and showed that *OsOSCA1.1* mediated OICI and $SICI_{cyt}$ in rice roots under hyperosmolality and salt stress [32]. These results further proved that the *OSCA* family members are functionally conservative and specific.

Soybean, which provides more than one-half of global oilseed production and one-quarter of the world's protein for humans and livestock, is susceptible to osmotic stress such as drought and salinity. Soybean-growing regions and production are seriously limited due to the two major threats of drought and salt stresses [35,36]. Therefore, cloning genes related to osmotic stress tolerance and studying their underlying molecular mechanisms are crucial for breeding new drought- or salt-tolerant soybean cultivars and improving soybean yield through molecular breeding techniques. In recent decades, especially after the release of the soybean genome, with the demand for agricultural production and the rapid development of molecular biology technology, some genes involved in the osmotic stress regulation in soybean have been cloned and reported. These genes are mainly transcription factors regulating the expression of downstream stress genes or stress-responsive genes themselves, which are downstream of the signal transduction pathway. In this study, we first systematically studied and experimentally demonstrated the functions of *GmOSCA* members, which may involve osmotic stress perception as osmosensors in soybean. A total of 20 *GmOSCA* members in four subgroups were identified in the soybean genome (Figure 1). The number of *OSCA* genes in soybean was significantly higher than that in diploid *Arabidopsis* (15) and rice (11), which is consistent with the fact that soybean is paleo-polyploid. In addition, our collinearity analysis showed that all of the *GmOSCA* genes, apart from *GmOSCA1.5* and *GmOSCA2.7*, are duplicated genes (Supplementary Materials Figure S2), which could provide a chance for gene evolution. On the one hand, soybean has more *OSCA* members, and, on the other hand, most of these genes are duplicated genes, indicating that they may have greater functional redundancy and diversity. The comparative analyses of homologs between soybean and *Arabidopsis* on gene structures and protein domains and structures suggest that the *GmOSCA* members in the same cluster may have similar functions, and *GmOSCA* genes in clusters III and IV may have generated divergent functions (Figure 1 and Supplementary Materials Figure S3). Our gene-expression data confirmed that *GmOSCA* members produced nonfunctionalization and subfunctionalization at gene expression. For instance, the spatial- and temporal-expression patterns of soybean *GmOSCAs* genes are very different, and some genes are even pseudogenes (Figure 2 and Supplementary Materials Figure S4). In addition, they also have different responses to drought and salt stress (Figure 5 and Supplementary Materials Figure S5). Finally, our complementary experiments in *osca1* showed that the *GmOSCA* family also produced functional differentiation at the protein level (Figures 4 and 6).

Here, the presented data preliminarily proved that GmOSCA1.1, GmOSCA1.2, GmOSCA1.3, GmOSCA1.4, and GmOSCA1.5 are likely calcium-permeable osmosensors in soybean by heterologous expression of *GmOSCA* members in *osca1* (Figure 4), and the overexpression of *GmOSCA1.1*, *GmOSCA1.2*, *GmOSCA1.3*, *GmOSCA1.4*, *GmOSCA1.5*, *GmOSCA3.1*, and *GmOSCA3.2* conferred enhanced drought tolerance to transgenic *Arabidopsis* (Figure 6). Unexpectedly, it was found that these *GmOSCA* genes did not affect salt tolerance, due to the similar survival rates and above-ground growth status of wildtype, *osca1*, and *OSCA* transgenic plants under hydroponic NaCl (50 mM and 80 mM) treatment (data not shown). In fact, when the hydroponic NaCl concentration is below 50 mM or above 80 mM, the *Arabidopsis* seedlings will survive or die accordingly. One of the most likely explanations for this phenomenon is that these *OSCA* genes may act primarily on osmotic stress. Although high salinity can increase osmotic stress, it also causes ion toxicity to plants. The accumulation of Na^+ and Cl^- in cells affects the absorption and transport of mineral elements and inhibits the activity of intracellular enzymes [37]. Recently, using a similar calcium-imaging-based genetic screen, we identified that *MOCA1*, encoding a glucuronosyltransferase for glycosyl inositol phosphorylceramide (GIPC) sphingolipids in PM, is involved in sensing salt-associated ionic stress [38]. The increase in $[Ca^{2+}]_i$, found to be induced by 200 mM NaCl, was lower in the *moca1* mutant, while the rise in $[Ca^{2+}]_i$ caused by 400 mM sorbitol was similar between *moca1* and the wildtype. These data distinguished the ionic effect from the osmotic effect of salt stress. Thus, when the concentration of NaCl is low, it is mainly ion stress that plays a role. With the increase in NaCl concentration until the *Arabidopsis* seedlings die due to ion toxicity, osmotic stress does not play a significant role; thereby, highly concentrated NaCl cannot reflect the function of *OSCA* genes to enhance osmotic stress tolerance. However, to elucidate the natural biological function of *GmOSCA* genes, we still need further study of these genes, by overexpressing or silencing them in soybean. Moreover, since *GmOSCA* members may be functionally redundant, it may be necessary to knock out multiple genes simultaneously to study their function.

Fortunately, CRISPR/Cas9-mediated gene-editing technology provides a solution to this problem. In addition, nearly 1000 soybean resequencing materials have been released so far. The association analysis between *GmOSCAs* genotypes and drought or salinity tolerance phenotypes using these materials can not only further support whether they are involved in the regulation of drought stress but also screen out the excellent alleles of *GmOSCAs* for breeding. Despite these problems, our study is the first to explore the process of osmotic stress perception in soybean, so our data will lay the foundation for further study of this process and provide valuable genetic resources for the development of drought-tolerant soybean cultivars using genetic engineering and molecular breeding.

4. Materials and Methods

4.1. Identification and General Characterization of GmOSCA Family Members in Soybean

The 15 *Arabidopsis* OSCA protein sequences were used to blast in the soybean genome (Wm82.a2.v1) in Phytozome (https://phytozome.jgi.doe.gov/pz/portal.html, accessed on 15 August 2022). The Pfam tool (http://pfam.xfam.org/) was then used to identify the conserved domains of candidates, including the typical RSN1_7TM domain in OSCA proteins [39]. Finally, 20 GmOSCA candidates with high sequence identity and RSN1_7TM domain were screened out in assembled soybean chromosomes. Protparam (https://web.expasy.org/protparam/) was adopted to calculate the relative molecular masses and isoelectric points of GmOSCA proteins. A neighbor-joining phylogenetic tree of OSCA proteins was constructed by MEGA 6.0 with the Poisson model, 1000 bootstrap replications, and a complete deletion treatment for gaps/missing data [40]. The *OSCA* gene structures were drawn using GSDS 2.0 software (http://gsds.cbi.pku.edu.cn/) [41]. The three-dimensional structures of OSCA proteins were predicted using Phyre2 (http://www.sbg.bio.ic.ac.uk/phyre2/html/) [42]. The collinear blocks carrying *GmOSCA* genes in soybean were identified with the MCScanX toolkit [43], and the collinear relationships were drawn by Circos [44]. Transmembrane helices and subcellular localizations of GmOSCA proteins were predicted using MemPype (https://mu2py.biocomp.unibo.it/mempype/default/index) [45].

4.2. RNA-Seq Data Analysis

The spatiotemporal expression profiles of *GmOSCA* genes in 25 samples were obtained by mining the Illumina RNA-seq raw data released by Shen et al. [24]. The expression patterns of *GmOSCA* genes against dehydration and salinity treatments were detected by reanalysis of the Illumina RNA-seq raw data from Belamkar et al. [26] and Liu et al. [25], respectively. The short reads were mapped and aligned with the soybean reference genome Wm82.a2.v1 using HISAT [46]. The assembly and expression calculations of these transcripts were achieved by StringTie [47]. The mean fragments per kilobase of exon per million fragments mapped (FPKM) value was regarded as the gene expression value. The heat maps were visualized utilizing Heml, with the FPKM values as input data [48].

4.3. Plant Materials and Abiotic Stress Treatments

The soybean cultivar Williams 82 was used in this study. Soybean plants for gene cloning were grown outdoors during the sowing season. For abiotic stress treatments, soybean seeds with uniform size and harvest time were firstly germinated on moist sterile filter paper for four days in the dark at 25 °C. They then were transferred to half-strength Hoagland-modified nutrient solution (Coolaber Biotech, Beijing, China) in a growth chamber with a 12 h light/12 h dark photoperiod, 65%~75% relative humidity, and an ambient temperature of 28 °C. The nutrient solution was renewed every two days. When the unifoliolate leaves were fully opened, seedlings were transferred into half-strength Hoagland-modified nutrient solution supplemented with 10% (*w*/*w*) PEG and 0.9% (*w*/*w*; ~150 mM) NaCl, respectively. 10% PEG hydroponics can cause plant dehydration and thus simulate soil drought. The unifoliolate leaf and root tissues were harvested at 0, 1, 4, and 10 h after treatments, and two biological replicates and 10 plants per time point were maintained for each treatment.

The *Arabidopsis* WT (Col-0 constitutively expressing intracellular Ca^{2+} indicator aequorin, from M. Knight) [49], *osca1*, and all the transgenic plants in *osca1* background were grown in a greenhouse with a 16 h light/8 h dark photoperiod at 20~23 °C. The sterilized seeds were vernalized at 4 °C for three days, before being sown on a half-strength MS medium. For the drought treatment in soil, the WT, *osca1*, and homozygous transgenic seedlings were first grown on half-strength MS medium for about one week and then transferred into the weighed soil for two weeks. The remaining water in the pot's base was poured out and then withheld until the *osca1* plants developed the wilting phenotypes. The survival rates were surveyed after re-watering with three biological replicates [50].

4.4. RNA Extraction, cDNA Synthesis, and RT-qPCR

Total RNA was extracted using an EasyPure Plant RNA Kit (TransGen Biotech, Beijing, China). cDNA synthesis was performed with a kit of TransScript One-Step gDNA Removal and cDNA Synthesis SuperMix (TransGen Biotech, Beijing, China) or HiScript III All-in-one RT SuperMix Perfect for qPCR (Vazyme Biotech, Nanjing, China). Real-time quantitative PCR (RT-qPCR) was performed using the BIO-RAD C1000 Touch Thermal Cycler PCR system and Applied Biosystems PowerUp SYBR Green Master Mix Kit (Thermo Fisher Scientific, Shanghai, China), in accordance with the instructions of the manufacturer, with a slight modification. The relative gene expression levels of *GmOSCA* genes were calculated from three or four replicates according to the $2^{-\Delta CT}$ method, with a reference gene ACTIN. The primers used for RT-qPCR are listed in Supplementary Materials Table S2.

4.5. Subcellular Localization of GmOSCA proteins

The full-length CDS of *GmOSCAs* and *AtOSCA1.1* were amplified from the soybean and *Arabidopsis* cDNA, respectively. Then, CDS was cloned into the PucGFP vector without stop codons for fusion with a green fluorescent protein (GFP) tag at the C-terminus through enzymatic digestion and ligation. The empty PucGFP vector and these recombinant plasmids were transformed into the digested *Arabidopsis* protoplasts, as previously described [51]. The GFP fluorescence was detected by laser confocal microscopy. The amplified primers for PucGFP are listed in Supplementary Materials Table S3.

4.6. Construction of OSCA Transgenic Arabidopsis Lines

The full-length CDS of *GmOSCAs* and *AtOSCA1.1* were cloned into the pfgc5941 vector for genetic complementation assay through enzymatic digestion and ligation. The empty pfgc5941 vector and these recombinant plasmids were transformed into the *osca1* mutant, using the floral dip method with *Agrobacterium tumefaciens* GV3101 strain. Transgenic *Arabidopsis* lines were screened by basta spraying (50 mg/L) and PCR test, and the homozygous lines with single copy insertion were used for experimental analysis. The amplified primers for pfgc5941 are listed in Supplementary Materials Table S3.

4.7. Aequorin Bioluminescence-Based Ca^{2+} Imaging

$[Ca^{2+}]_i$ was detected using *Arabidopsis* plants expressing aequorin, as described previously [15]. Nine-day-old *Arabidopsis* seedlings were evenly sprayed with 6 mL of 10 μM coelenterazine (Prolume) per Petri dish (15 cm in diameter) and then placed in the dark at 22 °C for 12 h before imaging. The aequorin bioluminescence-based Ca^{2+} imaging was conducted using a Lumazone Pylon1300B system (Roper Scientific, Tuscon, AZ, USA) equipped with a cooled CCD camera in a light-tight box. A liquid nitrogen autofiller was connected to this system to maintain constant cooling. The camera was controlled by WinView/32 (Roper Scientific, Tuscon, USA) software. The plate was treated with 90 mL 600 mM sorbitol (Sigma-Aldrich, Shanghai, China), and the recording of luminescence (L) was started 10 s before treatment and collected for 5 min. The total aequorin luminescence (L_{max}) was recorded for 3 min by discharging with 0.9 M $CaCl_2$ in 10% (v/v) ethanol. The bioluminescence images were analyzed using ImageJ software. The $[Ca^{2+}]_i$ was measured according to the formula (pCa = 0.6747 × (−log L/Lmax) + 5.3177), and calculated from four replicates [15,38].

5. Conclusions

In this study, we identified 20 *GmOSCA* members in soybean and systematically compared their phylogenetic relationships, gene structures, protein domains and structures, spatiotemporal and osmotic-stressed expression profiles, and protein functions in transgenic *Arabidopsis*. The results showed that 20 *GmOSCA* members were divided into four clades and that the members in the same cluster may have more similar functions. In addition, *GmOSCA1.1*, *GmOSCA1.2*, *GmOSCA1.3*, *GmOSCA1.4*, and *GmOSCA1.5* from cluster I might function in the osmotic stress sensing of soybean. Furthermore, *GmOSCA1.1*, *GmOSCA1.2*, *GmOSCA1.3*, *GmOSCA1.4*, *GmOSCA1.5*, *GmOSCA3.1*, and *GmOSCA3.2* might confer enhanced drought tolerance in soybean. These results greatly promote the research progress of the *GmOSCA* family, enrich the molecular mechanism of how soybean responds to osmotic stress, and lay a foundation for improving soybean drought tolerance.

Supplementary Materials: The following supporting information can be downloaded at: https://www.mdpi.com/article/10.3390/ijms231810570/s1.

Author Contributions: Project administration and supervision, Z.H., Z.-M.P. and Q.L.; funding acquisition, Z.-M.P. and Q.L.; investigation, C.L., H.W., Y.Z., H.C. and Q.L.; data curation, C.L., H.W. and Q.L.; writing—original draft, C.L. and Q.L.; writing—review and editing, Q.L. All authors have read and agreed to the published version of the manuscript.

Funding: This research was supported by the Shenzhen Peacock Innovation Team Plan (Grant No. KQTD2017032715165926), the China Postdoctoral Science Foundation (Grant No. 2017M622751), the Ph.D. Start-up Fund of Guangdong Natural Science Foundation (Grant No. 2018A030310455), and the Central Public-interest Institution Basal Research Fund (CPSIBRF-CNRRI-202110).

Institutional Review Board Statement: Not applicable.

Informed Consent Statement: Not applicable.

Data Availability Statement: Not applicable.

Conflicts of Interest: The authors declare no conflict of interest.

References

1. Zhu, J.K. Abiotic stress signaling and responses in plants. *Cell* **2016**, *167*, 313–324. [CrossRef] [PubMed]
2. Daryanto, S.; Wang, L.X.; Jacinthe, P.A. Global synthesis of drought effects on maize and wheat production. *PLoS ONE* **2016**, *11*, e0156362. [CrossRef] [PubMed]
3. Zhao, C.Z.; Zhang, H.; Song, C.P.; Zhu, J.K.; Shabala, S. Mechanisms of plant responses and adaptation to soil salinity. *Innovation* **2020**, *1*, 100017. [CrossRef] [PubMed]
4. Zhang, H.M.; Zhu, J.H.; Gong, Z.Z.; Zhu, J.K. Abiotic stress responses in plants. *Nat. Rev. Genet.* **2022**, *23*, 104–119. [CrossRef] [PubMed]
5. Zhu, J.K. Salt and drought stress signal transduction in plants. *Annu. Rev. Plant Biol.* **2002**, *53*, 247–273. [CrossRef]
6. Huang, G.T.; Ma, S.L.; Bai, L.P.; Zhang, L.; Ma, H.; Jia, P.; Liu, J.; Zhong, M.; Guo, Z.F. Signal transduction during cold, salt, and drought stresses in plants. *Mol. Biol. Rep.* **2012**, *39*, 969–987. [CrossRef]
7. van Zelm, E.; Zhang, Y.X.; Testerink, C. Salt tolerance mechanisms of plants. *Annu. Rev. Plant Biol.* **2020**, *71*, 403–433. [CrossRef]
8. Yoshida, T.; Mogami, J.; Yamaguchi-Shinozaki, K. ABA-dependent and ABA-independent signaling in response to osmotic stress in plants. *Curr. Opin. Plant Biol.* **2014**, *21*, 133–139. [CrossRef]
9. Fujii, H.; Chinnusamy, V.; Rodrigues, A.; Rubio, S.; Antoni, R.; Park, S.Y.; Cutler, S.R.; Sheen, J.; Rodriguez, P.L.; Zhu, J.K. In vitro reconstitution of an abscisic acid signalling pathway. *Nature* **2009**, *462*, 660–664. [CrossRef]
10. Fujii, H.; Zhu, J.K. Arabidopsis mutant deficient in 3 abscisic acid-activated protein kinases reveals critical roles in growth, reproduction, and stress. *Proc. Natl. Acad. Sci. USA* **2009**, *106*, 8380–8385. [CrossRef]
11. Fujita, Y.; Yoshida, T.; Yamaguchi-Shinozaki, K. Pivotal role of the AREB/ABF-SnRK2 pathway in ABRE-mediated transcription in response to osmotic stress in plants. *Physiol. Plant.* **2013**, *147*, 15–27. [CrossRef] [PubMed]

12. Sakuma, Y.; Maruyama, K.; Osakabe, Y.; Qin, F.; Seki, M.; Shinozaki, K.; Yamaguchi-Shinozaki, K. Functional analysis of an Arabidopsis transcription factor, DREB2A, involved in drought-responsive gene expression. *Plant Cell* **2006**, *18*, 1292–1309. [CrossRef]
13. Sakuma, Y.; Maruyama, K.; Qin, F.; Osakabe, Y.; Shinozaki, K.; Yamaguchi-Shinozaki, K. Dual function of an Arabidopsis transcription factor DREB2A in water-stress-responsive and heat-stress-responsive gene expression. *Proc. Natl. Acad. Sci. USA* **2006**, *103*, 18822–18827. [CrossRef] [PubMed]
14. Singh, D.; Laxmi, A. Transcriptional regulation of drought response: A tortuous network of transcriptional factors. *Front. Plant Sci.* **2015**, *6*, 895. [CrossRef] [PubMed]
15. Yuan, F.; Yang, H.M.; Xue, Y.; Kong, D.D.; Ye, R.; Li, C.J.; Zhang, J.Y.; Theprungsirikul, L.; Shrift, T.; Krichilsky, B.; et al. OSCA1 mediates osmotic-stress-evoked Ca^{2+} increases vital for osmosensing in Arabidopsis. *Nature* **2014**, *514*, 367–371. [CrossRef]
16. Hou, C.C.; Tian, W.; Kleist, T.; He, K.; Garcia, V.; Bai, F.L.; Hao, Y.L.; Luan, S.; Li, L.G. DUF221 proteins are a family of osmosensitive calcium-permeable cation channels conserved across eukaryotes. *Cell Res.* **2014**, *24*, 632–635. [CrossRef]
17. Thor, K.; Jiang, S.S.; Michard, E.; George, J.; Scherzer, S.; Huang, S.G.; Dindas, J.; Derbyshire, P.; Leitao, N.; DeFalco, T.A.; et al. The calcium-permeable channel OSCA1.3 regulates plant stomatal immunity. *Nature* **2020**, *585*, 569–573. [CrossRef]
18. Li, Y.S.; Yuan, F.; Wen, Z.H.; Li, Y.H.; Wang, F.; Zhu, T.; Zhuo, W.Q.; Jin, X.; Wang, Y.D.; Zhao, H.P.; et al. Genome-wide survey and expression analysis of the *OSCA* gene family in rice. *BMC Plant Biol.* **2015**, *15*, 261. [CrossRef]
19. Ding, S.C.; Feng, X.; Du, H.W.; Wang, H.W. Genome-wide analysis of maize *OSCA* family members and their involvement in drought stress. *PeerJ* **2019**, *7*, e6765. [CrossRef]
20. Cao, L.R.; Zhang, P.Y.; Lu, X.M.; Wang, G.R.; Wang, Z.H.; Zhang, Q.J.; Zhang, X.; Wei, X.; Mei, F.J.; Wei, L.; et al. Systematic analysis of the maize *OSCA* genes revealing *ZmOSCA* family members involved in osmotic Stress and *ZmOSCA2.4* confers enhanced drought tolerance in transgenic Arabidopsis. *Int. J. Mol. Sci.* **2020**, *21*, 351. [CrossRef]
21. Tong, K.; Wu, X.Y.; He, L.; Qiu, S.Y.; Liu, S.; Cai, L.N.; Rao, S.F.; Chen, J.P. Genome-wide identification and expression profile of *OSCA* gene family members in *Triticum aestivum* L. *Int. J. Mol. Sci.* **2022**, *23*, 469. [CrossRef] [PubMed]
22. Schmutz, J.; Cannon, S.B.; Schlueter, J.; Ma, J.X.; Mitros, T.; Nelson, W.; Hyten, D.L.; Song, Q.J.; Thelen, J.J.; Cheng, J.L.; et al. Genome sequence of the palaeopolyploid soybean. *Nature* **2010**, *463*, 178–183. [CrossRef] [PubMed]
23. Xu, G.X.; Guo, C.C.; Shan, H.Y.; Kong, H.Z. Divergence of duplicate genes in exon-intron structure. *Proc. Natl. Acad. Sci. USA* **2012**, *109*, 1187–1192. [CrossRef] [PubMed]
24. Shen, Y.T.; Zhou, Z.K.; Wang, Z.; Li, W.Y.; Fang, C.; Wu, M.; Ma, Y.M.; Liu, T.F.; Kong, L.A.; Peng, D.L.; et al. Global dissection of alternative splicing in paleopolyploid soybean. *Plant Cell* **2014**, *26*, 996–1008. [CrossRef] [PubMed]
25. Liu, A.L.; Xiao, Z.X.; Li, M.W.; Wong, F.L.; Yung, W.S.; Ku, Y.S.; Wang, Q.W.; Wang, X.; Xie, M.; Yim, A.K.Y.; et al. Transcriptomic reprogramming in soybean seedlings under salt stress. *Plant Cell Environ.* **2019**, *42*, 98–114. [CrossRef] [PubMed]
26. Belamkar, V.; Weeks, N.T.; Bharti, A.K.; Farmer, A.D.; Graham, M.A.; Cannon, S.B. Comprehensive characterization and RNA-Seq profiling of the *HD-Zip* transcription factor family in soybean (*Glycine max*) during dehydration and salt stress. *BMC Genom.* **2014**, *15*, 950. [CrossRef] [PubMed]
27. Maity, K.; Heumann, J.M.; McGrath, A.P.; Kopcho, N.J.; Hsu, P.K.; Lee, C.W.; Mapes, J.H.; Garza, D.; Krishnan, S.; Morgan, G.P.; et al. Cryo-EM structure of OSCA1.2 from *Oryza sativa* elucidates the mechanical basis of potential membrane hyperosmolality gating. *Proc. Natl. Acad. Sci. USA* **2019**, *116*, 14309–14318. [CrossRef] [PubMed]
28. Zhang, M.F.; Wang, D.L.; Kang, Y.L.; Wu, J.X.; Yao, F.Q.; Pan, C.F.; Yan, Z.Q.; Song, C.; Chen, L. Structure of the mechanosensitive OSCA channels. *Nat. Struct. Mol. Biol.* **2018**, *25*, 850–858. [CrossRef]
29. Liu, X.; Wang, J.W.; Sun, L.F. Structure of the hyperosmolality-gated calcium-permeable channel OSCA1.2. *Nat. Commun.* **2018**, *9*, 5060. [CrossRef]
30. Jojoa-Cruz, S.; Saotome, K.; Murthy, S.E.; Tsui, C.C.A.; Sansom, M.S.P.; Patapoutian, A.; Ward, A.B. Cryo-EM structure of the mechanically activated ion channel OSCA1.2. *Elife* **2018**, *7*, e41845. [CrossRef]
31. Murthy, S.E.; Dubin, A.E.; Whitwam, T.; Jojoa-Cruz, S.; Cahalan, S.M.; Mousavi, S.A.R.; Ward, A.B.; Patapoutian, A. OSCA/TMEM63 are an evolutionarily conserved family of mechanically activated ion channels. *Elife* **2018**, *7*, e41844. [CrossRef] [PubMed]
32. Han, Y.; Wang, Y.X.; Zhai, Y.J.; Wen, Z.H.; Liu, J.; Xi, C.; Zhao, H.P.; Wang, Y.D.; Han, S.C. *OsOSCA1.1* mediates hyperosmolality and salt stress sensing in *Oryza sativa*. *Biology* **2022**, *11*, 678. [CrossRef] [PubMed]
33. Zhai, Y.J.; Wen, Z.H.; Fang, W.Q.; Wang, Y.X.; Xi, C.; Liu, J.; Zhao, H.P.; Wang, Y.D.A.; Han, S.C. Functional analysis of rice *OSCA* genes overexpressed in the arabidopsis *osca1* mutant due to drought and salt stresses. *Transgenic Res.* **2021**, *30*, 811–820. [CrossRef] [PubMed]
34. Zhai, Y.J.; Wen, Z.H.; Han, Y.; Zhuo, W.Q.; Wang, F.; Xi, C.; Liu, J.; Gao, P.; Zhao, H.P.; Wang, Y.D.A.; et al. Heterogeneous expression of plasma-membrane-localised OsOSCA1.4 complements osmotic sensing based on hyperosmolality and salt stress in Arabidopsis *osca1* mutant. *Cell Calcium.* **2020**, *91*, e102261. [CrossRef]
35. Valliyodan, B.; Ye, H.; Song, L.; Murphy, M.; Shannon, J.G.; Nguyen, H.T. Genetic diversity and genomic strategies for improving drought and waterlogging tolerance in soybeans. *J. Exp. Bot.* **2017**, *68*, 1835–1849. [CrossRef]
36. Phang, T.H.; Shao, G.H.; Lam, H.M. Salt tolerance in soybean. *J. Integr. Plant Biol.* **2008**, *50*, 1196–1212. [CrossRef]
37. Dai, L.; Li, P.; Li, Q.; Leng, Y.; Zeng, D.; Qian, Q. Integrated multi-omics perspective to strengthen the understanding of salt tolerance in rice. *Int. J. Mol. Sci.* **2022**, *23*, 236. [CrossRef]

38. Jiang, Z.H.; Zhou, X.P.; Tao, M.; Yuan, F.; Liu, L.L.; Wu, F.H.; Wu, X.M.; Xiang, Y.; Niu, Y.; Liu, F.; et al. Plant cell-surface GIPC sphingolipids sense salt to trigger Ca^{2+} influx. *Nature* **2019**, *572*, 341–346. [CrossRef]
39. Finn, R.D.; Coggill, P.; Eberhardt, R.Y.; Eddy, S.R.; Mistry, J.; Mitchell, A.L.; Potter, S.C.; Punta, M.; Qureshi, M.; Sangrador-Vegas, A.; et al. The Pfam protein families database: Towards a more sustainable future. *Nucleic Acids Res.* **2016**, *44*, D279–D285. [CrossRef]
40. Tamura, K.; Stecher, G.; Peterson, D.; Filipski, A.; Kumar, S. MEGA6: Molecular evolutionary genetics analysis version 6.0. *Mol. Biol. Evol.* **2013**, *30*, 2725–2729. [CrossRef]
41. Hu, B.; Jin, J.P.; Guo, A.Y.; Zhang, H.; Luo, J.C.; Gao, G. GSDS 2.0: An upgraded gene feature visualization server. *Bioinformatics* **2015**, *31*, 1296–1297. [CrossRef] [PubMed]
42. Kelley, L.A.; Mezulis, S.; Yates, C.M.; Wass, M.N.; Sternberg, M.J.E. The Phyre2 web portal for protein modeling, prediction and analysis. *Nat. Protoc.* **2015**, *10*, 845–858. [CrossRef] [PubMed]
43. Wang, Y.P.; Tang, H.B.; DeBarry, J.D.; Tan, X.; Li, J.P.; Wang, X.Y.; Lee, T.H.; Jin, H.Z.; Marler, B.; Guo, H.; et al. MCScanX: A toolkit for detection and evolutionary analysis of gene synteny and collinearity. *Nucleic Acids Res.* **2012**, *40*, e49. [CrossRef] [PubMed]
44. Krzywinski, M.; Schein, J.; Birol, I.; Connors, J.; Gascoyne, R.; Horsman, D.; Jones, S.J.; Marra, M.A. Circos: An information aesthetic for comparative genomics. *Genome Res.* **2009**, *19*, 1639–1645. [CrossRef] [PubMed]
45. Pierleoni, A.; Indio, V.; Savojardo, C.; Fariselli, P.; Martelli, P.L.; Casadio, R. MemPype: A pipeline for the annotation of eukaryotic membrane proteins. *Nucleic Acids Res.* **2011**, *39*, W375–W380. [CrossRef]
46. Kim, D.; Landmead, B.; Salzberg, S.L. HISAT: A fast spliced aligner with low memory requirements. *Nat. Methods* **2015**, *12*, 357–360. [CrossRef]
47. Pertea, M.; Pertea, G.M.; Antonescu, C.M.; Chang, T.C.; Mendell, J.T.; Salzberg, S.L. StringTie enables improved reconstruction of a transcriptome from RNA-seq reads. *Nat. Biotechnol.* **2015**, *33*, 290–295. [CrossRef]
48. Deng, W.K.; Wang, Y.B.; Liu, Z.X.; Cheng, H.; Xue, Y. HemI: A toolkit for illustrating heatmaps. *PLoS ONE* **2014**, *9*, e111988.
49. Knight, M.R.; Campbell, A.K.; Smith, S.M.; Trewavas, A.J. Transgenic plant aequorin reports the effects of touch and cold-shock and elicitors on cytoplasmic calcium. *Nature* **1991**, *352*, 524–526. [CrossRef]
50. Chen, J.N.; Nolan, T.M.; Ye, H.X.; Zhang, M.C.; Tong, H.N.; Xin, P.Y.; Chu, J.F.; Chu, C.C.; Li, Z.H.; Yin, Y.H. Arabidopsis WRKY46, WRKY54, and WRKY70 transcription factors are involved in brassinosteroid-regulated plant growth and drought responses. *Plant Cell* **2017**, *29*, 1425–1439. [CrossRef]
51. Li, Q.; Fang, C.; Duan, Z.B.; Liu, Y.C.; Qin, H.; Zhang, J.X.; Sun, P.; Li, W.B.; Wang, G.D.; Tian, Z.X. Functional conservation and divergence of *GmCHLI* genes in polyploid soybean. *Plant J.* **2016**, *88*, 584–596. [CrossRef] [PubMed]

International Journal of
Molecular Sciences

MDPI

Article

Responsiveness of Candidate Genes on *CoPv01*^{CDRK}/*PhgPv01*^{CDRK} Loci in Common Bean Challenged by Anthracnose and Angular Leaf Spot Pathogens

Maike Lovatto [1], Maria Celeste Gonçalves-Vidigal [1,*], Mariana Vaz Bisneta [1], Alexandre Catto Calvi [1], Josmar Mazucheli [2], Pedro Soares Vidigal Filho [1], Eduardo Gomes Rosa Miranda [1] and Maeli Melotto [3]

1 Departamento de Agronomia, Universidade Estadual de Maringá, Maringá 87020-900, Brazil
2 Departamento de Estatística, Universidade Estadual de Maringá, Maringá 87020-900, Brazil
3 Department of Plant Sciences, University of California, Davis, CA 95616, USA
* Correspondence: mcgvidigal@uem.br

Abstract: Anthracnose (ANT) and angular leaf spot (ALS) are significant diseases in common bean, leading to considerable yield losses under specific environmental conditions. The California Dark Red Kidney (CDRK) bean cultivar is known for its resistance to multiple races of both pathogens. Previous studies have identified the *CoPv01*^{CDRK}/*PhgPv01*^{CDRK} resistance loci on chromosome Pv01. Here, we evaluated the expression levels of ten candidate genes near the *CoPv01*^{CDRK}/*PhgPv01*^{CDRK} loci and plant defense genes using quantitative real-time PCR in CDRK cultivar inoculated with races 73 of *Colletotrichum lindemuthianum* and 63-39 of *Pseudocercospora griseola*. Gene expression analysis revealed that the *Phvul.001G246300* gene exhibited the most elevated levels, showing remarkable 7.8-fold and 8.5-fold increases for ANT and ALS, respectively. The *Phvul.001G246300* gene encodes an abscisic acid (ABA) receptor with pyrabactin resistance, PYR1-like (PYL) protein, which plays a central role in the crosstalk between ABA and jasmonic acid responses. Interestingly, our results also showed that the other defense genes were initially activated. These findings provide critical insights into the molecular mechanisms underlying plant defense against these diseases and could contribute to the development of more effective disease management strategies in the future.

Keywords: *CoPv01*^{CDRK}/*PhgPv01*^{CDRK} loci; candidate gene expression; common bean–anthracnose interaction; common bean–angular leaf spot interaction; plant defense genes

1. Introduction

Colletotrichum lindemuthianum (Sacc. & Magnus) Briosi & Cavara is a highly destructive pathogen that causes anthracnose (ANT) in common beans. In conditions of low temperature and high humidity, ANT can result in up to 100% yield losses. Furthermore, the pathogenic variability of *C. lindemuthianum* and the emergence of new races have resulted in the reduction or total loss of yield of previously resistant cultivars [1]. Angular leaf spot (ALS) is another disease that impacts common beans globally, caused by *Pseudocercospora griseola* (Sacc.) Crous & U. Braun. This disease can lead to yield losses of up to 70% [2–4]. Adopting genetically resistant cultivars provides a cost-effective, user-friendly, and environmentally conscious strategy for managing *C. lindemuthianum* and *P. griseola* infections in common beans [4]. Consequently, identifying and molecularly characterizing resistance genes is crucial for enhancing resistance efficacy and durability [5–8].

The *Co* and *Phg* loci associated with ANT and ALS, respectively, are often found in disease resistance clusters on various chromosomes. Although several independent genes confer resistance to *C. lindemuthianum*, most resistance genes found in Andean cultivars have been mapped to the common bean chromosome Pv01. The *Co-1*, *Co-1*^{2}, *Co-1*^{3}, *Co-1*^{4}, *Co-1*^{5}, and *Co-1*^{HY} alleles of the *Co-1* genes are present in the cultivars of Michigan Dark Red Kidney (MDRK), Kaboon, Perry Marrow, AND 277, Widusa, and Hongyundou, respectively [9–12]. Other resistance genes were mapped to the end of Pv01: *Co-x* in Jalo EEP558 and *Co-AC* in Amendoim Cavalo [13,14]. Moreover, the $CoPv01^{CDRK}$/$PhgPv01^{CDRK}$ gene in CDRK, which confers resistance to *C. lindemuthianum* races 73, 2047, and 3481, as well as race 63-39 of *P. griseola*, was recently fine-mapped to Pv01 [15].

A single dominant resistance gene primarily confers resistance to the ALS pathogen; however, recent studies have also identified quantitative resistance loci (QRLs) [16,17]. To date, five resistance loci have been mapped, including three independent loci, *Phg-1*, *Phg-2*, and *Phg-3*, located on chromosomes Pv01, Pv08, and Pv04, respectively [9,18,19]. Additionally, two major QRLs, *Phg-4* and *Phg-5*, have been found on Pv04 and Pv10 [16,17,20,21].

Candidate genes for ANT resistance loci have been analyzed through gene expression analysis to infer functionality in resistant cultivars [11,22,23]. Generally, examining the expression of the candidate and disease resistance genes can reveal their roles and interactions, thereby contributing to our understanding of how these genes collaborate in effective resistance responses. Chen [11] evaluated the expression analysis of four candidate genes at the $Co\text{-}1^{HY}$ allele in the Hongyundou cultivar inoculated with *C. lindemuthianum* race 81. The authors observed significant induction of all genes at an early stage. However, expression levels decreased at 24 h post-inoculation (hpi) and beyond. In the susceptible cultivar, high expression was only observed at 120 hpi, suggesting that delayed gene expression might facilitate pathogen penetration and proliferation, ultimately leading to disease development.

To elucidate the precise timing and magnitude of expression related to resistance against *C. lindemuthianum* race 73, Mahiya-Farooq [22] analyzed the expression of four candidate genes within the *Co-1* locus using near-isogenic lines that differ in the presence of the $Co\text{-}1^2$ resistance allele. They observed that the *Phvul.001G243800* gene exhibited substantially higher expression levels, nearly 144-fold, in the resistant near-isogenic line. The molecular basis of the ANT resistance *Co-x* locus was established by sequencing a 58-kb target region in the Jalo EEP558 cultivar. The *KTR2/3* gene was identified as an additional gene within a CRINKLY4 kinase cluster between the candidate genes *Phvul.001G243600* and *Phvul.001G243700*. Gene expression analysis demonstrated that *KTR2/3* was upregulated in Jalo EEP558 at 24 hpi in plants inoculated with strain 100 of *C. lindemuthianum* [13,23]. In summary, gene expression studies regarding the ANT resistance locus have identified the candidate genes *Phvul.001G243800* for $Co\text{-}1^2$, *KTR2/3* for *Co-x* and *Phvul.001G243600* and *Phvul.001G243700* for $Co\text{-}1^{HY}$, which showed high expression levels under inoculation conditions [11,13,22,23].

Phaseolus vulgaris CDRK from the breeding program at the University of California Davis is a landrace collected around Sacramento, CA, USA [24]. CDRK is resistant to Andean races 2, 39, and 55 and Mesoamerican races 9, 64, 65, 73, 89, 1545, and 2047, and race 3481 of *C. lindemuthianum* [15]. Allelism test revealed that $CoPv01^{CDRK}$/$PhgPv01^{CDRK}$ gene is not allelic to *Co-1* [15]. Through fine mapping, $CoPv01^{CDRK}$/$PhgPv01^{CDRK}$ was delimited in a genomic region of 33 Kb on chromosome Pv01, wherein the physical distances between $CoPv01^{CDRK}$/$PhgPv01^{CDRK}$ and *Co-1*, *Co-x*, and $Co\text{-}1^{HY}$ are 211 kb, 193 kb, and 181 kb, respectively. This previous study detected five candidate genes: *Phvul.001G246000* (ATP-dependent RNA helicase), *Phvul.001G246100* (cation-dependent mannose-6-phosphate receptor), *Phvul.001G246200* (protein trichome birefringence-like 33), *Phvul.001G246300* (abscisic acid (ABA) receptor PYL5), and *Phvul.001G246400* (SNF2 domain-containing protein class 1-related). Additionally, the candidate genes *Phvul.001G245300* and *Phvul.001G246800*, which encode putative leucine-rich repeat protein kinases, are close to the $CoPv01^{CDRK}$/$PhgPv01^{CDRK}$ loci [15].

The present study hypothesizes that each of the candidate genes that overlap with the $CoPv01^{CDRK}$/$PhgPv01^{CDRK}$ loci on Pv01 exhibits distinct expression patterns in response to inoculations with race 73 of *C. lindemuthianum* and race 63-39 of *P. griseola* in the California Dark Red Kidney cultivar. The objective of this study was to investigate the expression patterns of the $CoPv01^{CDRK}$/$PhgPv01^{CDRK}$ candidate genes (*Phvul.001G246000*, *Phvul.001G246100*, *Phvul.001G246200*, *Phvul.001G246300*, and *Phvul.001G245300*) in the

CDRK cultivar, not only in response to *C. lindemuthianum* race 73 but also to *P. griseola* race 63-39, using gene expression analysis employing quantitative real-time PCR. Specifically, we seek to gain insights into their potential roles in the plant's defense mechanisms against these pathogens, contributing to a deeper understanding of disease resistance in common beans.

2. Results

2.1. Phenotypic Evaluation of Cultivars

The results of inoculation of *C. lindemuthianum* race 73 at 120 h post-inoculation (hpi) onto the resistant cultivar CDRK and also the susceptible control cultivar Yolano are displayed in Figure 1. The disease in Yolano appears as small water-soaked lesions on the underside of the leaf and small sunken lesions on the stem and eventually leads to plant death. In contrast, no symptoms or hypersensitive response was observed in the resistant cultivar CDRK. Inoculations results of *Pseudocercospora griseola* race 63-39 onto CDRK resistant cultivar and Yolano susceptible cultivar 216 hpi are also displayed in Figure 1. The symptoms observed in Yolano cultivar were angular lesions in leaf area leading to early defoliation and plant death.

Figure 1. (**A**) Disease reaction of susceptible Yolano cultivar and no reaction in resistant California Dark Red Kidney cultivar at 120 h post-inoculation (hpi) with *C. lindemuthianum* race 73. (**B**) Disease reaction of susceptible Yolano cultivar and no reaction in resistant California Dark Red Kidney cultivar at 216 h post-inoculation with *P. griseola* race 63-39.

2.2. Differential Expression of Candidate and Defense Genes in the CDRK Cultivar Inoculated with Race 73 of C. lindemuthianum

We conducted an in-depth exploration into the expression patterns of the following candidate genes: *KTR2/3*, *Phvul.001G243800*, *Phvul.001G244300*, *Phvul.001G244400*, *Phvul.001G244500*, *Phvul.001G245300*, *Phvul.001G246000*, *Phvul.001G246100*, *Phvul.001G246200*, and *Phvul.001G246300*. These genes are located in regions that overlap with *Co-x*, $Co\text{-}1^2$, *Co-AC*, and $CoPv01^{CDRK}/PhgPv01^{CDRK}$ on the Pv01 chromosome. Additionally, defense genes *PR1a*, *PR1b*, and *PR2*, were evaluated to identify the molecular basis of CDRK resistance upon *C. lindemuthianum* race 73 inoculation.

Detailed information about the evaluated genes, accompanied by their functional annotations, can be found in Table S1. The magnitude of gene expression levels within the candidate and defense genes are shown in Table 1. Particularly, the gene *Phvul.001G246300* standout among all the candidate genes in the CDRK cultivar. This gene exhibited a remarkable 7.2-fold change in expression at 24 hpi in the resistant CDRK cultivar. As the interaction progressed to 72 hpi, the gene *Phvul.001G246300* showed significantly heightened expression of 7.8-fold. Moreover, the expression of *Phvul.001G246300* remained constant across multiple time points: 24, 48, 72, 96, and 120 hpi (Figures 2A and 3, and Table 1).

Table 1. Summary table of mean relative gene expression (Log_2(fold change)) of *CoPv01*^{CDRK}/*PhgPv01*^{CDRK} candidate genes and pathogenesis-related genes in response to ANT and ALS in CDRK cultivar.

Gene	Gene Model	*C. lindemuthianum* Race 73 (ANT)					*P. griseola* Race 63-39 (ALS)				
		24 hpi	48 hpi	72 hpi	96 hpi	120 hpi	24 hpi	72 hpi	120 hpi	168 hpi	216 hpi
Co-x	*KTR2/3*	1.4	1.3	3.1	0.5	0.5	0.2	0.8	0.8	−0.1	−1.4
Co-1	*Phvul.001G243800*	1.4	0.9	2.0	1.8	0.6	−1.1	−0.2	0.1	−0.9	−0.3
Co-AC	*Phvul.001G244300*	−0.2	−0.9	−0.2	−1.4	−1.3	−0.8	−0.1	0.2	0.2	0.0
	Phvul.001G244400	0.0	0.0	−0.7	−0.4	−0.2	−0.2	−0.2	−0.3	−0.7	−1.0
	Phvul.001G244500	0.4	0.3	−0.4	−0.5	0.8	−2.4	−2.5	−2.3	−2.3	−2.4
CoPv01^{CDRK}/*PhgPv01*^{CDRK}	*Phvul.001G245300*	3.7	2.5	3.4	3.3	3.5	−1.4	0.0	0.4	−1.7	0.7
	Phvul.001G246000	0.9	0.8	0.8	0.7	0.3	−0.5	0.4	0.5	−0.1	−0.2
	Phvul.001G246100	0.6	1.5	1.2	0.7	1.3	0.0	0.3	0.1	0.4	0.0
	Phvul.001G246200	2.7	2.6	2.6	0.5	1.6	0.0	1.5	0.1	−0.4	−0.2
	Phvul.001G246300	7.2	7.3	7.8	7.4	6.3	1.7	8.5	6.5	1.3	6.1
Pathogenesis-related genes	*PR1a*	6.6	6.2	8.0	8.1	6.3	1.5	3.6	4.7	1.0	5.8
	PR1b	13.6	13.9	14.1	14.2	16.7	−1.3	0.9	3.3	−1.1	4.3
	PR2	10.2	10.1	11.3	11.4	11.0	2.5	4.5	5.3	1.4	5.5

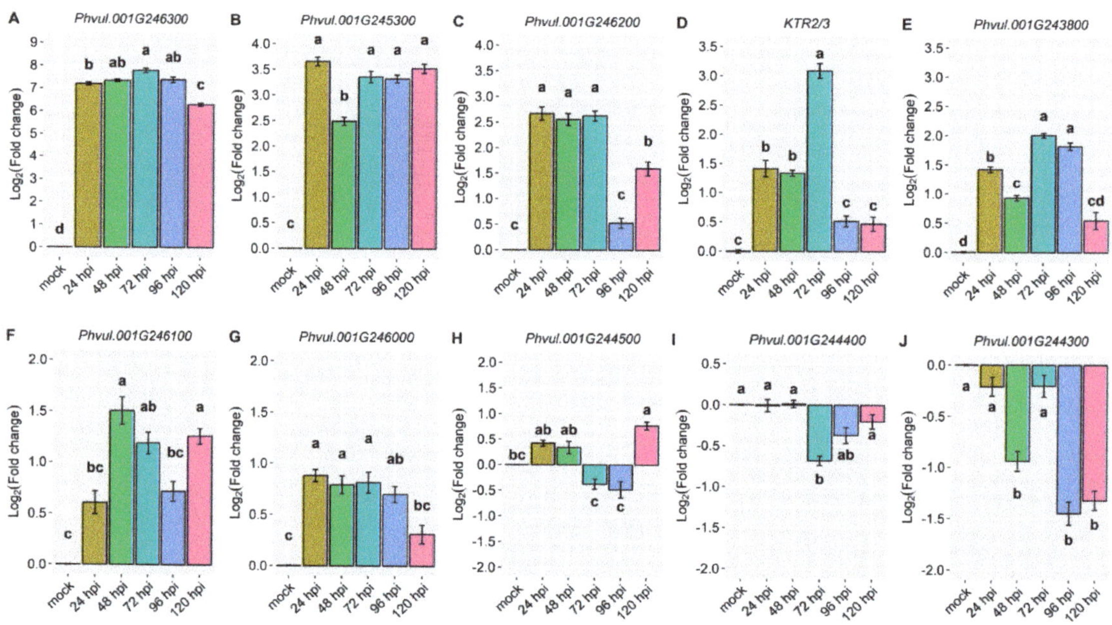

Figure 2. Relative expression of candidate genes: (**A**) *Phvul.001G246300*; (**B**) *Phvul.001G245300*; (**C**) *Phvul.001G246200*; (**D**) *KTR2/3*; (**E**) *Phvul.001G243800*; (**F**) *Phvul.001G246000*; (**G**) Phvul.001G246100; (**H**) *Phvul.001G244500*; (**I**) *Phvul.001G244400*; and (**J**) *Phvul.001G244300* in California Dark Red Kidney at 24, 48, 72, 96, and 120 h post-inoculation (hpi) with race 73 of *C. lindemuthianum* and a mock. The results are presented as logarithmic base 2 of the fold change of gene expression. Means with the same letter, for each gene, are not significantly different at the 5% significance level, using the Alexander-Govern test.

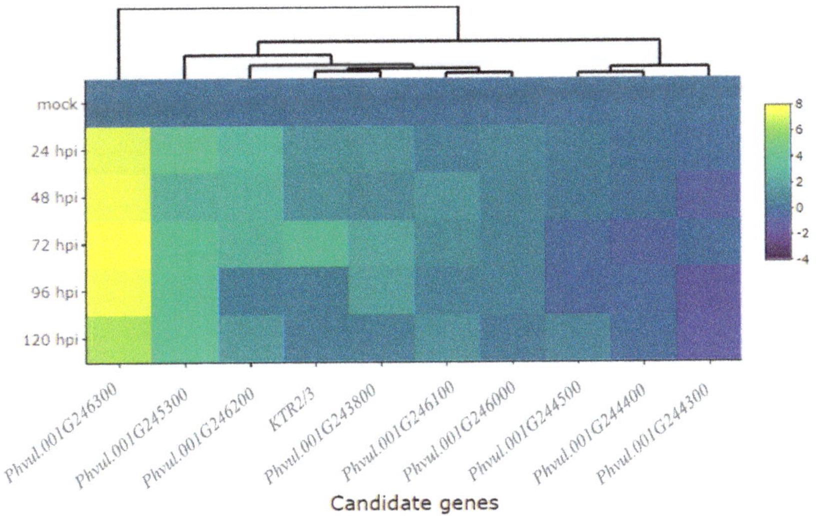

Figure 3. Heatmap of relative expression of candidate genes for the *CoPv01^{CDRK}/PhgPv01^{CDRK}* and genes proximal to these loci in California Dark Red Kidney at 24, 48, 72, 96, and 120 h post-inoculation (hpi) with race 73 of *C. lindemuthianum* and a mock. The genes evaluated were *Phvul.001G246300, Phvul.001G245300, Phvul.001G246200, KTR2/3, Phvul.001G243800, Phvul.001G246100, Phvul.001G246000, Phvul.001G244500, Phvul.001G244400*, and *Phvul.001G244300*. Yellow shading indicates higher expression and dark blue shading lower expression.

Expression of the candidate gene *Phvul.001G245300* revealed an approximate 3.7-fold increase at 24, 72, 96, and 120 hpi (Figure 2B), and *Phvul.001G246200* expression increased 2.7-fold at 24, 48, and 72 hpi (Figure 2C and Table 1). KTR2/3 expression exhibits higher expression only at 72 hpi, a 3.1-fold change (Figure 2D and Table 1). Although these genes exhibited some different expression levels in relation to the mock, they did not show large differences, as observed in *Phvul.001G246300* or in pathogen-related genes (Figures 2 and 3, and Table 1).

The candidate genes *Phvul.001G243800, Phvul.001G246100*, and *Phvul.001G246000* had the lowest levels of relative expression. The *Phvul.001G243800* gene exhibited 2.0-fold change in relative expression at 72 and 96 hpi (Figure 2E and Table 1). *Phvul.001G246100* and *Phvul.001G246000* displayed an average of 1-fold change in relative expression (Figure 2F,G and Table 1). The candidate genes for resistance in Amendoim Cavalo, namely, *Phvul.001G244500, Phvul.001G244400*, and *Phvul.001G244300*, were not expressed in CDRK plants after inoculation of race 73 in relation to mock plants, indicating that these genes may not be involved in the *CoPv01^{CDRK}/PhgPv01^{CDRK}* response to *C. lindemuthianum* race 73 (Figure 2H–J and Figure 3, and Table 1).

Regarding the pathogenesis-related genes *PR1b* (*Phvul.006G196900*), *PR2* (*Phvul.009G256400*), and *PR1a* (*Phvul.003G109100*), their reaction to the pathogen was high, with their expression surging by more than 6-fold and, in some cases, even up to 16.7-fold post-inoculation (Figures 4 and 5, and Table 1). Among them, *PR1b* (*Phvul.006G196900*) stood out as the most responsive to the pathogen, exhibiting a substantial 13.6-fold increase in expression from 24 to 96 hpi and a remarkable 16.7-fold increase at 120 hpi (Figures 4 and 5, and Table 1). The gene *PR2* (*Phvul.009G256400*) showed a moderate level of expression and response upon exposure to the pathogen. Remarkably, there was a 10-fold increase in expression observed at both 24 and 48 hpi, alongside an 11-fold increase at 72, 96, and 120 hpi (Figure 4B). Similarly, the gene *PR1a* (*Phvul.003G109100*) exhibited a significant increase from 6.2- to 8.1-fold in expression following inoculation (Figure 4C). These findings shed light on the distinct patterns of gene response to the pathogen, indicating that *Phvul.009G256400* and *Phvul.003G109100* may play crucial roles in plant defenses.

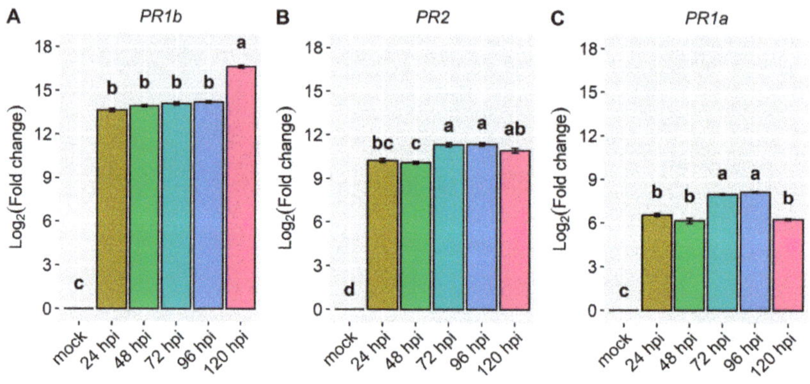

Figure 4. Relative expression of plant defense genes: (**A**) *Phvul.006G196900 (PR1b)*; (**B**) *Phvul.009G256400 (PR2)*; and (**C**) *Phvul.003G109100 (PR1a)* in the common bean cultivar of California Dark Red Kidney at 24, 48, 72, 96, and 120 h post-inoculation (hpi) with race 73 of *C. lindemuthianum* and a mock. The results are presented as logarithmic base 2 of the fold change of gene expression. Means with the same letter, for each gene, are not significantly different at the 5% significance level, using the Alexander-Govern test.

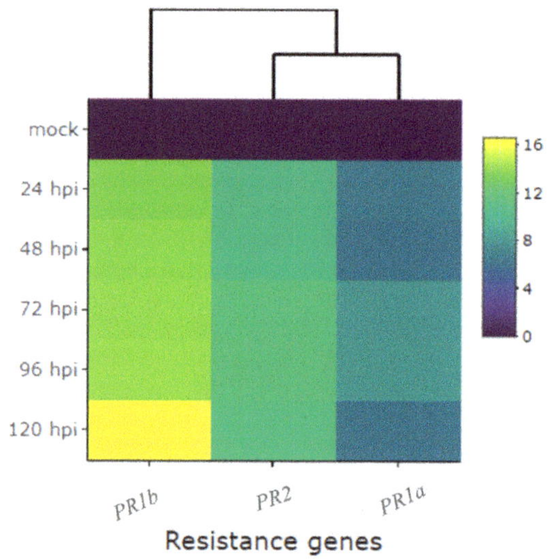

Figure 5. Heatmap of relative expression of plant defense genes *Phvul.006G196900* (*PR1b*), *Phvul.009G256400* (*PR2*), and *Phvul.003G109100* (*PR1a*) in the common bean cultivar of California Dark Red Kidney at 24, 48, 72, 96, and 120 h post-inoculation (hpi) with race 73 of *C. lindemuthianum* and mock. Yellow shading indicates higher expression, and dark blue indicates lower expression.

2.3. Differential Expression of Candidate and Defense Genes in CDRK Cultivar Inoculated with *P. griseola* Race 63-39

Upon inoculating the CDRK cultivar with race 63-39 of *P. griseola*, significant alterations in the relative expression levels of candidate genes for the $CoPv01^{CDRK}/PhgPv01^{CDRK}$ loci and resistance genes were observed. Among the candidate genes assessed at the $CoPv01^{CDRK}/PhgPv01^{CDRK}$ loci, the *Phvul.001G246300* gene showcased the utmost responsiveness to race 63-39 of *P. griseola* in the CDRK cultivar, as illustrated in Figures 6A and 7.

Notably, this gene exhibited a substantial increase in expression at 72 hpi, reaching a fold change of up to 8.5-fold (Figure 6A). Furthermore, at 120 and 216 hpi, the gene maintained a consistent mean fold change of 6.1-fold, indicating sustained response to the pathogen. These findings emphasize the potential significance of *Phvul.001G246300* in the response of the CDRK cultivar to *P. griseola* race 63-39.

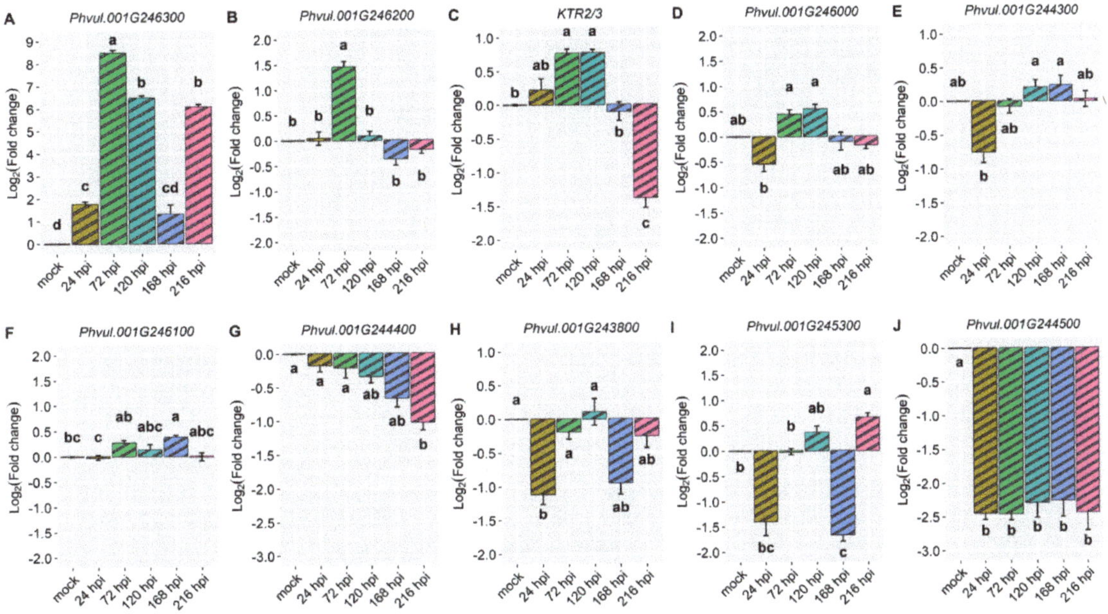

Figure 6. Relative expression of candidate genes: (**A**) *Phvul.001G246300*; (**B**) *Phvul.001G246200*; (**C**) *KTR2/3*; (**D**) *Phvul.001G246000*; (**E**) *Phvul.001G244300*; (**F**) *Phvul.001G246100*; (**G**) *Phvul.001G244400*; (**H**) *Phvul.001G243800*; (**I**) *Phvul.001G245300*; and (**J**) *Phvul.001G244500* in California Dark Red Kidney at 24, 72, 120, 168, and 216 h post-inoculation (hpi) with race 63-39 of *P. griseola* and a mock. The results are presented as logarithmic base 2 of the fold change of gene expression. Means with the same letter, for each gene, are not significantly different at the 5% significance level, using the Alexander-Govern test.

The gene *Phvul.001G246200* showed induction only at 72 hpi, with an increase of nearly 1.5-fold (Figure 6B). As for the KTR2/3 gene, it exhibited induction at 72 and 120 hpi, with a fold change increase of 0.8, but was subsequently downregulated at 216 hpi (Figure 6C). The gene *Phvul.001G246000* demonstrated minimal changes in expression levels at 72 and 120 hpi (Figure 6D), while *Phvul.001G244300* exhibited similar patterns with few changes at 120 and 168 hpi (Figure 6E). The *Phvul.001G246100* gene displayed induction from 72 to 168 hpi, albeit with a small increase of 0.3-fold (Figure 6F).

On the contrary, the *Phvul.001G244400* gene displayed downregulation at 216 hpi, with a reduction of 1-fold (Figure 6G). The *Phvul.001G243800* gene exhibited downregulation at 24 hpi, with a reduction of 1-fold (Figure 6H). *Phvul.001G245300* demonstrated downregulation at 24 and 168 hpi, followed by induction at 120 and 216 hpi, but with a slight increase of 0.4 and 0.7-fold (Figure 6I). The gene *Phvul.001G244500* consistently showed downregulation at all evaluated time points, with a significant reduction of approximately 2.3-fold in response to the pathogen (Figure 6J). These observations provide valuable insights into the dynamic expression patterns of these genes in response to the pathogen, shedding light on their potential roles in the defense mechanisms of CDRK plants.

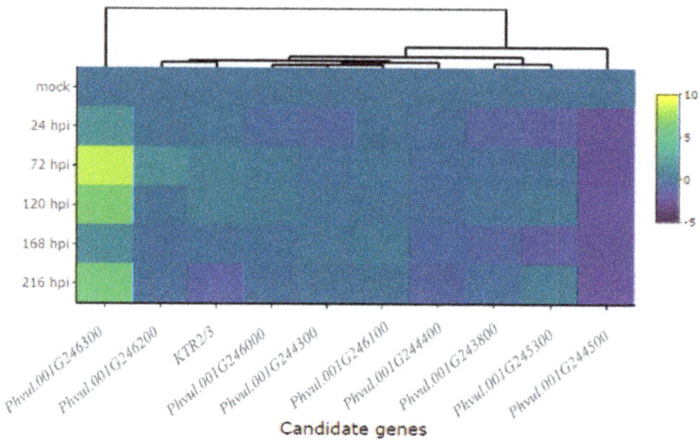

Figure 7. Heatmap of relative expression of candidate genes for the *CoPv01*^{CDRK}/*PhgPv01*^{CDRK} and genes proximal to these loci in California Dark Red Kidney at 24, 72, 120, 168, and 216 h post-inoculation (hpi) with race 63-39 of *P. griseola* and a mock. The genes evaluated were *Phvul.001G246300, Phvul.001G246200, KTR2/3, Phvul.001G246000, Phvul.001G244300, Phvul.001G246100, Phvul.001G244400, Phvul.001G243800, Phvul.001G245300,* and *Phvul.001G244500*. Yellow shading indicates higher expression and dark blue shading lower expression.

When the CDRK cultivar was inoculated with race 63-39 of *P. griseola*, pathogen-related genes exhibited high expression responses, as evidenced in Figures 8 and 9. The *PR2* gene displayed a significant increase in expression at 72 hpi, reaching peak induction levels at 120 and 216 hpi with fold increases of around 5.3. At 24 and 168 hpi, there was a slight increase in the expression level of this gene, with an average fold change above 1.9 (Figure 8A and Table 1). The *PR1a* gene demonstrated an increase in gene expression starting from 24 hpi and peaking at 216 hpi, with a fold increase of nearly six in response to the pathogen (Figure 8B). The *PR1b* gene exhibited induction at 120 and 216 hpi, with fold changes above three and four, respectively. However, at 24 and 168 hpi, the expression level of this gene decreased compared to 72, 120, and 216 hpi, but no differences were observed compared to the mock (Figure 8C). Overall, the *PR1a* and *PR2* genes showed similar expression patterns and were more responsive to race 63-39 of *P. griseola* than the *PR1b* gene (Figure 9).

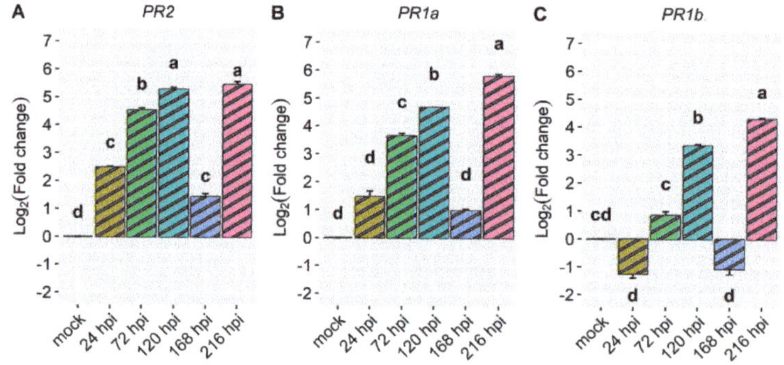

Figure 8. Relative expression of plant defense genes: (**A**) *Phvul.009G256400* (*PR2*); (**B**) *Phvul.003G109100* (*PR1a*); and (**C**) *Phvul.006G196900* (*PR1b*) in the common bean cultivar of California Dark Red Kidney at 24, 72, 120, 168, and 216 h post-inoculation (hpi) with race 63-39 of *P. griseola* and mock. The results are presented as logarithmic base 2 of the fold change of gene expression. Means with the same letter, for each gene, are not significantly different at the 5% significance level, using the Alexander-Govern test.

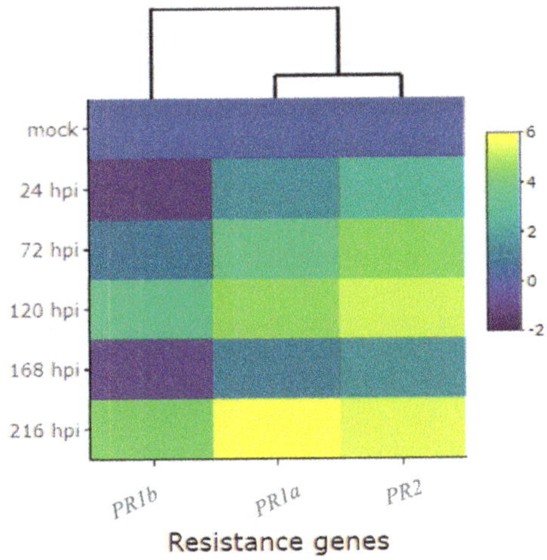

Figure 9. Heatmap of relative expression of plant defense genes *Phvul.006G196900* (*PR1b*), *Phvul.003G109100* (*PR1a*) and *Phvul.009G256400* (*PR2*) in the common bean cultivar of California Dark Red Kidney at 24, 72, 120, 168, and 216 h post-inoculation (hpi) with race 63-39 of *P. griseola* and a mock. Yellow shading indicates higher expression and dark blue has lower expression.

3. Discussion

This study discusses the cellular mechanisms employed by plants to combat unfavorable conditions caused by biotic factors. These responses are complex networks that involve changes in gene expression, regulation of metabolic processes, reinforcement of the plant cell wall, and hormone signaling pathways. In particular, the majority of resistance genes identified encode NBS-LRR proteins, which consist of an amino-terminal signaling domain, a nucleotide-binding site (NBS), and carboxy-terminal leucine-rich repeats (LRRs). These NBS-LRR proteins are capable of recognizing pathogen effectors through protein–protein interactions, subsequently triggering effector-triggered immunity [25,26]. However, it is noteworthy that the common bean chromosome Pv01 exhibits a low abundance of NBS-LRR genes. Instead, this chromosome harbors genes that encode other proteins involved in the resistance response, such as kinases functioning as pattern-recognition receptors (PRRs) that recognize pathogen-associated molecular patterns (PAMPs) and initiate PAMP-triggered immunity (PTI).

In this research, we assessed the expression levels of candidate genes near the *CoPv01^{CDRK}*/*PhgPv01^{CDRK}* loci and plant defense genes in CDRK cultivar. Through this analysis, we identified genes that consistently manifested high expression in resistant plants under pathogen inoculation. We observed that the candidate gene *Phvul.001G246300* as a potential candidate for the *CoPv01^{CDRK}*/*PhgPv01^{CDRK}* resistance loci in CDRK plants inoculated with *C. lindemuthianum* race 73, as it demonstrated the highest relative expression among the candidate genes. Among the tested candidate genes, the expression of *Phvul.001G246300* demonstrated a significant 7.8-fold increase at 24, 48, 72, 96, and 120 hpi of race 73 of *C. lindemuthianum*. Moreover, it exhibited an 8.5-fold change at 72 hpi and over a 6-fold change at 120 and 216 hpi with race 63-39 of *P. griseola*, indicating its heightened responsiveness to both pathogens compared to the other candidate genes examined. This particular gene encodes an abscisic acid (ABA) receptor, PYL5 protein, which is known to play a crucial role in mediating the crosstalk between ABA and jasmonic acid (JA) responses [27]. ABA has been implicated in plant defense against pathogens and shows synergistic interactions with the ethylene (ET) signaling pathway. The upregulation of the PYR receptor during biotic stress suggests its involvement in perceiving ABA and initiating downstream signaling mediated by kinases [28].

The *Phvul.001G245300* gene ranked second in terms of induction among the candidate genes for *CoPv01*^{CDRK}/*PhgPv01*^{CDRK} loci in plants inoculated with race 73 of *C. lindemuthianum*. However, its expression was only half that of *Phvul.001G246300*, and when the plants were inoculated with race 63-39 of *P. griseola*, this gene showed a less pronounced response, with higher expression observed only at 216 hpi. Hypothetically, this gene encodes a protein belonging to the protein kinase superfamily, which is predominantly composed of catalytic domains of serine/threonine-specific and tyrosine-specific protein kinases. Furthermore, the protein contains a leucine-rich repeat (LRR) domain. Proteins containing LRRs, including tyrosine kinase receptors, are involved in diverse biological processes such as signal transduction, cell adhesion, DNA repair, recombination, transcription, RNA processing, apoptosis, disease resistance, and immune responses [29,30].

The *Phvul.001G246200* gene ranked third among the upregulated genes, mainly between 24 and 72 hpi with race 73 of *C. lindemuthianum*, but its induction was three times lower than that of *Phvul.001G246300*. When plants were inoculated with race 63-39 of *P. griseola*, this gene exhibited a modest upregulation at 72 hpi, five times less than *Phvul.001G246300*. In Arabidopsis thaliana, a homologous gene plays a crucial role in xylan acetylation and the proper deposition of secondary walls [31]. Acetylation of wall polymers is important for cell wall strength and disease resistance, as evidenced by several Arabidopsis mutants and overexpression lines [31,32]. Therefore, the *Phvul.001G246200* gene may contribute to the resistance response in the CDRK cultivar by modifying cell wall strength, thereby impeding pathogen infection and/or colonization of plant tissues during the biotrophic life stage.

The majority of the ANT resistance genes identified in Andean cultivars are located in a resistance cluster on the common bean chromosome Pv01, including *Co-1*, *Co-1*^{2}, *Co-1*^{3}, *Co-1*^{4}, *Co-1*^{5}, *Co-1*^{HY}, *Co-1*^{X}, *Co-x*, *Co-AC*, and *CoPv01*^{CDRK}/*PhgPv01*^{CDRK} (Figure 10). Interestingly, the resistance gene from the Jalo EEP558 cultivar, *Co-x*, conferring resistance against race 3993 of *C. lindemuthianum*, is located in close proximity to *CoPv01*^{CDRK}/*PhgPv01*^{CDRK} loci. The protein KTR2/3 has been identified as the controlling factor for resistance of *Co-x* in Jalo EEP558 cultivar [23]. In our study with the CDRK cultivar, we observed that the expression of the *KTR2/3* gene was 2.6 times lower than that of *Phvul.001G246300* at 72 hpi in response to race 73 of *C. lindemuthianum*. Furthermore, in response to race 63-39 of *P. griseola*, the *KTR2/3* gene exhibited a remarkably lower expression level, approximately 10.6 times lower, compared to *Phvul.001G246300*. Taking together, the physical position where *CoPv01*^{CDRK}/*PhgPv01*^{CDRK} and *Co-x* were mapped, and different candidate genes significantly upregulated after inoculation in each cultivar, *KTR2/3* for Jalo EEP558 and *Phvul.001G246300* for CDRK, and its possible to conclude that different genetic resistances are involved in each cultivar against the same pathogen.

Using near-isogenic lines with differing resistance alleles, expression analysis of candidate genes for the *Co-1*^{2} allele against *C. lindemuthianum* race 73 resistance revealed high levels of *Phvul.001G243800* in the resistant NIL [22]. Furthermore, based on transcriptional analysis, it was observed that *Phvul.001G243700*, located near the *Co-1* locus, was differentially upregulated in the resistant NIL at 72 and 96 hpi after race 73 inoculation [33]. In the present study, significant differences in the expression of the *Phvul.001G246300* and *Phvul.001G243800* genes in response to race 73 of *C. lindemuthianum* and race 63-39 of *P. griseola* were observed in CDRK cultivar. At 72 hpi with race 73 of *C. lindemuthianum*, we observed a notable expression of the *Phvul.001G246300* gene, which was approximately three times higher compared to *Phvul.001G243800*. Remarkably, within the context of *P. griseola* race 63-39, we observed that *Phvul.001G243800* gene was repressed, and on the other hand, the expression of *Phvul.001G246300* gene exhibited a substantial increase of more than 8-fold.

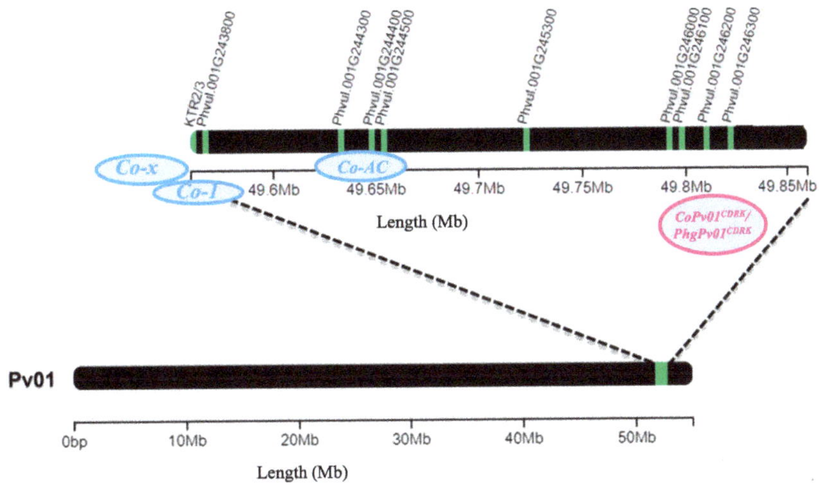

Figure 10. Common bean chromosome Pv01 containing candidate genes for anthracnose resistance genes *Co-x* (*KTR2/3*), *Co-1* (*Phvul.001G243800*), *Co-AC* (*Phvul.001G244300*, *Phvul.001G244400*, and *Phvul.001G244500*), and *CoPv01*^{CDRK}/*PhgPv01*^{CDRK} (*Phvul.001G245300*, *Phvul.001G246000*, *Phvul.001G246100*, *Phvul.001G246200*, and *Phvul.001G246300*).

Overall, our findings underscore the contrasting expression patterns of the *Phvul.001G246300* and *Phvul.001G243800* genes within the CDRK cultivar, shedding explanation on their potential roles in the defense mechanisms against *C. lindemuthianum* race 73 and *P. griseola* race 63-39. It is worth noting that the *Co-1* and *CoPv01*^{CDRK}/*PhgPv01*^{CDRK} loci were accurately mapped to separate regions towards the terminal end of the common bean chromosome Pv01, positioned 211 kb apart. Consequently, it is of paramount importance to emphasize that the CDRK cultivar harbors a distinct and independent gene from the *Co-1* locus.

The pathogenesis-related defense genes *PR1a*, *PR1b*, and *PR2* exhibited significant responsiveness to the pathogen. Notably, *PR1b* displayed the highest level of responsiveness to pathogen race 73, with a pronounced increase in expression at 120 hpi, reaching up to 16.7-fold higher than the control. Both *PR2* and *PR1a* displayed elevated expression levels, exhibiting 11.4-fold and 8.1-fold increases, respectively, compared to the mock. Notably, *PR2* maintained consistently high expression levels from 24 hpi to 120 hpi in response to the pathogen. Mahiya-Farooq et al. [22] reported early expression of plant defense genes in the resistant NIL, with *PR1b* and *PR2* showing accumulation at 24 hpi of *C. lindemuthianum* race 73. Similarly, Shams et al. [34] observed higher expression of *PR2* in the Naz-resistant bean cultivar upon inoculation with *C. lindemuthianum* race 2. Although to a lesser extent, these pathogenesis-related defense genes were also upregulated in plants inoculated with race 63-39 of *P. griseola*. *PR1a* and *PR1b* showed higher induction at 216 hpi, with nearly a 6-fold increase and over a 4-fold increase in gene expression, respectively. The *PR2* gene exhibited a more than 5-fold increase at 120 and 216 hpi. Interestingly, at 168 hpi, all the genes displayed reduced or no responsiveness to the pathogen.

These results reveal that the ANT and ALS resistance genes in the CDRK cultivar are controlled independently from those previously identified at the *Co-1* locus. This indicates that the robust resistance against ANT and ALS in CDRK is manifested through the heightened response of the candidate gene *Phvul.001G246300* to the respective pathogens. These findings point out the complex nature of plant–pathogen interactions, emphasizing the significance of comprehending gene expression mechanisms. Our findings might contribute to an enhanced comprehension of novel and efficient strategies for the development of cultivars resistant to angular leaf spot and ANT.

4. Materials and Methods

4.1. Plant Material, Growth Conditions, and Experimental Design

Two experiments were conducted using a completely randomized design. In Experiment I, seedlings from both the resistant (R) CDRK and the susceptible (S) Yolano cultivars were inoculated with race 73 of *C. lindemuthianum*. The isoline derived from the CDRK × Yolano cross, namely, CY 19, was used as a positive control for disease scoring to ensure that the fungus was virulent on them and that the absence of disease on the resistant genotypes was not attributed to the fungus lacking virulence. Additionally, the CY 70 isoline was used as a resistant control. The relative expression levels of 13 specific genes only in CDRK cultivar were evaluated at multiple time points: 24, 48, 72, 96, and 120 h post-inoculation (hpi), as well as in the mock. In Experiment II, a parallel approach was adopted, involving the same genotypes as in Experiment I. However, the genotypes were subjected to inoculation with race 63-39 of *P. griseola*. Similarly, the relative expression of the same 13 genes only in the CDRK cultivar were evaluated at 24, 72, 120, 168, and 216 hpi, in addition to the mock condition. Each experimental condition was replicated across three separate biological replicates (plants). Within each biological replicate, the assessment was further reinforced by performing three technical replicates of quantitative polymerase chain reaction (qPCR) for each experiment.

The experiments were conducted at the Núcleo de Pesquisa Aplicada à Agricultura (Nupagri) at the Universidade Estadual de Maringá (UEM) in Maringá, Paraná, Brazil (latitude 23° 26′8″ S, longitude 51° 53′42″ W). Briefly, seeds were planted in plastic trays filled with a commercial substrate, MecPlant (MEC PREC—Ind. Com Ltd.a, Telemaco Borba, Brazil), that had been previously sterilized and fertilized. The seedlings were grown in greenhouses under natural light at a temperature of 25 °C until the first trifoliate leaf growth stage [35].

4.2. Pathogenesis Assays

Monosporic cultures of *C. lindemuthianum* and *P. griseola* were prepared according to the methodologies described by Mathur et al. [36] and Sanglard et al. [37], respectively. Inocula of the ANT pathogen races were produced by incubating them on a young green common bean pod medium [38] at 22 °C for 14 days. The inoculum for the ALS pathogens was multiplied in Petri dishes with tomato medium [37,39] containing 1.61% agar (m/v), 0.25% calcium carbonate (m/v), 61.94% distilled water (v/v), and 36.2% V8® vegetable juice (Campbell's company soup (v/v)) and maintained in a bio-oxygen demand incubator at 24 °C for 18 days. The concentration of fungal spores for the ANT pathogen was adjusted to 1.2×10^6 [38]. For the ALS pathogen, it was adjusted to 1.2×10^4 conidia ml^{-1} [40] using a hemacytometer (1/400 mm^2, Hausser Scientific, Horsham, PA, USA). Fourteen-day-old seedlings were inoculated with each race of the pathogen on the underside of their leaves. The inoculation process was carried out by using a manual pressurized pump sprayer for spraying. For the mock treatment, which served as the negative control, the seedlings were sprayed with only distilled water and Tween 20® (0.01%). After inoculation, the plants were maintained in a mist chamber at >95% relative humidity, at a temperature of 20 ± 2 °C, with 12 h of daily light (680 lux) for 72. Following the inoculation process, the seedlings were moved to benches and placed under the same conditions as before, except for a high-humidity environment. This environment was maintained until the end of the experiment when all the samples were collected. Anthracnose and angular leaf spot symptoms were evaluated using 1-to-9 disease severity scales proposed by Pastor-Corrales et al. [35] and Inglis et al. [41]. Plants with disease reaction scores between 1 and 3 were considered resistant, whereas plants with scores from 4 to 9 were considered susceptible.

4.3. Total RNA Extraction and cDNA Synthesis

Leaf samples were collected from CDRK plants before inoculation (mock) and during incompatible reactions with race 73 of *C. lindemuthianum*, as well as with race 63-39 of *P. griseola*. Sampling was performed at specific time points critical for pathogens development: 24, 48, 72, 96, and 120 hpi for *C. lindemuthianum*, and 24, 72, 120, 168, and 216 hpi for *P. griseola*. Samples were obtained from three biological replicates for each pathogen. To ensure RNA integrity, the leaf samples were promptly frozen in liquid nitrogen for subsequent extraction [42].

Total RNA was extracted from 100 mg frozen and purified using GeneJet Plant RNA Purification Kit (Thermo Fisher Scientific, Waltham, MA, USA) following the manufacturer's instructions. The integrity of the total RNA was assessed with electrophoresis on a 1% m/v agarose gel, run for 80 min at 80 volts, at 5 °C, and in the absence of light. For the assessment of both the quality and quantity of total RNA, a spectrophotometer (FEMTO 700S^{TM}) was employed to measure absorbance at specific wavelengths: 230 nm, 240 nm, 260 nm, and 280 nm. The criteria for RNA purity were determined based on the following absorbance ratios: A_{260}/A_{230} ranging from 1.9 to 2.4, A_{260}/A_{240} of at least 1.4, and A_{260}/A_{280} between 1.8 and 2.2. To compute the concentration of total RNA, the formula [RNA] (ng μL^{-1}) = $A_{260\ nm} \times 40 \times 100$, as outlined by Farrell [43], was utilized. Total RNA samples that met the purity and integrity criteria were treated with DNase I^{TM} (Invitrogen^{TM}, Waltham, MA, USA) to eliminate any possible genomic DNA contamination. The purification reaction involved 1 μg of total RNA, following the manufacturer's instructions.

To synthesize cDNA, the "Superscript® IV First-Strand Synthesis System" kit (Invitrogen^{TM}, Waltham, MA, USA) was used according to the manufacturer's instructions. The cDNA synthesis reaction was prepared with a total volume of 20 μL, containing the following constituents: 1 μg of total RNA, primer-oligo d(T) at a concentration of 2.5 μM, dNTP mix with each nucleotide at 0.5 mM, First-Strand Buffer at 1× concentration, DL-dithiothreitol at 5 mM, ribonuclease inhibitor with a concentration of 2 U μL^{-1}, MMLV-RT at 10 U μL^{-1}, and RNase-free water. The procedure was initiated by combining total RNA, primer-oligo d(T), dNTP mix, and RNase-free water to reach a cumulative volume of 13 μL. The samples were incubated in a thermocycler (Applied Biosystems® Veriti® 96-Well Fast Thermal Cycler, Waltham, MA, USA) at 65 °C for 5 min, followed by 4 °C for 1 min. Then, the First-Strand Buffer, DL-dithiothreitol, ribonuclease inhibitor, and 1 MMLV-RT were added to the reaction. The samples were incubated at 55 °C for 10 min for cDNA synthesis activation, followed by 80 °C for 10 min to inactivate the reaction. To remove residual RNA after cDNA synthesis, 1 μL of Escherichia coli RNase H was added, and the samples were incubated at 37 °C for 20 min. The cDNA synthesis product (20 μL) was diluted 1:100 for qPCR analysis. To assess the cDNA synthesis efficiency, positive control was included, in which HeLa-S3 RNA (10 ng) was used instead of total RNA.

To verify the quality of the cDNA synthesis, a PCR reaction was performed on the positive control and a negative control (containing water instead of cDNA) using the following reaction mix: 5 μL PCR buffer (10×), 2 μL MgCl (50 mM), 1 μL dNTP Mix (10 mM), 1 μL sense primer (10 μM), 1 μL antisense primer (10 μM), 2 μL of cDNA for the positive control, and 2 μL of ultrapure H_2O for the negative control, 0.2 μL of Taq PlatinumTM DNA polymerase (Invitrogen^{TM}, Waltham, USA), and 37.8 μL of ultrapure H_2O. The PCR reaction was performed for 35 cycles, with an initial denaturation at 94 °C for 2 min, denaturation at 94 °C for 15 s, annealing at 55 °C for 30 s, and synthesis at 68 °C for 1 min. The PCR products were analyzed using electrophoresis on a 1.5% *m/v* agarose gel. The positive control showed a single band of approximately 353 bp, while no product was observed in the negative control, confirming the efficiency of the cDNA synthesis.

4.4. Target Genes and Primer Design

Ten candidate genes were selected for expression analysis in CDRK plants inoculated with race 73 of *C. lindemuthianum* and race 63-39 of *P. griseola*. The genes evaluated in the *CoPv01^{CDRK}/PhgPv01^{CDRK}* resistance loci were *Phvul.001G246000*, *Phvul.001G246100*, *Phvul.001G246200*, and *Phvul.001G246300*. Additionally, the gene *Phvul.001G245300* located near these loci was also assessed [15]. The genes proposed for the *Co-AC* locus, namely, *Phvul.001G244300*, *Phvul.001G244400*, and *Phvul.001G244500* [14], were also tested. The gene *Phvul.001G243800* was evaluated as it was induced in the near isogenic line

T9576R, which carries the *Co-1*^{2} resistance allele, when inoculated with race 73 of *C. lindemuthianum* [22]. The *KTR2/3* gene, a candidate gene for *Co-x* in Jalo EEP558, which was induced in response to strain 100 of *C. lindemuthianum*, was also included [23]. Finally, three known plant defense genes, namely, *Phvul.003G109100* (*PR1a*), *Phvul.006G196900* (*PR1b*), and *Phvul.009G256400* (*PR2*), were evaluated [16,22,42]. To standardize gene expression levels, we employed the reference genes *Phvul.008G011000* (actin–ACT) and *Phvul.001G133200* (insulin-degrading enzyme–IDE) [44]. ACT had previously been validated for quantifying the relative expression of candidate genes in studies conducted by Mahiya-Farooq et al. [22] and Shams et al. [34], while IDE's validation was previously established by Oblessuc et al. [20]. Both genes were utilized as reference genes for quantifying the relative expression of resistance genes against ANT in studies conducted by Borges et al. [44] and Richard et al. [23].

To design primers for qPCR, the coding sequences (CDS) and DNA sequences of the target genes were downloaded from the common bean (*P. vulgaris* L.) genome available at Phytozome 12 [45]. The "Primer-BLAST web tool" [46] was used to design primers that met the following specifications for efficient qPCR: primer size 18-24 bp, melting temperature between 59–61 °C, amplicon size between 80-160 bp, and whenever possible, at least one intron on the corresponding genomic DNA sequence was included between the primer pair. The primers for the *KTR2/3* gene were obtained from Richard et al. [23].

To ensure the specificity and efficiency of the primers, dimers and secondary structures were checked using Gene Runner software (version 6.5.52), the "Multiple Prime Analyzer" web tool (Thermo Fisher Scientific: https://bit.ly/34kZpnP (accessed on 29 May 2020)), and "The Sequence Manipulation Suite" web tool [47]. The amplicon secondary structure was also verified using the "The Mfold Web Server" platform [48] with coding sequences downloaded from Phytozome 12. All procedures used for primer design and in silico validation followed literature recommendations [49,50]. The primer sequences for each candidate gene evaluated are listed in Table 2.

4.5. Quantitative PCR (qPCR) and Data Analysis

The determination of PCR efficiency for each primer involved establishing a standard curve through a fivefold serial dilution, utilizing the cDNA pool as the template. This process incorporated three replicates at every dilution point, following the methodologies outlined by Svec et al. [51] and Rasmussen [52]. The amplification efficiency was computed employing the equation $E = [10^{(-1/\text{slope})}] - 1$ [52], utilizing the slope values derived from linear regression analysis. This analysis encompassed the $\log_{10}$-transformed cDNA concentrations on the x-axis and corresponding Cq values on the y-axis. The calculated amplification efficiency for each primer pair ranged from 0.92 to 1.09, while maintaining a coefficient of determination (R^2) for the linear regression of at least 0.97 (Table 2).

The cDNA quantification reactions were conducted in the StepOnePlus™ real-time PCR system (Applied Biosystems™; StepOnePlus™ Real-Time PCR Systems) using 96-well microplates (MicroAmp™ Fast 96 -well Reaction Plate (0.1 mL)) sealed with MicroAmp™ Optical Adhesive Film. The total reaction volume was 10 µL, consisting of 3.4 µL of cDNA, 1.6 µL of forward and reverse primer mix (800 nM), and 5 µL of PowerUp™ SYBR™ Green Master Mix (Applied Biosystems™, Waltham, MA, USA). The thermocycling conditions included 50 °C for 2 min, 95 °C for 2 min, 40 cycles of 15 s at 95 °C, and 30 s at 60 °C.

After completion of the cDNA quantitation reaction, a dissociation curve was performed to verify target specificity using the manufacturer's standard continuous melt curve setup, and only samples exhibiting specificity based on the dissociation curve were used. Cq (quantification cycle) values were obtained using StepOnePlus™ Software v2.3 (Applied Biosystems™), with the baseline determined automatically and the threshold was determined manually in the exponential phase of amplification (0.7707 for all cDNA quantification reactions).

Table 2. Target genes, primers used, qPCR product size (amplicon), primer melting temperature (Tm), amplification efficiency (E,) and coefficient of determination of linear regression (R^2).

Gene Model [a]	Reference Genes	Primers Forward (F) and Reverse (R) (5′-3′)	Amplicon (bp)	Tm (°C)	E [b]	R^2 [c]
Phvul.001G133200	*IDE*	F: AAGCAGGTATCTTGGCCATCTC R: AAAGCAAACTCCAAGCTCCAATC	126	60.16 59.99	1.04	0.99
Phvul.008G011000	*ACT*	F: GAAGTTCTCTTCCAACCATCC R: TTTCCT TGCTCATTCTGTCCG	154	59.67 58.38	1	0.98
Phvul.001G243800	*Co-1*	F: CCTCAAGGTGGGGCTTTTGAG R: TCACCGAGAAACTCCCATTGC	118	61.16 60.61	1.04	0.99
KTR2/3	*Co-x*	F: ATGCACAGGGGAATGGGATG R: GCCATAGCGAGTGAGAGTGCG	279	60.11 63.42	1.04	0.98
Phvul.001G244300	*Co-AC*	F: GAAACGTCTCCGCAGAATAGTG R: GTCTTGTGTTGTTCCTTGGAGTTG	150	59.4 60.44	1.03	0.99
Phvul.001G244400	*Co-AC*	F: TACAGCAAGAGAGCGGTTAAAGG R: CCCTTTGTCACTTTGTTTTGAAGC	121	60.62 59.67	1.01	0.99
Phvul.001G244500	*Co-AC*	F: CAATGCACAGCTCGCAACTC R: GGAACTGTGAAAGCTCTGCTAAC	141	60.45 59.81	1.09	0.97
Phvul.001G245300	*CoPv01^CDRK^/ PhgPv01^CDRK^*	F: TCTGCTGGAAGGGTGGTAGTC R: GGACGTTATGTGAACAAGGTTTGC	93	61.17 61.08	1.07	0.98
Phvul.001G246000	*CoPv01^CDRK^/ PhgPv01^CDRK^*	F: ATGAAGCGGATGGATGTCTTG R: TCTACGAAGCTTAGGCAATTGAG	132	58.43 58.57	1.01	0.97
Phvul.001G246100	*CoPv01^CDRK^/ PhgPv01^CDRK^*	F: CACGGTATCCTCAGCGAAGAC R: CAGCAGTCAGCACATACTGGAG	119	60.53 60.99	1.05	0.99
Phvul.001G246200	*CoPv01^CDRK^/ PhgPv01^CDRK^*	F: GAAGGAGGCTGTGACGTGTTC R: CCATCGCCACCGTTGATACTC	150	61.2 61.13	1.04	0.99
Phvul.001G246300	*CoPv01^CDRK^/ PhgPv01^CDRK^*	F: CTTCTTCCCTTCACTTCGATACC R: GTTGAGAGTGTTTGTGGCAGT	87	58.57 58.98	1.09	0.99
Phvul.003G109100	*PR1a*	F: GTCCTAACGGAGGATCACTCA R: CAGGGATTGGCCAGAAGGTAT	148	58.62 59.5	1.06	0.99
Phvul.006G196900	*PR1b*	F: GGTTTGCCTATGATCCCAATGC R: TGTTGTGAGCGTTGAGGAAGTC	115	59.96 61.06	1.03	0.99
Phvul.009G256400	*PR2*	F: CAGAGGTTCTCATTTGCTGCTTTC R: ATGCCATAACACACCCCGATTTG	98	60.62 61.75	1.07	0.99

[a] Based on the *Phaseolus vulgaris* genome available on the Phytozome 12 platform: https://phytozome.jgi.doe.gov/pz/portal.html# (accessed on 29 May 2020). [b] Amplification efficiency obtained from Equation $E = [10^{(-1/\text{slope})}] - 1$ [52]. [c] Coefficient of determination of linear regression.

The genes *Phvul.001G133200* (*ACT*) and *Phvul.008G011000* (*IDE*) were used as reference genes [44], with the arithmetic mean of C_T values [53] calculated in each experimental condition evaluated. Relative expression levels were determined as follows by normalizing the C_T values with reference genes using the $2^{-\Delta\Delta CT}$ method [54,55]:

$$\Delta\Delta C_T = [(C_T \text{ gene of interest} - \text{arithmetic mean of } C_T \text{ of reference genes}) \text{ at time x} - (C_T \text{ gene of interest} - \text{arithmetic mean of } C_T \text{ of reference genes}) \text{ at mock}].$$

The mean C_T values for each gene under each experimental condition were calculated based on three biological replicates and three technical replicates ($n = 3 \times 3$).

The relative expression of candidate genes to the *CoPv01^CDRK^/PhgPv01^CDRK^* loci and known disease resistance genes were investigated in response to race 73 of *C. lindemuthianum* at 24, 48, 72, 96, and 120 hpi (Experiment I), as well as in response to race 63-39 of *P. griseola* at 24, 72, 120, 168, and 216 hpi (Experiment II) in CDRK cultivar. The calibrator condition for each gene was the relative expression in the mock (control, without pathogen). To analyze the data and show results, logarithmic base 2 transformation of $\Delta\Delta C_T$ values were performed before statistical analysis. The expression levels among experimental conditions were compared using the Alexander-Govern test with a significance level of 5%. Pairwise comparisons of relative expression mean among time points for each gene were assessed, and the significance level was adjusted using Bonferroni correction ($p \leq 0.05$). These statistical analyses were performed using the "oneway-test" [56] and "companion" R packages. All data wrangling and statistical analysis were performed using R software (version 4.0.3) (R Core Team) [57], with plots generated using the package ggplot2 [58] and R base. Error bars represent the standard deviation of the means from three biological and three technical replicates (3 × 3). Heatmaps were generated with mean C_T values using the "heatmaply" R package, and the dendrogram was based on the Euclidean distance measure and the average linkage function [59] among the relative expression of the genes.

5. Conclusions

In conclusion, our study provides valuable insights into the genomic organization of ANT and ALS resistance genes on Pv01, specifically in the context of the CDRK cultivar. We highlight the importance of closely situated candidate genes within resistance gene clusters, as they play a crucial role in conferring effective resistance against ANT and ALS pathogens. Furthermore, the $CoPv01^{CDRK}$ and $PhgPv01^{CDRK}$ resistance loci are concomitantly inherited and can be efficiently tracked using molecular markers. Our findings shed light on the fact that the *Phvul.001G246300* gene develops a pleiotropic mechanism, demonstrating the highest responsiveness to both pathogens: *C. lindemuthianum* race 73 and *P. griseola* race 63-39. These discoveries carry significant practical implications for breeding endeavors aimed at developing bean cultivars resistant to ANT and ALS, facilitated by the implementation of marker-assisted selection.

Supplementary Materials: The supporting information can be downloaded at: https://www.mdpi.com/article/10.3390/ijms242216023/s1.

Author Contributions: Conceptualization, M.L. and M.C.G.-V.; methodology, M.L., M.C.G.-V., A.C.C. and J.M.; data curation and formal analysis, M.L., A.C.C., E.G.R.M. and J.M.; funding acquisition and project administration, P.S.V.F. and M.C.G.-V.; supervision, M.C.G.-V., P.S.V.F. and M.V.B.; writing—original draft preparation, M.L.; writing—review and editing, M.L., M.C.G.-V., M.V.B. and M.M. All authors have read and agreed to the published version of the manuscript.

Funding: This research received financial support from the National Council for Scientific and Technological Development (CNPq) Grant/Award Numbers: 141112/2019-0, 301861/2017-0, 311599/2017-6, and from the Coordination for the Improvement of Higher Education Personnel (Capes) Grant/Award Number: BEX 88887.470337/2019-00.

Institutional Review Board Statement: Not applicable.

Informed Consent Statement: Not applicable.

Data Availability Statement: Data sharing not applicable to this article.

Conflicts of Interest: The authors declare no conflict of interest.

References

1. Padder, B.A.; Sharma, P.N.; Awale, H.E.; Kelly, J.D. *Colletotrichum lindemuthianum*, the causal agent of bean anthracnose. *J. Plant Pathol.* **2017**, *99*, 317–330. [CrossRef]
2. Crous, P.W.; Liebenberg, M.M.; Braun, U.; Johannes, Z.; Groenewald, J.Z. Re-evaluating the taxonomic status of *Phaeoisariopsis griseola*, the causal agent of angular leaf spot of bean. *Stud. Mycol.* **2006**, *55*, 163–173. [CrossRef] [PubMed]
3. Miklas, P.N.; Kelly, J.D.; Beebe, S.E.; Blair, M.W. Common bean breeding for resistance against biotic and abiotic stresses: From classical to MAS breeding. *Euphytica* **2006**, *147*, 105–131. [CrossRef]
4. Singh, S.P.; Schwartz, H.F. Breeding common bean for resistance to diseases: A review. *Crop Sci.* **2010**, *50*, 2199–2223. [CrossRef]
5. Guzman, P. Characterization of variability in the fungus *Phaeoisariopsis griseola* suggests coevolution with the common bean (*Phaseolus vulgaris*). *Phytopathology* **1995**, *85*, 600. [CrossRef]
6. Pastor-Corrales, M.A.; Jara, C.E. La evolución de *Phaeoisariopsis griseola* con el frijol común en América Latina. *Fitop Colomb.* **1995**, *19*, 15–24.
7. Pastor-Corrales, M.A.; Jara, C.; Singh, S.P. Pathogenic variation in, sources of, and breeding for resistance to *Phaeoisariopsis griseola* causing angular leaf spot in common bean. *Euphytica* **1998**, *103*, 161–171. [CrossRef]
8. Stenglein, S.A.; Balatti, P.A. Genetic diversity of *Phaeoisariopsis griseola* in Argentina as revealed by pathogenic and molecular markers. *Physiol. Mol. Plant Pathol.* **2006**, *68*, 158–167. [CrossRef]

9. Gonçalves-Vidigal, M.C.; Cruz, A.S.; Garcia, A.; Kami, J.; Vidigal Filho, P.S.; Sousa, L.L.; McClean, P.; Gepts, P.; Pastor-Corrales, M.A. Linkage mapping of the *Phg-1* and *Co-1*^{4} genes for resistance to angular leaf spot and anthracnose in the common bean cultivar AND 277. *Theor. Appl. Genet.* **2011**, *122*, 893–903. [CrossRef]
10. Zuiderveen, G.H.; Padder, B.A.; Kamfwa, K.; Song, Q.; Kelly, J.D. Genome-wide association study of anthracnose resistance in Andean beans (*Phaseolus vulgaris*). *PLoS ONE* **2016**, *11*, e0156391. [CrossRef]
11. Chen, M.; Wu, J.; Wang, L.; Mantri, N.; Zhang, X.; Zhu, Z.; Wang, S. Mapping and genetic structure analysis of the anthracnose resistance *locus Co-1*^{HY} in the common bean (*Phaseolus vulgaris* L.). *PLoS ONE* **2017**, *12*, e0169954. [CrossRef] [PubMed]
12. Lima, L.R.L.; Gonçalves-Vidigal, M.C.; Vaz Bisneta, M.; Valentini, G.; Vidigal Filho, P.S.; Martins, V.S.R.; Souza, T.L.P.O. Genetic fine-mapping of anthracnose disease-resistance allele *Co-1*^{4} present in the Andean common bean cultivar AND 277. *Crop Sci.* **2023**, *63*, 750–763. [CrossRef]
13. Richard, M.M.S.; Pflieger, S.; Sevignac, M.; Thareau, V.; Blanchet, S.; Li, Y.; Jackson, S.A.; Jackson, S.A.; Geffroy, V. Fine mapping of *Co-x*, an anthracnose resistance gene to a highly virulent strain of *Colletotrichum lindemuthianum* in common bean. *Theor. Appl. Genet.* **2014**, *127*, 1653–1666. [CrossRef] [PubMed]
14. Gilio, T.A.S.; Hurtado-Gonzales, O.P.; Gonçalves-Vidigal, M.C.; Valentini, G.; Elias, J.C.F.; Song, Q.; Pastor-Corrales, M.A. Fine mapping of an anthracnose-resistance *locus* in Andean common bean cultivar Amendoim Cavalo. *PLoS ONE* **2020**, *15*, e0239763. [CrossRef] [PubMed]
15. Gonçalves-Vidigal, M.C.; Gilio, T.A.S.; Valentini, G.; Vaz Bisneta, M.; Vidigal Filho, P.S.; Song, Q.; Oblessuc, P.R.; Melotto, M. New Andean source of resistance to anthracnose and angular leaf spot: Fine-mapping of disease-resistance genes in California Dark Red Kidney common bean cultivar. *PLoS ONE* **2020**, *15*, e0235215. [CrossRef] [PubMed]
16. Oblessuc, P.R.; Baroni, R.M.; Garcia, A.A.F.; Chioratto, A.F.; Carbonell, S.A.M.; Camargo, L.E.A.; Benchimol, L.L. Mapping of angular leaf spot resistance QTL in common bean (*Phaseolus vulgaris* L.) under different environments. *BMC Genet.* **2012**, *13*, 50. [CrossRef] [PubMed]
17. Keller, B.; Manzanares, C.; Jara, C.; Lobaton, J.D.; Studer, B.; Raatz, B. Fine-mapping of a major QTL controlling angular leaf spot resistance in common bean (*Phaseolus vulgaris* L.). *Theor. Appl. Genet.* **2015**, *128*, 813–826. [CrossRef] [PubMed]
18. Sartorato, A.; Nietsche, S.; Barros, E.G.; Moreira, M.A. Inheritance of angular leaf spot resistance and RAPD markers linked to disease resistance gene in common beans. *Ann. Rep. Bean Improv. Coop.* **1999**, *42*, 21–22.
19. Gonçalves-Vidigal, M.C.; Cruz, A.S.; Lacanallo, G.F.; Vidigal Filho, P.S.; Sousa, L.L.; Pacheco, C.M.N.A.; McClean, P.; Gepts, P.; Pastor-Corrales, M.A. Co-segregation analysis and mapping of the anthracnose *Co-10* and angular leaf spot *Phg-ON* disease-resistance genes in the common bean cultivar Ouro Negro. *Theor. Appl. Genet.* **2013**, *126*, 2245–2255. [CrossRef]
20. Oblessuc, P.R.; Perseguini, J.M.K.C.; Baroni, R.M.; Chiorato, A.F.; Carbonell, S.A.M.; Mondego, J.M.C.; Vidal, R.O.; Camargo, L.E.A.; Benchimol-Reis, L.L. Increasing the density of markers around a major QTL controlling resistance to angular leaf spot in common bean. *Theor. Appl. Genet.* **2013**, *126*, 2451–2465. [CrossRef]
21. Nay, M.M.; Souza, T.L.P.O.; Raatz, B.; Mukankusi, C.M.; Gonçalves-Vidigal, M.C.; Abreu, A.F.B.; Melo, L.C.; Pastor-Corrales, M.A. A review of angular leaf spot resistance in common bean. *Crop Sci.* **2019**, *59*, 1376. [CrossRef]
22. Mahiya-Farooq; Padder, B.A.; Bhat, N.N.; Shah, M.D.; Shikari, A.B.; Awale, H.E.; Kelly, J.D. Temporal expression of candidate genes at the *Co-1* locus and their interaction with other defense related genes in common bean. *Physiol. Mol. Plant Pathol.* **2019**, *108*, 101424. [CrossRef]
23. Richard, M.M.S.; Gratias, A.; Diaz, J.C.A.; Thareau, V.; Pflieger, S.; Meziadi, C.; Blanchet, S.; Marande, W.; Bitocchi, E.; Papa, R.; et al. A common bean truncated CRINKLY4 kinase controls gene-for-gene resistance to the fungus *Colletotrichum lindemuthianum*. *J. Exp. Bot.* **2021**, *72*, 3569–3581. [CrossRef] [PubMed]
24. McClean, P.E.; Myers, J.R.; Hammond, J.J. Coefficient of parentage and cluster analysis of North American dry bean cultivars. *Crop Sci.* **1993**, *33*, 190–197. [CrossRef]
25. Boller, T.; Felix, G. A renaissance of elicitors, perception of microbe-associated molecular patterns and danger signals by pattern-recognition receptors. *Annu. Rev. Plant Biol.* **2009**, *60*, 379–406. [CrossRef]
26. Vaz Bisneta, M.; Gonçalves-Vidigal, M.C. Integration of anthracnose resistance loci and RLK and NBS-LRR-encoding genes in the *Phaseolus vulgaris* L. genome. *Crop Sci.* **2020**, *60*, 2901–2918. [CrossRef]
27. Aleman, F.; Yazaki, J.; Lee, M.; Takahashi, Y.; Kim, A.Y.; Li, Z.; Kinoshita, T.; Ecker, J.R.; Schroeder, J.I. ABA-increased interaction of the PYL6 ABA receptor with MYC2 transcription factor: A putative link of ABA and JA signaling. *Sci. Rep.* **2016**, *6*, 28941. [CrossRef]
28. García-Andrade, J.; González, B.; Gonzalez-Guzman, M.; Rodriguez, P.L.; Vera, P. The role of ABA in plant immunity is mediated through the PYR1 receptor. *Int. J. Mol. Sci.* **2020**, *21*, 5852. [CrossRef] [PubMed]
29. Ng, A.; Xavier, R.J. Leucine-rich repeat (LRR) proteins: Integrators of pattern recognition and signaling in immunity. *Autophagy* **2011**, *7*, 1082–1084. [CrossRef] [PubMed]
30. Li, X.; Kapos, P.; Zhang, Y. NLRs in plants. *Curr. Opin. Immunol.* **2015**, *32*, 114–121. [CrossRef] [PubMed]
31. Yuan, Y.; Teng, Q.; Zhong, R.; Haghighat, M.; Richardson, E.A.; Ye, Z.-H. Mutations of *Arabidopsis TBL32* and *TBL33* affect Xylan acetylation and secondary wall deposition. *PLoS ONE* **2016**, *11*, e0146460. [CrossRef] [PubMed]
32. Yuan, Y.; Teng, Q.; Zhong, R.; Ye, Z.-H. The *Arabidopsis DUF231* Domain-Containing Protein ESK1 Mediates 2-O- and 3-O-Acetylation of Xylosyl Residues in Xylan. *Plant Cell Physiol.* **2013**, *54*, 1186–1199. [CrossRef] [PubMed]

33. Padder, B.A.; Kamfwa, K.; Awale, H.E.; Kelly, J.D. Transcriptome profiling of the *Phaseolus vulgaris—Colletotrichum lindemuthianum* pathosystem. *PLoS ONE* **2016**, *11*, e0165823. [CrossRef] [PubMed]
34. Shams, E.; Javan-Nikkhah, M.; Mirzadi Gohari, A. Dissecting molecular events and gene expression signatures involved in *Colletotrichum lindemuthianum-Phaseolus vulgaris* pathosystem in compatible and incompatible interactions. *Eur. J. Plant Pathol.* **2020**, *156*, 925–937. [CrossRef]
35. Pastor-Corrales, M.A.; Otoya, M.M.; Molina, A.; Singh, S.P. Resistance to *Colletotrichum lindemuthianum* isolates from middle America and Andean South America in different common bean races. *Plant Dis.* **1995**, *79*, 63–67. [CrossRef]
36. Mathur, R.S.; Barnett, H.l.; Lilly, V.G. Sporulation of *Colletotrichum lindemuthianum* in culture. *Phytopathology* **1950**, *40*, 104–114.
37. Sanglard, D.A.; Balbi, B.P.; Barros, E.G.; Moreira, M.A. An efficient protocol for isolation, sporulation and maintenance of *Pseudocercospora griseola*. *Ann. Rep. Bean Improv. Coop.* **2009**, *52*, 62–63.
38. Cárdenas, F.; Adams, M.W.; Andersen, A. The genetic system for reaction of field beans (*Phaseolus vulgaris* L.) to infection by three physiologic races of *Colletotrichum lindemuthianum*. *Euphytica* **1964**, *13*, 178–186. [CrossRef]
39. de Almeida, C.P.; Arruda, N.; de Carvalho Paulino, J.F.; de Freitas, G.M.; Bonfante, G.F.J.; Bajay, M.M.; Deus, B.C.; Patrício, F.R.A.; Carbonell, S.A.M.; Chiorato, A.F.; et al. Genetic diversity of *Pseudocercospora griseola* resistance loci in common beans. *Trop. Plant Pathol.* **2021**, *46*, 129–138. [CrossRef]
40. Castellanos, G.; Jara, C.; Mosquera, G. *Bean Pathogens: Practical Guide for Lab and Greenhouse Work*; Centro Internacional de Agricultura Tropical (CIAT): Cali, Colombia, 2015.
41. Inglis, D.A.; Hagedorn, J.; Rand, R.E. Use of dry inoculum to evaluate beans for resistance to anthracnose and angular leaf spot. *Plant Dis.* **1988**, *72*, 771–774. [CrossRef]
42. Borges, A.; Melotto, M.; Tsai, S.M.; Caldas, D.G.G. Changes in spatial and temporal gene expression during incompatible interaction between common bean and anthracnose pathogen. *J. Plant Physiol.* **2012**, *169*, 1216–1220. [CrossRef]
43. Farrell, R.E. *RNA Methodologies: Laboratory Guide for Isolation and Characterization*, 5th ed.; Elsevier: Amsterdam, The Netherlands, 2017; pp. 1–855.
44. Borges, A.; Tsai, S.M.; Caldas, D.G.G. Validation of reference genes for RT-qPCR normalization in common bean during biotic and abiotic stresses. *Plant Cell Rep.* **2012**, *31*, 827–838. [CrossRef] [PubMed]
45. Goodstein, D.M.; Shu, S.; Howson, R.; Neupane, R.; Hayes, R.D.; Fazo, J.; Mitros, T.; Dirks, W.; Hellsten, U.; Putnam, N.; et al. Phytozome: A comparative platform for green plant genomics. *Nucleic. Acids Res.* **2012**, *40*, D1178–D1186. [CrossRef] [PubMed]
46. Ye, J.; Coulouris, G.; Zaretskaya, I.; Cutcutache, I.; Rozen, S.; Madden, T.L. Primer-BLAST: A tool to design target-specific primers for polymerase chain reaction. *BMC Bioinform.* **2012**, *13*, 134. [CrossRef] [PubMed]
47. Stothard, P. The sequence manipulation suite: JavaScript programs for analyzing and formatting protein and DNA sequences. *BioTechniques* **2000**, *28*, 1102–1104. [CrossRef]
48. Zuker, M. Mfold web server for nucleic acid folding and hybridization prediction. *Nucleic Acids Res.* **2003**, *31*, 3406–3415. [CrossRef] [PubMed]
49. Bustin, S.A.; Benes, V.; Garson, J.A.; Hellemans, J.; Huggett, J.; Kubista, M. The MIQE Guidelines: Minimum information for publication of quantitative real-time PCR Experiments. *Clin. Chem.* **2009**, *55*, 611–622. [CrossRef]
50. Bustin, S.; Huggett, J. qPCR primer design revisited. *Biomol. Detect. Quantif.* **2017**, *14*, 19–28. [CrossRef]
51. Svec, D.; Tichopad, A.; Novosadova, V.; Pfaffl, M.W.; Kubista, M. How good is a PCR efficiency estimate: Recommendations for precise and robust qPCR efficiency assessments. *Biomol. Detect. Quantif.* **2015**, *3*, 9–16. [CrossRef]
52. Rasmussen, R. Quantification on the LightCycler. In *Rapid Cycle Real-Time PCR*; Meuer, S., Wittwer, C., Nakagawara, K.-I., Eds.; Springer: Heidelberg/Berlin, Germany, 2001; pp. 21–34. [CrossRef]
53. Riedel, G.; Rüdrich, U.; Fekete-Drimusz, N.; Manns, M.P.; Vondran, F.W.R.; Bock, M. An extended ΔCT-method facilitating normalisation with multiple reference genes suited for quantitative RT-PCR analyses of human hepatocyte-like cells. *PLoS ONE* **2014**, *9*, e93031. [CrossRef]
54. Livak, K.J.; Schmittgen, T.D. Analysis of relative gene expression data using real-time quantitative PCR and the $2^{-\Delta\Delta CT}$ method. *Methods* **2001**, *25*, 402–408. [CrossRef] [PubMed]
55. Schmittgen, T.D.; Livak, K.J. Analyzing real-time PCR data by the comparative CT method. *Nat. Protoc.* **2008**, *3*, 1101–1108. [CrossRef] [PubMed]
56. Dag, O.; Dolgun, A.; Konar, N.M. onewaytests: An R Package for One-Way Tests in Independent Groups Designs. *R J.* **2018**, *10*, 175–199. [CrossRef]
57. R Core Team. *R: A Language and Environment for Statistical Computing*; R Foundation for Statistics Computing: Vienna, Austria, 2020.
58. Wickham, H. *Ggplot2: Elegant Graphics for Data Analysis*, 2nd ed.; Use R!; Springer International Publishing: Cham, Switzerland, 2016; ISBN 978-3-319-24277-4.
59. Galili, T.; O'Callaghan, A.; Sidi, J.; Sievert, C. Heatmaply: An R package for creating interactive cluster heatmaps for online publishing. *Bioinformatics* **2018**, *34*, 1600–1602. [CrossRef] [PubMed]

International Journal of
Molecular Sciences

MDPI

Article

Effect of Triazole Fungicides Titul Duo and Vintage on the Development of Pea (*Pisum sativum* L.) Symbiotic Nodules

Artemii P. Gorshkov [1], Pyotr G. Kusakin [1], Yaroslav G. Borisov [2], Anna V. Tsyganova [1] and Viktor E. Tsyganov [1,3,*]

1 Laboratory of Molecular and Cell Biology, All-Russia Research Institute for Agricultural Microbiology, Saint Petersburg 196608, Russia; a.gorshkov@arriam.ru (A.P.G.); pyotr.kusakin@arriam.ru (P.G.K.); avtsyganova@arriam.ru (A.V.T.)

2 Research Resource Centre "Molecular and Cell Technologies", Saint Petersburg State University, Saint Petersburg 199034, Russia; gis.ib88@gmail.com

3 Saint Petersburg Scientific Center RAS, Universitetskaya Embankment 5, Saint Petersburg 199034, Russia

* Correspondence: vetsyganov@arriam.ru; Tel.: +7-812-4705100

Abstract: Triazole fungicides are widely used in agricultural production for plant protection, including pea (*Pisum sativum* L.). The use of fungicides can negatively affect the legume-*Rhizobium* symbiosis. In this study, the effects of triazole fungicides Vintage and Titul Duo on nodule formation and, in particular, on nodule morphology, were studied. Both fungicides at the highest concentration decreased the nodule number and dry weight of the roots 20 days after inoculation. Transmission electron microscopy revealed the following ultrastructural changes in nodules: modifications in the cell walls (their clearing and thinning), thickening of the infection thread walls with the formation of outgrowths, accumulation of poly-β-hydroxybutyrates in bacteroids, expansion of the peribacteroid space, and fusion of symbiosomes. Fungicides Vintage and Titul Duo negatively affect the composition of cell walls, leading to a decrease in the activity of synthesis of cellulose microfibrils and an increase in the number of matrix polysaccharides of cell walls. The results obtained coincide well with the data of transcriptomic analysis, which revealed an increase in the expression levels of genes that control cell wall modification and defense reactions. The data obtained indicate the need for further research on the effects of pesticides on the legume-*Rhizobium* symbiosis in order to optimize their use.

Keywords: *Pisum sativum* L.; symbiotic nodule; symbiosome; bacteroid; cell wall; fungicide

1. Introduction

The biotic stresses (e.g., pests, diseases, and weeds) are important factors limiting plant growth and agricultural production. Among them, fungal diseases have reduced the world's crop yields by almost 20% [1]. Root rot, seedling rot, rust, and powdery mildew lead to partial or complete crop loss [2–6]. The crops in Russia are most susceptible to Fusarium root rot, the losses are up to 50%, and in some years, the crop may die completely as a result of the disease [7]. The application of chemical fungicides is considered the main method of protecting crops from many diseases due to their convenience and low costs [8].

Combined triazole fungicides Titul Duo (propiconazole 200 g/L + tebuconazole 200 g/L) and Vintage (difenoconazole 65 g/L + flutriafol 25 g/L) are used in complex protection of legume crops in Russia [9]. When spraying, the preparations are sorbed by leaves and stems, penetrating the plant tissues. Triazoles, upon penetration into phytopathogenic fungi, inhibit the enzyme lanosterol 14α-demethylase, which is necessary for the biosynthesis of ergosterol, the main sterol in many fungal species [10,11], which leads to their impaired growth and death [12,13].

Different triazole compounds, such as triadimefon, propiconazole, hexaconazole, and paclobutrazol, are widely used as fungicides, and they can influence plant growth [14–17]. Triazoles act as plant growth regulators and affect hormonal balance, photosynthesis rate, enzyme activity, lipid peroxidation, and yield in various crops [14,18–20]. In particular, triazoles inhibit cytochrome P450-mediated oxidative demethylation, as well as the conversion of kaurene to kaurenoic acid, in the gibberellin biosynthetic pathway [14]. Triazoles cause morphological (stimulation of root growth and inhibition of shoot elongation) and biochemical (increased cytokinin synthesis and a temporary increase in abscisic acid) changes [14,18]. In addition, due to their inherent induction of an efficient free radical scavenging system that allows plants to detoxify reactive oxygen species (ROS), triazole compounds are sometimes used as stress protectants [21–26].

Pea (*Pisum sativum* L.) is one of the main legume crops in the world [27]. Like other legumes, pea forms a symbiotic relationship with *Rhizobium leguminosarum* bv. *viciae*. Rhizobia can contribute to overcoming the negative effects of pesticides. They secrete siderophores [28,29], produce 1-aminocyclopropane-1-carboxylate (ACC) deaminase that catalyzes ACC (precursor of ethylene) degradation [30,31], and solubilize insoluble phosphorus [32]. Therefore, the effects of fungicide treatment should be considered in terms of legume-*Rhizobium* symbiosis [33,34].

Despite the visible positive effects of triazoles on various crops [35–40], phytotoxic effects were reported [41–45]. Particularly, strong effects were manifested during the development of legume-*Rhizobium* symbiosis [41,42,46]. The use of tebuconazole caused a decrease in the number and weight of nodules, as well as the dry weight of the roots and shoots of pea plants [42]. In another study, tebuconazole significantly reduced the biomass of roots and shoots in pea, lentil (*Lens esculenta* Moench), mungbean (*Vigna radiate* L. (R) Wilczek), and chickpea (*Cicer arietinum* L.) plants by an average of 30% compared with controls, and also reduced the nodule number, with a maximum decrease of 67% in pea [41].

In this study, the effects of two triazole fungicides (Titul Duo and Vintage) on morphological and transcriptomic changes in pea symbiotic nodules were investigated. As a result, the dose-dependent negative effects of Titul Duo and Vintage treatment on legume-*Rhizobium* symbiosis were revealed. The influence of the stage of plant development when fungicides were applied was also shown.

2. Results

2.1. Nodulation and Plant Growth Parameters

Both fungicides affected the plant growth of the pea cultivars 'Finale' and 'Frisson'. In the 20- and 30-day-old plants, the shoot height of the treated plants decreased, they became thinner, and the leaves turned yellow in a dose-dependent manner (Figure 1A–H). Treatment at 10 days after inoculation (DAI) with both the double- and tenfold-concentrated solutions of the fungicide Titul Duo caused the strong inhibition of plant growth in the cv. 'Frisson' (Figure 1E).

Nodules of plants treated with both fungicides did not differ from untreated ones by color (Figure 2). However, plants of cv. 'Frisson' formed a decreased number of nodules or did not form nodules at all when treated at 10 DAI with the double- and tenfold-concentrated solutions of fungicide Vintage (Figure 2).

Growth and nodule parameters were measured for plants of the cv. 'Frisson' treated with both fungicides. Nodule numbers followed this trend: untreated $>1\times > 2\times > 10\times$, except for Vintage at 10 DAI, where the trend was untreated $>2\times > 1\times > 10\times$. However, only the difference between the untreated and the $10\times$ was statistically different (Figure 3). When treated with various concentrations of fungicides, no statistically significant differences were found in the dry mass of shoots at both treatment time points (Supplementary Materials Figure S1A–D). Only treatment with fungicides at the highest concentration at 10 DAI reduced the dry weight of the roots (Supplementary Materials Figure S2A–D).

Figure 1. Phenotypes of pea plants (*Pisum sativum* L.) of the cultivars 'Finale' (**A–D**) and 'Frisson' (**E–H**). Untreated plants and plants treated with recommended by the manufacturer (1×), double-(2×), and tenfold-concentrated (10×) solutions of Titul Duo (**A**,**C**,**E**,**G**) and Vintage (**B**,**D**,**F**,**H**). (**A**,**B**,**E**,**F**) Plants treated at 10 DAI. (**C**,**D**,**G**,**H**) Plants treated at 20 DAI.

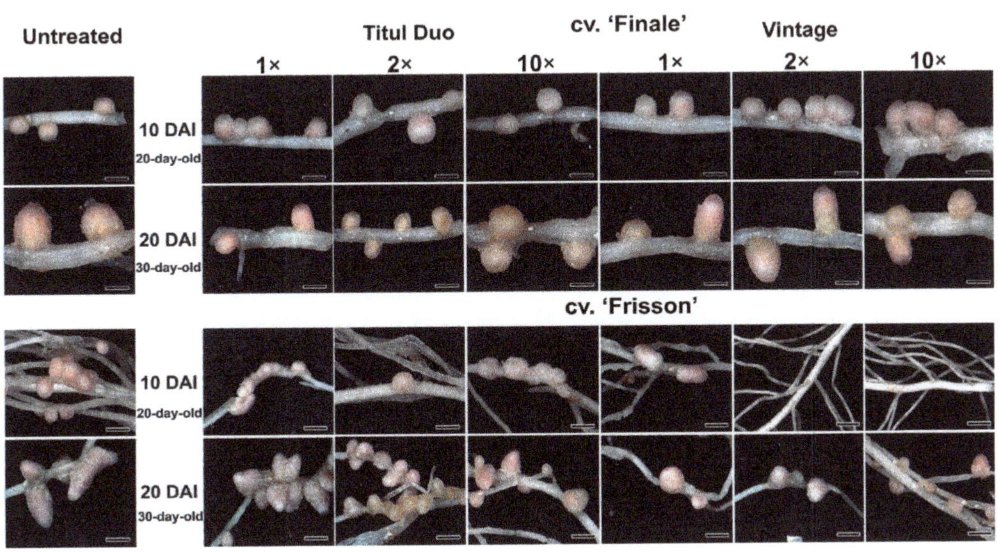

Figure 2. Nodule phenotypes of pea plants (*Pisum sativum* L.) of the cultivars 'Finale' and 'Frisson'. Untreated plants and plants treated with recommended by the manufacturer (1×), double-(2×), and tenfold-concentrated (10×) solutions of Titul Duo and Vintage. Bars = 1 mm.

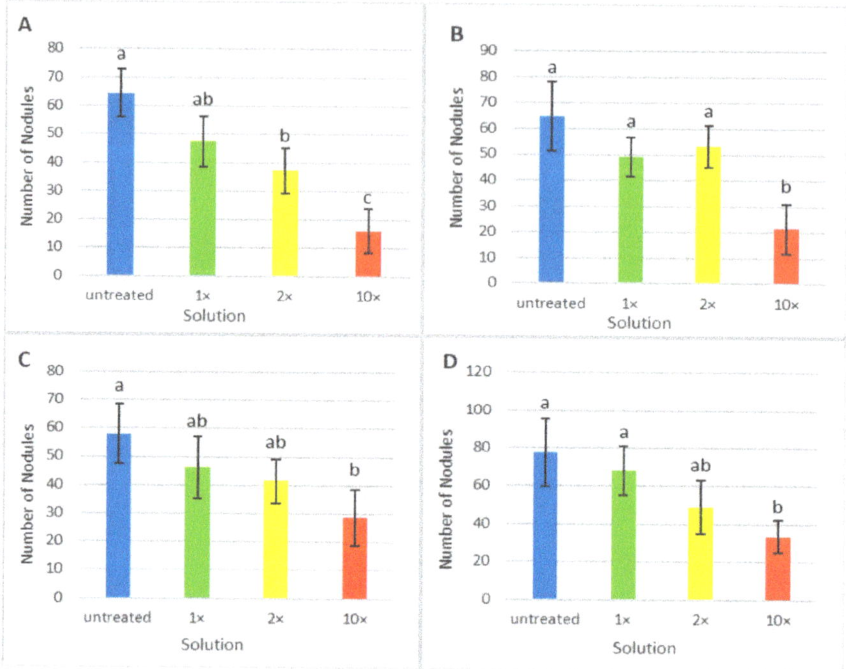

Figure 3. Mean nodule number per plant of pea (*Pisum sativum* L.) cv. 'Frisson' treated with recommended by the manufacturer (1×), double-(2×), and tenfold-concentrated (10×) solutions of fungicides Titul Duo (**A**,**C**) and Vintage (**B**,**D**). (**A**,**B**) Fungicide treatment at 10 DAI. (**C**,**D**) Fungicide treatment at 20 DAI. Different letters indicate groups with a significant difference according to the least significant difference test ($p < 0.05$; $n = 20$). Vertical bars represent standard deviation.

2.2. Nodule Histological Organization

The detailed analysis of both 20- and 30-day-old nodules of untreated plants of the cv. 'Frisson' (Figure 4A) showed a histological organization typical for an indeterminate nodule. Meristematic cells had numerous small vacuoles, a large nucleus with a nucleolus, and an electron-dense cytoplasm (Figure 4B). Metaphase plates were often seen (Figure 4B). Numerous infection threads and droplets were present in the infection zone, and a few juvenile bacteroids were located along the cell periphery (Figure 4C). Mature nitrogen-fixing cells with a central vacuole were filled with numerous pleomorphic bacteroids (Figure 4D).

Plants of cv. 'Frisson' treated with the different concentrations of fungicide Titul Duo at 10 DAI demonstrated clearly visible abnormalities in the histological zones of nodules (Figure 5A–C). Meristematic cells had a folded cell surface; small vacuoles merged into large vacuoles (Figure 5D–F). At the highest concentration of fungicide in the vacuoles of meristematic cells, dark inclusions, presumably, phenolic compounds, were observed (Figure 5F). The cell walls in the meristem and the infection zone were curved and sometimes thinned; in such places, it was difficult to distinguish cell boundaries (Figure 5D–I). These effects of Titul Duo intensified with increasing concentrations. In the nitrogen fixation zone, the boundaries of infected cells sometimes were barely visible (Figure 5K,L); however, such cells were less common when compared with the meristem and the infection zone. In addition, numerous degenerating cells were seen (Figure 5K). At the highest concentration of fungicide, cells in the nitrogen fixation zone contained an increased amount of starch granules (Figure 5L). Even more, a senescence zone was formed at the base of the nodule, which occupied more than half of the nodule (Figure 5C).

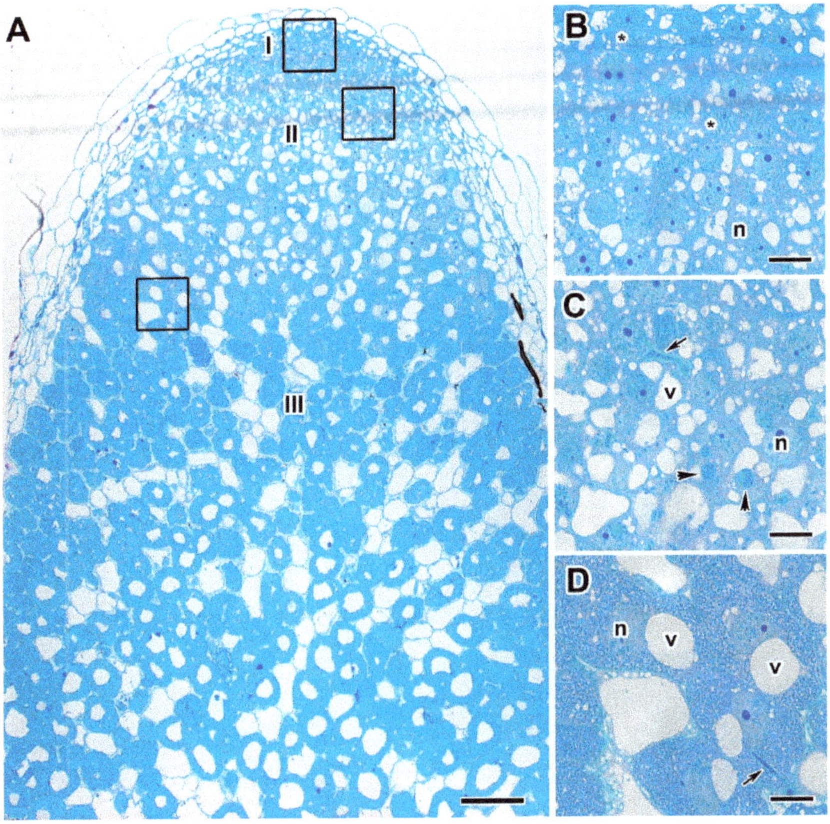

Figure 4. Histological organization of nodules of 20-day-old untreated plants of pea (*Pisum sativum* L.) cv. 'Frisson'. (**A**) Longitudinal section of a nodule. (**B–D**) High magnification of the boxed area in (**A**). (**B**) Nodule meristematic cells. (**C**) Cells in the infection zone. (**D**) Infected cells in the nitrogen fixation zone. I, meristem; II, infection zone; III, nitrogen fixation zone; n, nucleus; v, vacuole; *, metaphase plate. Arrows indicate infection threads; arrowheads indicate infection droplets. Bars (**A**) = 100 μm, (**B–D**) = 10 μm.

Treatment with the fungicide Vintage at 10 DAI caused similar abnormalities to those induced with Titul Duo treatment (Supplementary Materials Figure S3). At the highest fungicide concentration, there were nodules where the senescence zone occupied the entire tissue of the nodule (Supplementary Materials Figure S3C).

There were no pronounced differences in the types of abnormalities caused by the treatment with fungicides Titul Duo and Vintage at 10 and 20 DAI. However, the histological structure of nodules of the 30-day-old plants treated with both fungicides was less damaged (Supplementary Materials Figure S4A–C). Meristematic cells also had folded edges and small vacuoles fused into large vacuoles (Supplementary Materials Figure S4D–F). Degenerating cells were seen in the meristem (Supplementary Materials Figure S4D–F), as well as in the nitrogen fixation zone (Supplementary Materials Figure S4J–L). The cells of the meristem and the infection zone had a folded cell surface, but to a lesser extent in comparison with the treatment with fungicides at 10 DAI (Supplementary Materials Figure S4D–I). In some cells in the nitrogen fixation zone, the tonoplast was destroyed (Supplementary Materials Figure S4J–L). The cell walls in all zones in individual cells were thinned; the cell boundaries were barely distinguishable (Supplementary Materials Figure S4). The dose-dependent differences in abnormalities caused by fungicide treatment at 20 DAI were not revealed.

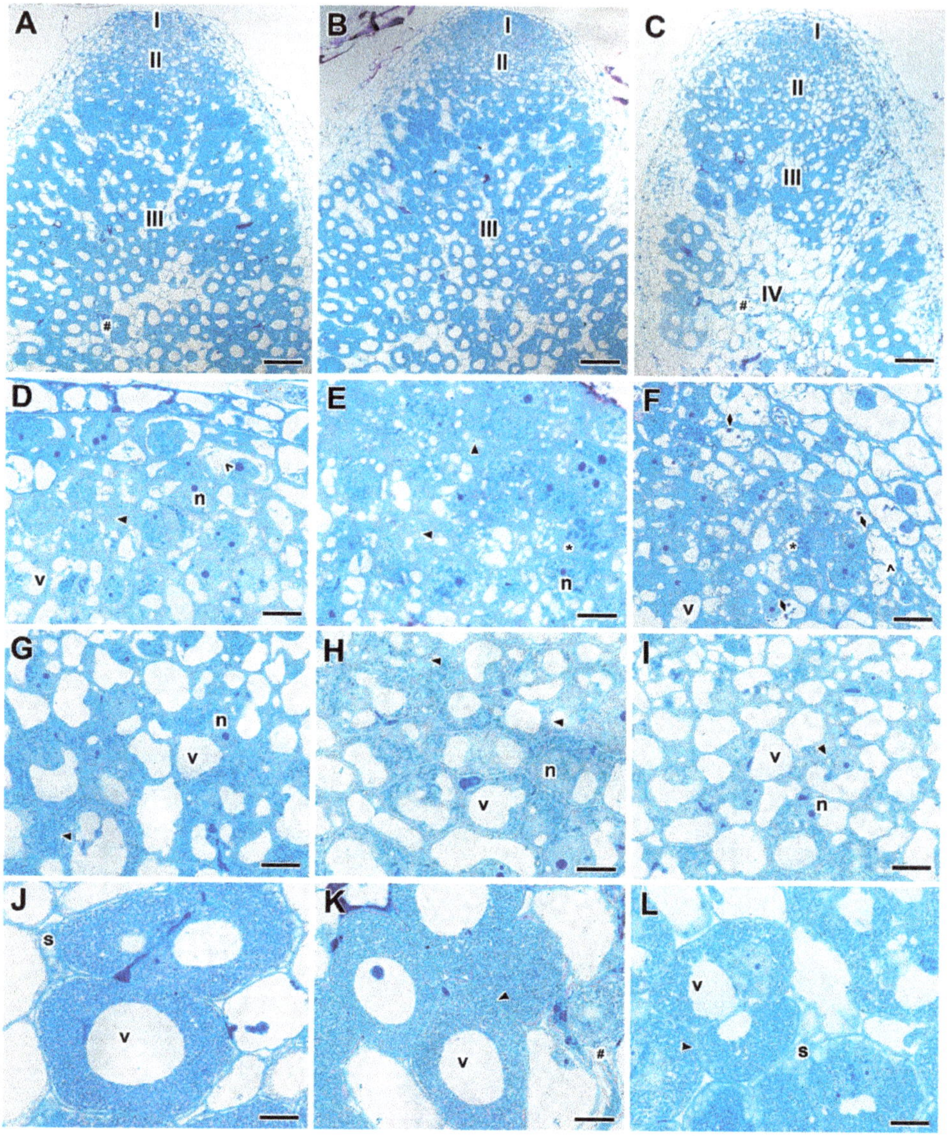

Figure 5. Histological organization of the nodules of pea (*Pisum sativum* L.) cv. 'Frisson' treated with fungicide Titul Duo at 10 DAI. (**A**,**D**,**G**,**J**) Treatment with fungicide at the concentration recommended by the manufacturer. (**B**,**E**,**H**,**K**) Treatment with a double-concentrated solution of fungicide. (**C**,**F**,**I**,**L**) Treatment with a tenfold-concentrated solution of fungicide. (**A–C**) Longitudinal section of a nodule. (**D–F**) Nodule meristematic cells. (**G–I**) Cells in the infection zone. (**J–L**) Infected cells in the nitrogen fixation zone. I, meristem; II, infection zone; III, nitrogen fixation zone; IV, senescence zone; n, nucleus; v, vacuole; *, metaphase plate; #, degrading cell; s, abnormal accumulation of starch in infected cells in the nitrogen fixation zone. Triangles indicate a barely visible cell wall between infected cells; rhombi indicate inclusions in vacuoles; empty arrowheads indicate vacuole fusion. Bars (**A–C**) = 100 μm, (**D–L**) = 10 μm.

2.3. Ultrastructure of Nodules

A comparative analysis of the ultrastructure of nodules of the two pea genotypes was carried out. Nodules of 20- and 30-day-old plants of cultivars 'Finale' and 'Frisson' grown without fungicide treatment had a similar ultrastructural organization characteristic of indeterminate nodules [47,48]. In the nitrogen fixation zone, in infected cells, numerous symbiosomes containing a single pleomorphic bacteroid were observed (Figure 6A). The cell walls in the entire tissue of the nodule were smooth and had a pronounced middle lamella (Figure 6A).

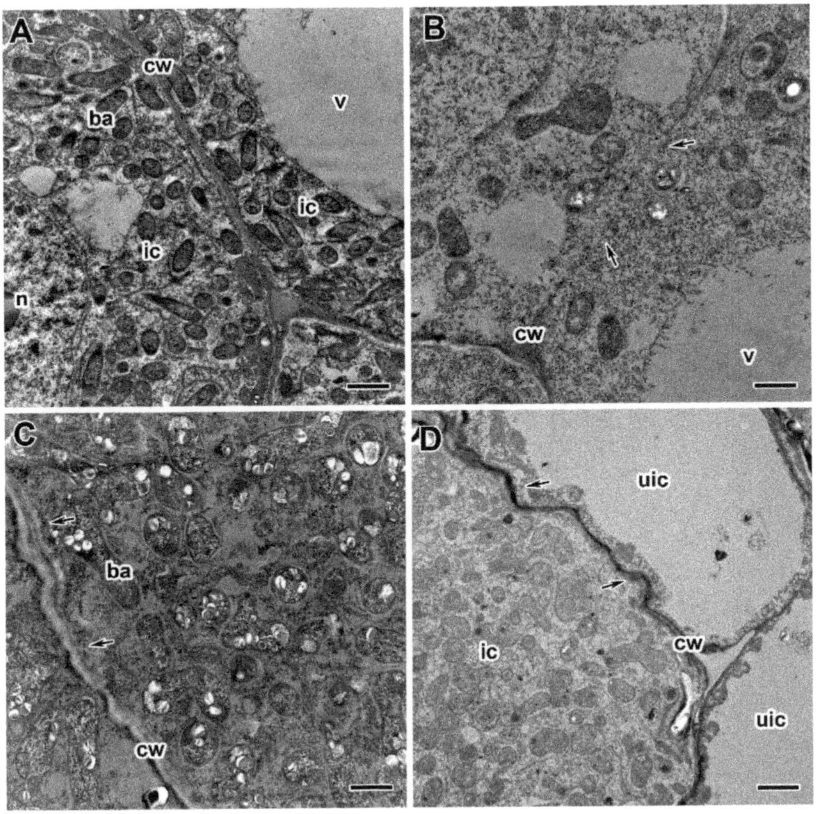

Figure 6. Ultrastructural organization of cell walls in nodules of pea (*Pisum sativum* L.) cultivars 'Finale' (**A–C**) and 'Frisson' (**D**). (**A**) Section of a nodule of an untreated 30-day-old plant. Treatment with fungicides Titul Duo (**B**,**D**) and Vintage (**C**) at 20 DAI at a concentration recommended by the manufacturer (**B**,**D**) and with a double-concentrated solution of fungicide (**C**). n, nucleus; v, vacuole; ic, infected cell; cw, cell wall; ba, bacteroid; uic, uninfected cell. Arrows indicate cell wall abnormalities. Bars (**D**) = 5 μm, (**A**,**C**) = 2 μm, (**B**) = 1 μm.

Treatment with fungicides Titul Duo and Vintage led to significant changes in the ultrastructure of nodules of both pea genotypes, and the severity of the changes depended on the concentration of fungicides and the time of treatment. The genotype-specific differences in abnormalities in nodule ultrastructure caused by the fungicide treatment were not revealed.

The most striking abnormalities in the nodule ultrastructure after treatment of plants with fungicides were changes in the cell wall structure (Figure 6B–D). The cell walls in the meristem and the infection zone were most susceptible to the negative effect of fungicides.

When treated with fungicides at the recommended concentration, the cell walls were curved and could also be less electron-dense (Figure 6C) or vice versa electron-dense (Figure 6D). These negative effects were more pronounced at the highest fungicide concentrations. However, in some cells, the cell wall became thinner, resulting in the cell borders being barely visible (Figure 6B).

Electron microscopy analysis revealed significant differences in the infection thread structure of nodules between untreated (Figure 7A) and fungicide-treated plants (Figure 7B–F). In nodules of fungicide-treated plants, the infection thread wall became less electron-dense (Figure 7B,C), and its outgrowths propagated in the cytoplasm (Figure 7D,E). In some infection threads, the wall was thickened and swelled (Figure 7F). Fungicide treatment at 20 DAI led to the formation of numerous fibrous layers in the infection thread wall (Figure 7F). Only at the highest concentrations of bacteria inside the infection threads underwent degenerative changes (Figure 7C,E).

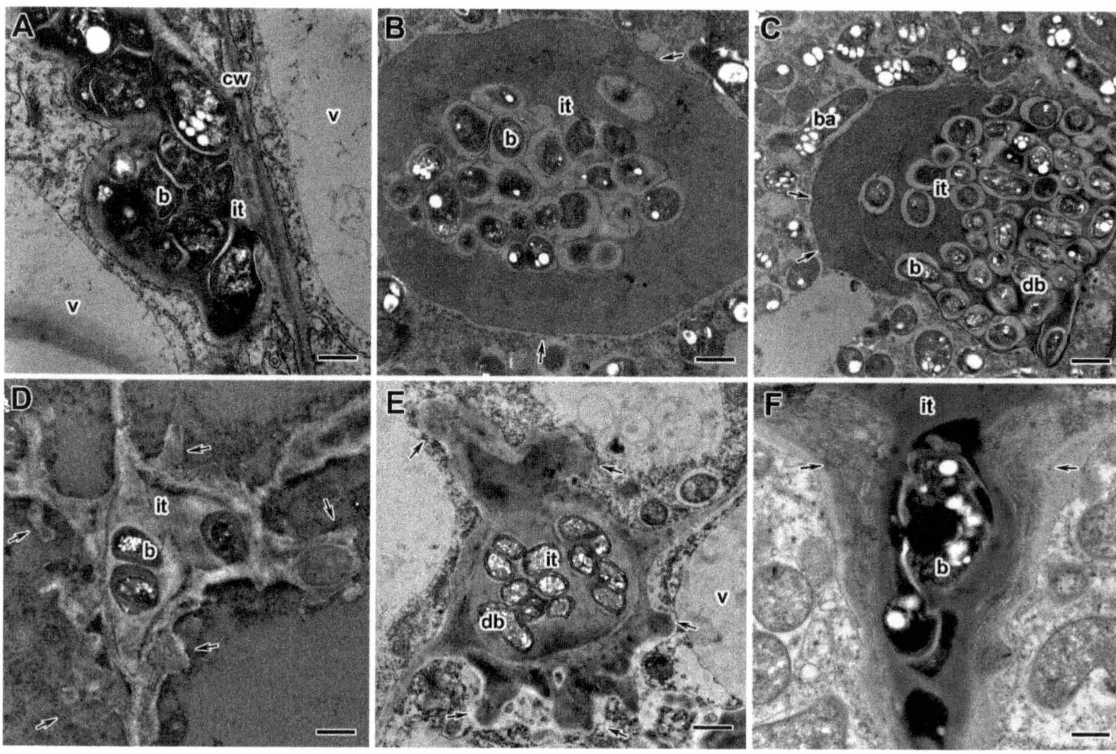

Figure 7. Ultrastructural organization of infection threads in nodules of the pea (*Pisum sativum* L.) cultivars 'Frisson' (**A**,**D**–**F**) and 'Finale' (**B**,**C**). (**A**) Section of a nodule of an untreated 30-day-old plant. Fungicide treatment with Titul Duo (**B**–**D**) and Vintage (**E**,**F**) at 10 DAI (**B**–**E**) and 20 DAI (**F**) at concentrations recommended by the manufacturer (**D**,**F**), and a tenfold-concentrated solution (**B**,**C**,**E**). it, infection thread; v, vacuole; cw, cell wall; ba, bacteroid; b, bacterium; db, degenerative bacterium. Arrows indicate infection thread wall abnormalities. Bars (**A**,**C**) = 2 μm, (**B**,**D**–**F**) = 1 μm.

Ultrastructural analysis revealed various morphological changes in bacteroids in the infected cells in nodules of fungicide-treated pea plants (Figure 8B–F) in comparison to untreated plants (Figure 8A). At the concentration recommended by the manufacturer, the accumulation of poly-β-hydroxybutyrates (PHB) was observed in bacteroids (Figure 8C,F). In some cells at these concentrations, the expansion of the peribacteroid space was visible

(Figure 8C–F). In some cells, symbiosomes, as a result of membrane fusion, contained several bacteroids at different stages of degeneration (Figure 8C,D). In addition, the symbiosome transformation into lytic compartments appeared (Figure 8D,E). Treatment with fungicides caused the degeneration of infected cells filled with "ghosts" of bacteroids in the nitrogen fixation zone of nodules (Figure 8F).

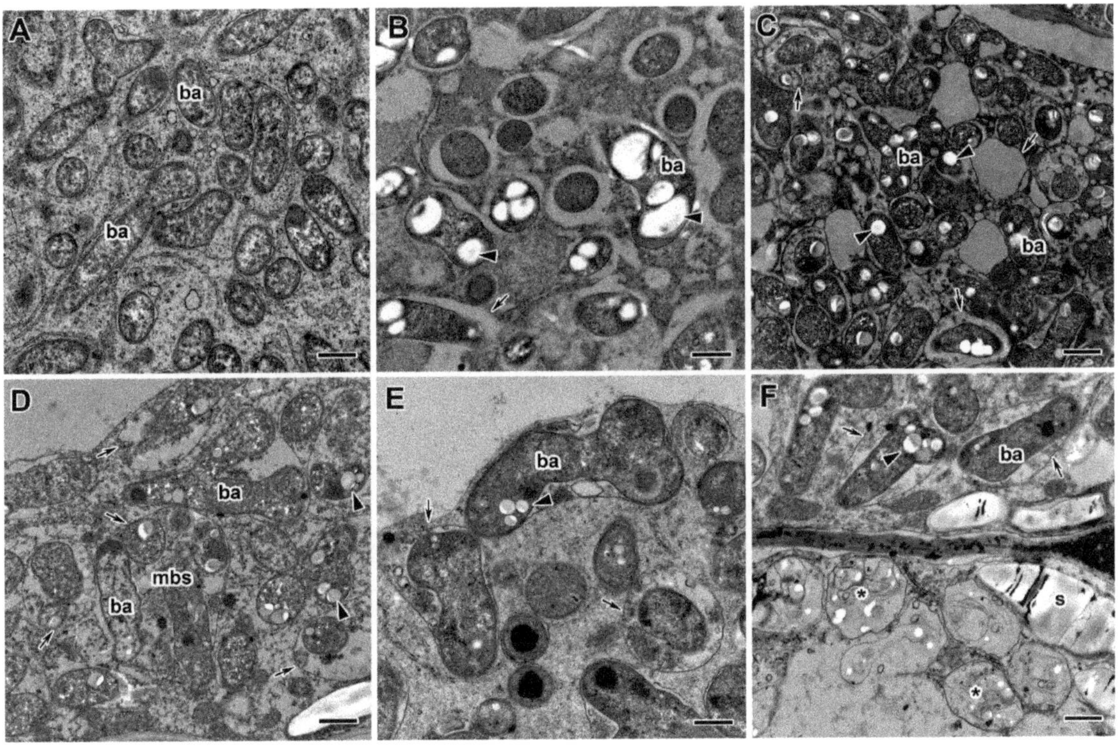

Figure 8. Ultrastructural organization of bacteroids in infected cells in nodules of the pea (*Pisum sativum* L.) cultivars 'Frisson' (**A**,**D**–**F**) and 'Finale' (**B**,**C**). (**A**) Untreated 30-day-old plants. Fungicide treatment of Titul Duo (**E**,**F**) and Vintage (**B**–**D**) at 10 DAI (**B**) and 20 DAI (**C**–**F**) at concentrations recommended by the manufacturer (**C**,**F**), with a double-concentrated solution (**E**) and a tenfold-concentrated solution (**B**,**D**). ba, bacteroid; *, "ghost" bacteroid; mbs, multibacteroid symbiosome; s, starch. Arrows indicate changes in the symbiosome membrane; arrowheads indicate poly-β-hydroxybutyrate drops in bacteroids. Bars (**C**) = 2 μm, (**A**,**B**,**D**–**F**) = 1 μm.

Fungicide treatment of pea plants led to the formation of abnormalities associated with the vacuole (Figure 9B–D) in comparison to untreated plants (Figure 9A). In meristematic cells, numerous small vacuoles merge into one (Figure 9B). In some cells, the tonoplast formed numerous invaginations and vesicles into the central vacuole, which led to the appearance of multivesicular bodies of various sizes (Figure 9D). Some vacuoles contained inclusions of unclear composition (Figure 9C). At high concentrations of fungicides, multivesicular bodies were observed in almost every cell in the infection and nitrogen fixation zones.

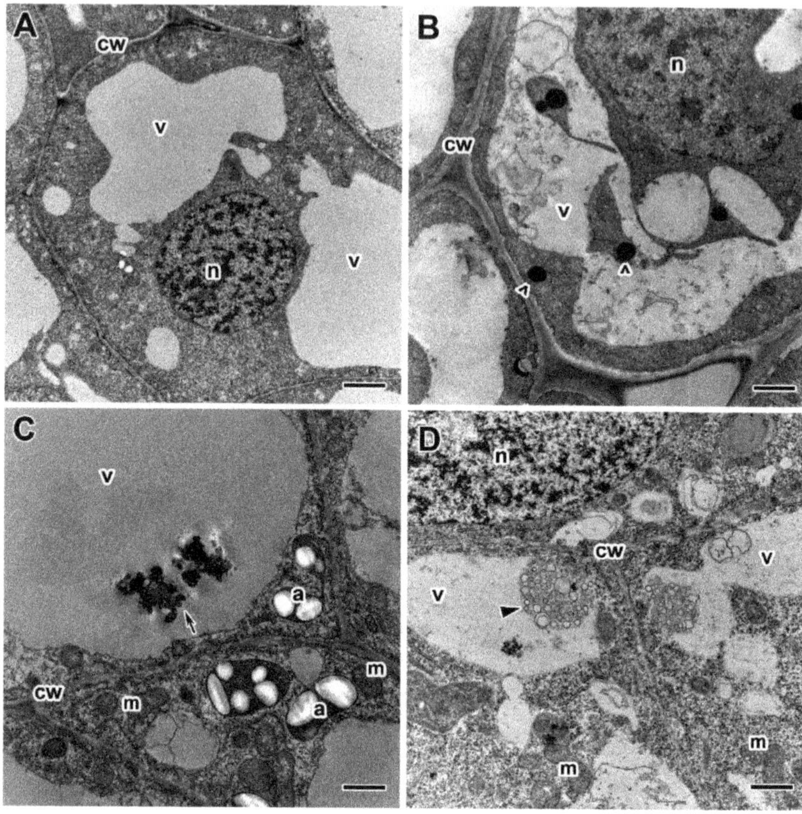

Figure 9. Ultrastructural organization of vacuoles in the nodules of the pea (*Pisum sativum* L.) cultivars 'Frisson' (**A**,**D**) and 'Finale' (**B**,**C**). (**A**) Section of a nodule of an untreated 20-day-old plant. Treatment with recommended by the manufacturer concentrations of fungicides Titul Duo (**D**) and Vintage (**B**,**C**) at 20 DAI. n, nucleus; v, vacuole; cw, cell wall; a, amyloplast; m, mitochondrion. Arrows indicate inclusions in vacuoles, arrowheads indicate multivesicular bodies, empty arrowheads indicate inclusions in meristematic cells, presumably of a phenolic nature. Bars = 2 μm.

In spite of observed abnormalities, fungicides did not affect the ultrastructure of plastids and mitochondria (Figure 9C,D). In addition, the use of fungicides resulted in earlier and more abundant starch accumulation, which is an indicator of an ineffective symbiosis (Figure 8F). Meristematic cells accumulated inclusions of presumably phenolic nature (Figure 9B).

2.4. *Immunocytochemical and Histochemical Analyses*

For a more detailed study of the composition of cell walls, the immunocytochemical and histochemical analyses were performed using monoclonal antibodies (MAbs) to various components of cell walls: pectins (homogalacturonans: 2F4 (Figure 10A,E,I) and LM20 (Figure 10B,F,J)), hemicelluloses (fucosylated xyloglucan: CCRC-M1 (Figure 10C,G,K)), and a fluorescent dye to cellulose SCRI Renaissance Stain 2200 (Figure 10D,H,L).

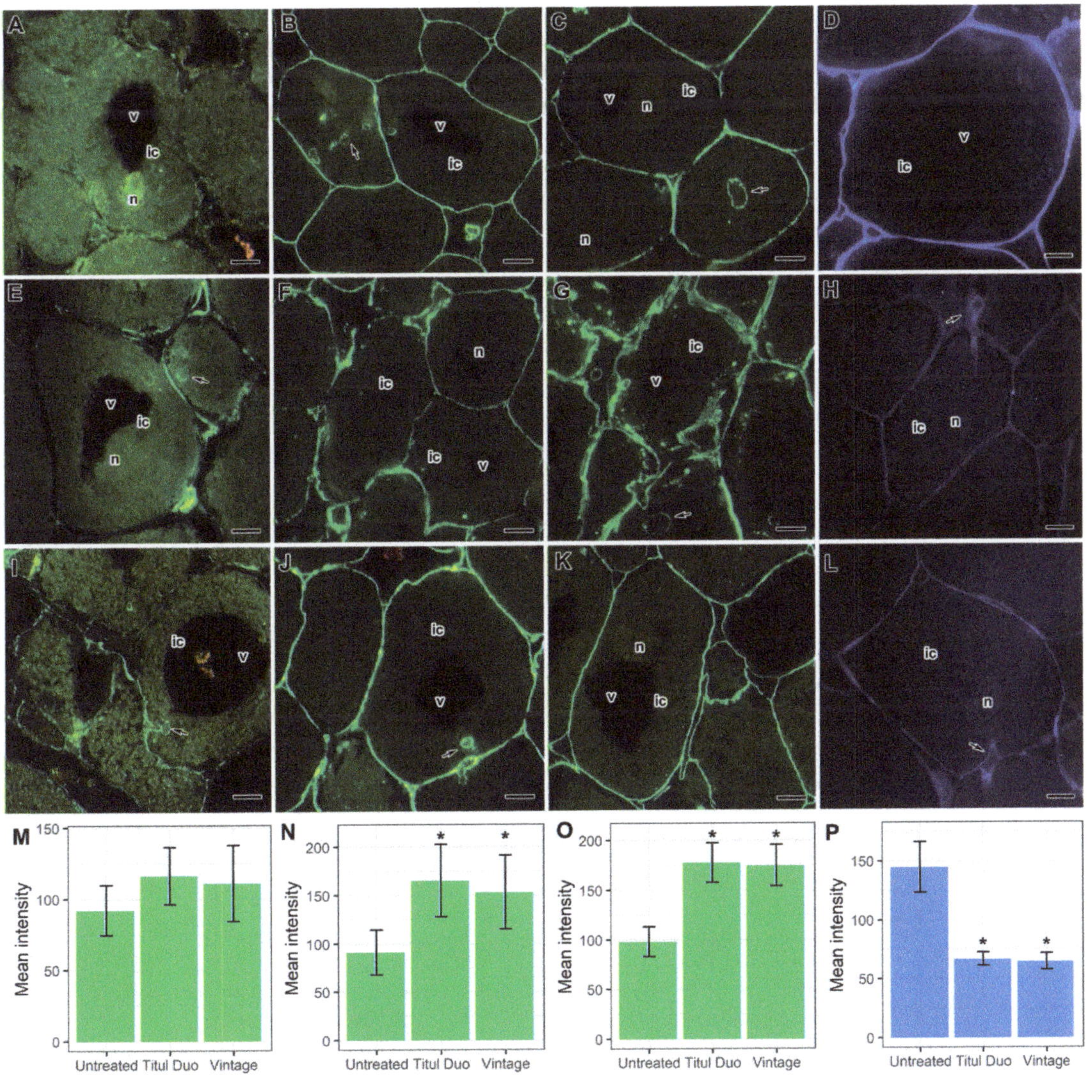

Figure 10. Effect of the fungicide treatment of pea (*Pisum sativum* L.) cv. 'Frisson' plants on the cell wall composition in nodule cells. (**A–D**) Section of nodules of untreated plants. Treatment with Titul Duo (**E–H**) and Vintage (**I–L**) with double-concentrated solutions at 10 DAI. (**M–P**) Mean fluorescence intensity. (**A**,**E**,**I**,**M**) Immunolocalization of homogalacturonan bound by Ca^{2+} labeled with 2F4 MAb, (**B**,**F**,**J**,**N**) highly methylesterified homogalacturonan labeled with LM20 MAb, (**C**,**G**,**K**,**O**) fucosylated xyloglucan labeled with CCRC-M1 MAb. (**D**,**H**,**L**,**P**) Histochemical staining of cellulose with SCRI Renaissance Stain 2200. The secondary antibodies used were goat anti-mouse (**A**,**C**,**E**,**G**,**I**,**K**) and anti-rat (**B**,**F**,**J**) IgG MAb conjugated with Alexa Fluor 488. ic, infected cell; n, nucleus; v, vacuole. Arrows indicate infection threads. Asterisks indicate statistically significant differences from untreated plants (Tukey's HSD test, p-value < 0.05; n = 15), vertical bars represent standard deviation. Bars = 10 μm.

Histochemical analyses showed that in nodules of treated plants, the intensity of fluorescence associated with cellulose microfibrils decreased (Figure 10D,H,L,P), but at the same time, the intensity of epitope labels for highly methylesterified homogalacturonan

(Figure 10B,F,J,N), and fucosylated xyloglucan (Figure 10C,G,K,O) was increased. Moreover, especially significant changes were observed when pea plants were treated with the fungicide Titul Duo (Figure 10F,G). However, the intensity of the epitope label for homogalacturonan cross-linked with Ca^{2+} was not significantly increased (Figure 10A,E,I,M).

Thus, the fungicides Titul Duo and Vintage influence the composition of cell walls. It is possible that the increase in the intensity of the label of matrix polysaccharides (pectins and hemicelluloses) in the cell walls of treated plants is associated with their unmasking as a result of a decrease in the number of cellulose microfibrils.

2.5. Transcriptome Analysis

Transcriptomic analysis was performed to unravel changes in gene expression associated with a fungicide treatment. Since the most distinct changes were observed in the cv. 'Frisson' plants at 20 DAI, and the development of the symbiotic nodule was not so critically disturbed when treated with the fungicide Titul Duo, nodules from such plants were selected for transcriptomic analysis. A total of 55 genes were identified as differentially expressed: 34 upregulated and 21 downregulated (Supplementary Table S1). Gene Ontology enrichment analysis was carried out for these genes, which showed as significantly enriched such "Biological process" terms as "sulfate assimilation", "polysaccharide catabolic process" and "reproduction" for upregulated genes, while "response to stress" and "polysaccharide biosynthetic process" were significantly enriched for downregulated genes.

3. Discussion

Treatment of plants with fungicides increases crop yield due to better plant survival [49–51]. Titul Duo and Vintage are systemic combined fungicides, the active ingredients of which belong to the triazole family, that are known to inhibit the synthesis of sterols [12]. Triazoles are broad-spectrum systemic fungicides. They are effective against various species of *Fusarium* [52–54], *Rhizoctonia, Alternaria, Pyricularia, Gibberella, Botrytis* [55], *Cercospora* [56], *Podosphaera, Erysiphe* [57], and *Colletotrichum* [58], inhibiting mycelial growth. Moreover, the role of triazoles in adaptive agriculture is unique since they not only have a pronounced fungicidal effect but also show growth-stimulating [14,59,60] and protective properties against various environmental stresses, such as high temperature [14], drought [23–26], salinity [22,61], and cooling [62]. Studies of the action of triazoles under various stresses revealed that plant resistance was enhanced due to an increase in the content of chlorophyll and the photosynthetic ability [63], regulation of the activity of enzymes involved in the metabolism of carbon and nitrogen, and changes in the level of endogenous hormones [25,64,65].

It is believed that triazoles are generally not phytotoxic. The most pronounced effect of triazoles on plants is a decrease in height, while the treated plants become greener and more compact [66,67]. Treatment of soybean (*Glycine max* (L.) Merill) plants with uniconazole promoted the accumulation and availability of sucrose and starch content in pods and seeds, thereby increasing the rate of pod setting and soybean yield [60]. However, the application of three different commercial fungicides based on triazole, strobilurin, or carboxamide during the pre-flowering and flowering stage on healthy soybean plants did not affect the physiological traits, pollen grain germination, and yield [17]. In this study, a significant increase in the dry weight of pea shoots under the action of triazole fungicides was not observed (Supplementary Figure S1).

In addition, triazoles have been shown to affect root growth, although this effect can be either inhibitory or stimulatory, depending on the plant and the concentration of the triazole compound used [14]. Paclobutrazol treatment of pea primary roots inhibited root expansion but promoted radial cell expansion [68]. In this study, the dry weight of the roots of fungicide-treated plants decreased compared to the untreated plants when plants were treated at 10 DAI with the highest concentrations of both Titul Duo and Vintage (Supplementary Figure S2A,B). A decrease in the total root weight, a decrease in the number of lateral roots, and root deformation, accompanied by the death of rhizodermal cells and the outer layers of the primary cortex, were also observed under the action of tebuconazole in wheat (*Triticum aestivum* L.) plants [69], hexaconazole [31,33], kitazin [29], naproxen [70], and fluoranthene in pea plants [71].

Despite the positive effects of triazole fungicides on plants, the negative effect of these compounds on various plants, in particular legumes, was also shown [31]. It should be noted that legume-*Rhizobium* symbiosis is very sensitive to stress factors [72]. Treatment of legumes with different fungicides can lead to growth inhibition and chlorosis [48, 73–76]. Plant treatment with certain fungicides reduced or stopped the formation of nodules and disrupted nitrogen fixation. For example, hexaconazole is used to control phytopathogenic fungi [77] but negatively affects biological nitrogen fixation, ureide levels, nitrogen transformation, and the overall yield of legume crops [78,79]. Previously, it was shown that the use of the tetramethylthiuram disulfide (TMTD) fungicide reduced the nodule number in pea laboratory lines Sprint-2 and SGE, as well as in the cv. 'Finale' [48]. A decrease in the number of nodules was also reported in pea, lentil, chickpea, and mungbean treated with hexaconazole, with a maximum decrease of 67% compared to untreated plants in pea [33,41]. In the present study, the application of fungicides Titul Duo and Vintage reduced the number of nodules at the highest concentration of fungicides (Figure 3). Moreover, the plants of the pea cv. 'Frisson' treated at 10 DAI with double- and tenfold-concentrated solutions of the fungicide Vintage did not form nodules (Figure 2).

Previously, the existence of genotypic variability in the resistance of pea genotypes to the fungicide TMTD [48], alfalfa (*Medicago sativa* L.) cultivars to the fungicide pentachloronitrobenzene [80], and peanut (*Arachis hypogaea* L.) cultivars to various fungicides [81] was shown. On the other hand, the effects of fungicides on chickpea plants did not reveal genotypic differences [82]. In this study, the significant genotypic differences in the nodule ultrastructure between the pea cultivars 'Finale' and 'Frisson' when treated with fungicides Titul Duo and Vintage were not revealed.

Various changes in physiological parameters have been described when plants are treated with triazole fungicides. For example, hexaconazole reduced the formation of chlorophyll and carotenoids in legumes, such as common bean (*Phaseolus vulgaris* L.) [83]. These fungicides act as polyfunctional inhibitors and have various types of toxic effects on cells, such as the chelation of calcium ions and the formation of mixed disulfide bonds, thereby disrupting membrane transport [33]. The toxicity of fungicides to plants also causes oxidative stress and leads to the accumulation of ROS. However, treatment with paclobutrazol in peanut plants [67,84], uniconazole in soybean plants [65], hexaconazole in mung bean plants [31], and difenoconazole in wheat plants [85] induced an increase in the content of ascorbic acid, alpha-tocopherol, proline, and glutathione, an enhancement in activity of superoxide dismutase, ascorbate peroxidase, and catalase. These antioxidant activities were not enough to remove excess ROS, which then caused oxidative stress and subsequent growth inhibition.

Studies of the action of fungicides on the morphology of legume nodules are extremely scarce. Serious changes in the histological and ultrastructural organization of pea nodules were previously described under the action of high doses of the TMTD fungicide [48]. In the present study, the morphological changes in pea nodules under treatment with low concentrations of widely used foliar fungicides were studied. However, even the recommended concentrations of fungicides caused changes in the structure of pea symbiotic nodules. The histological organization was characterized by changes in the shape of cells in the meristem and infection zone (Figure 5D–I, Supplementary Materials Figure S3D–I, Supplementary Materials Figure S4D–I), as well as the appearance of a senescence zone when treated with fungicides at 10 DAI (Figure 5C, Supplementary Materials Figure S3A–C). Previously, it was shown that infected cells in mung bean nodules treated with hexaconazole were smaller and deformed [31]. The early appearance of the senescence zone was found in common bean and pea nodules under dark shock conditions, after treatment with exogenous nitrates [86], in pea nodules after treatment with cadmium [87,88] and fungicide

TMTD [48], as well as in nodules of barrel medic (*Medicago truncatula* Gaertn.) during drought [89].

At the ultrastructural level, the most significant changes were cell wall modifications in response to the treatment of plants with fungicides Titul Duo and Vintage (Figure 6B–D), namely swelling, clearing, and curvature in the meristem and infection zone (Figure 6C), thinning and curvature cell walls of infected cells in the nitrogen fixation zone (Figure 6B,D). With an increase in the concentration of fungicides, negative manifestations were aggravated. It is well known that the cell wall is a cell compartment that performs numerous functions and directly responds to various morphogenic and stress factors [90]. Various cell wall changes have been described in white lupine (*Lupinus albus* L.) nodules after exposure to copper [91] and glyphosate [92], salt [93], and in pea nodules after the fungicide TMTD treatment [48]. The immunocytochemical and histochemical analyses showed that in pea nodules treated with fungicides, the intensity of fluorescence associated with cellulose microfibrils was decreased (Figure 10D,H,L,P), but at the same time, the intensity of the labels to highly methyl esterified homogalacturonan (Figure 10B,F,J,N) and to fucosylated xyloglucan (Figure 10C,G,K,O) was increased. At the same time, especially significant changes were observed when pea plants were treated with the fungicide Titul Duo (Figure 10F,G). The accumulation of pectins and hemicellulose in nodules of yellow lupine (*Lupinus luteus* L.) exposed to drought [94], as well as in the nodules of birdsfoot trefoil (*Lotus corniculatus* L.) impacted with nickel, cobalt, and chromium [95] was previously shown. Application of the herbicide isoxaben reduced the amount of cellulose in callus cultures of common bean [96] and in suspension cell cultures of thale cress (*Arabidopsis thaliana* (L.) Heynh.) [97]. The performed transcriptomic analysis showed that the treatment of pea plants with the fungicide Titul Duo resulted in the downregulation of several genes involved in the cell wall modification (Supplementary Table S1). There were notable changes in the expression of genes for dirigent-like proteins (Psat5g216680, Psat5g216760, Psat7g248760), which are known players in lignin biosynthesis in plants [98]. Moreover, the Psat1g162120 gene, characterized as a coding plant invertase/pectin methylesterase inhibitor, was down-regulated. This family includes proteins that are able to inhibit the activity of two classes of plant carbohydrate enzymes: invertases (which are essential for cellulose biosynthesis, sugar metabolism, and osmotic stress adaptations) and pectin methylesterases (which are involved in the modulation of cell wall stiffness) [99]. It was shown that a high level of de-esterified homogalacturonans was associated with various stresses in nodules: boron deficiency [100], aluminum treatment [101], and inefficient interaction with rhizobia in symbiotic mutants of pea [102,103]. Interestingly, an upregulation of the Psat3g077960 gene with a methyltransferase domain signature was also observed, which also suggests a possible link between the observed alteration in homogalacturonan distribution in treated nodules (Figure 10B,F,J,N) and these expression changes. Downregulation of the Psat1g004960 gene encoding a cellulose synthase-like protein was consistent with the detected decrease in cellulose microfibrils-associated fluorescence (Figure 10D,H,L,P) and the overall thinness of cell walls in treated nodules (Figure 6B–D).

The fungicide treatment affected not only the cell walls but also the infection thread walls. They thickened and swelled (Figure 7B,C), and outgrowths of the wall appeared into the cytoplasm (Figure 7D,E). A fibrillar matrix was distinguished in the infection thread walls (Figure 7F). These abnormalities were more pronounced with an increase in the concentration of fungicides but did not depend on the duration of treatment. At the highest concentration of fungicides, the bacteria trapped in the matrix within the infection thread were degraded (Figure 7C,E). Similar modifications of the infection thread walls and matrix have been described after treatment with the TMTD fungicide [48]. Treatment of *M. truncatula* [104] and pea [105] plants with high concentrations of aluminum and common kidneyvetch (*Anthyllis vulneraria* L.) with zinc and lead [106] led to modifications in the infection thread walls in the form of thickening, swelling, and the appearance of a fibrillar matrix. In addition, the cadmium treatment of pea nodules caused the formation of lateral outgrowths of the infection thread [87]. Interestingly, in the nodules of pea mutant *sym33–2*, bacteria inside infection threads also underwent degradation [107].

In addition to cell walls and infection thread walls, the treatment of pea plants with triazole fungicides exhibited disturbances in the bacteroids and symbiosomes. In infected cells, accumulation of PHB (Figure 8B–D,F), expansion of the peribacteroid membrane (Figure 8B–F), fusion of symbiosomes into multibacteroid symbiosomes (Figure 8D), and transformation of symbiosomes into lytic compartments (Figure 8D–F) were found. When plants were treated with fungicides at the highest concentration, degenerating infected cells with "ghosts" of bacteroids appeared in the nitrogen fixation zone of the nodule (Figure 8F). The same "ghost" bacteroids were observed when lupine plants were treated with glyphosate [92] and mercury [108], *A. vulneraria* with zinc [106], and pea with cadmium [87,88] and TMTD [48]. The accumulation of PHB in the cell is a response to various stresses such as heat shock, ultraviolet radiation, oxidizing agents, and osmotic shock [109]. Previously, we showed the accumulation of PHB in bacteroids in pea plants in response to cadmium [87,88] and the fungicide TMTD [48]. Expansion of the peribacteroid membrane and fusion of symbiosomes in the nodule is a widespread response to various stresses [48,87,88,92,104–106,108].

Treatment of pea plants with fungicides Titul Duo and Vintage not only affected the cell walls and infection structures in nodule cells but also led to the appearance of various inclusions in the vacuoles (Figures 5F and 9B–D). Previously, it was shown that electron-dense inclusions, probably of phenolic compounds, appeared in the vacuoles of meristematic cells of pea nodules when plants were treated with high doses of TMTD [48]. Similar inclusions have been described in nodules of *L. corniculatus* under metal stress [95], in pea under saline conditions [110], and in *A. vulneraria* after treatment with zinc and lead [104]. Phenolic compounds were also found in nodules of Chinese liquorice (*Glycyrrhiza uralensis* Fisch. ex DC.) when inoculated with *Mesorhizobium* sp. RCAM3115 [111], as well as big trefoil (*Lotus pedunculatus* Cav.) when inoculated with a rapidly growing strain of rhizobia NZP2037 [112]. The appearance of multivesicular bodies in the vacuoles of nodule cells indicates an increase in the autophagy process in response to stress conditions [113,114]. In the present work, when pea plants were treated with triazole fungicides Titul Duo and Vintage, multivesicular bodies of various shapes were found in the vacuoles of nodule cells (Figure 9D). Such inclusions were described when pea plants were treated with the fungicide TMTD [48] and during the formation of soybean nodules by the citrate synthase mutant strain *Sinorhizobium fredii* USDA257 [115].

4. Materials and Methods

4.1. Plant Material and Bacterial Strain

The pea (*Pisum sativum* L.) commercial cultivars 'Finale' [116] and 'Frisson' [117] were used. Both cultivars have a determinate flowering habit composed of white flowers, and they are cultivated in many European countries. 'Finale' is a late-ripening cultivar, and 'Frisson' is a mid-ripening one. The streptomycin-resistant *Rhizobium leguminosarum* bv. *viciae* strain 3841 was used for inoculation [118]. Bacteria were grown on a solid TY medium [119] at 28 °C with the addition of streptomycin at a concentration of 600–800 μg/L.

4.2. Inoculation and Plant Growth Conditions

Pea seeds of each cultivar were sterilized with concentrated sulfuric acid for 30 min and washed with sterile water 10 times. The seeds were planted in pots with vermiculite immediately after sterilization, and then each seed was inoculated with 1 mL of an aqueous suspension of bacteria (10^7–10^8 cells). Plants were grown in vermiculite moistened with a nitrogen-free nutrient solution [120] in a growth chamber (MLR-352H, Sanyo Electric Co., Ltd., Moriguchi, Japan) under controlled conditions: day/night, 16/8; temperature 21 °C; humidity 75%; illumination 280 mol photons m^{-2} s^{-1}). An active solution of the fungicide Titul Duo contains 200 g/L of propiconazole and 200 g/L of tebuconazole; Vintage contains 65 g/L of difenoconazole and 25 g/L of flutriafol [9]. Fungicide treatment was carried out with a manual sprayer at 10 and 20 DAI with solutions diluted as follows: Titul Duo—1:500 (recommended by the manufacturer), 1:250 (double-concentrated solution), 1:50 (tenfold-concentrated solution); Vintage—1:200 (recommended by the manufacturer), 1:100 (double-concentrated solution), 1:20 (tenfold-concentrated solution). Plants were harvested 10 days after treatment with fungicides (20- and 30-day-old plants, respectively).

4.3. Phenotypic Analysis of Plants and Nodules

Growth and nodule formation parameters were analyzed only for the cv. 'Frisson'; 20 plants were analyzed. Nodules were counted immediately after washing the plants from vermiculite. For weight measurements, cotyledons were removed, shoots and roots were separated and then dried in a Memmert UF160 oven (Memmert GmbH, Schwabach, Germany) at 40 °C. Pea nodules were photographed using a SteREO Lumar.V12 stereo microscope equipped with an AxioCam MRc 5 camera (Carl Zeiss, Oberkochen, Germany).

4.4. Statistical Analysis

Statistical data analysis was carried out using the software STATISTICA version 10 (StatSoft, Tulsa, OK, USA). For phenotypic analysis of plants and nodules, statistically significant differences were assessed using one-way ANOVA ($p < 0.05$) and the least significant difference test ($p < 0.05$). For mean fluorescence intensity, statistically significant differences were assessed using one-way ANOVA ($p < 0.05$) and Tukey's HSD test ($p < 0.05$).

4.5. Electron and Light Microscopy

Nodules (10–15 nodules from 10 plants for each variant) after harvesting were transferred directly into a 2.5% aqueous solution of glutaraldehyde (Sigma-Aldrich, St. Louis, MO, USA) in 0.01 M phosphate buffer (2.48 g/L NaH_2PO_4, 21.36 g/L Na_2HPO_4, and 87.66 g/L NaCl, pH 7.2). A lateral cut was made on each nodule for better penetration of the fixative. The samples were placed under a vacuum to remove air from the intercellular space and left in a fixative overnight at 4 °C.

Then, the nodules were washed in the buffer four times for 15 min each and postfixed in 1% aqueous solution of osmium tetroxide in 0.1 M phosphate buffer for 1 h. The nodules were then dehydrated in a series of increasing concentrations of ethanol followed by two changes to 100% acetone, as described previously [121]. Dehydrated samples were gradually infiltrated with epoxy resin Eponate 12 (Ted Pella, Inc., Redding, CA, USA). All these procedures were performed in the EM TP Tissue Processor (Leica Microsystems, Vienna, Austria) at 21 °C. The samples were transferred for embedding to small plastic containers with fresh resin, which were polymerized at 60 °C for 48 h.

For light microscopy, semi-thin sections (1 µm) obtained on a Leica EM UC7 ultramicrotome (Leica Microsystems) were placed on slides and stained with methylene blue–azure II [122] at 60 °C for 20 min. Sections were then placed in a drop of xylene and embedded in the EUKITT® Mounting Medium (Electron Microscopy Sciences, Hatfield, PA, USA). Sections were analyzed using an Axio Imager.Z1 microscope (Carl Zeiss). Photographs were taken with an Axiocam 506 digital camera (Carl Zeiss).

For transmission electron microscopy, ultrathin sections (90–100 nm thick) were cut with a Leica EM UC7 ultramicrotome (Leica Microsystems) using a diamond knife (Diatome, Nidau, Switzerland). The sections were collected on copper grids coated with 4% formvar and carbon. Sections were counterstained with 2% aqueous uranyl acetate for 30 min followed by lead citrate for 1 min in the automatic contrasting system for ultrathin sections EM AC20 (Leica Microsystems) at 21 °C. All solutions were filtered before use, and filter-sterilized deionized water was used throughout the experiment. Nodule tissues were examined using a JEM-1400 EM transmission electron microscope (JEOL Ltd., Tokyo, Japan) at 80 kV. Electron micrographs were taken with a Veleta CCD camera (Olympus, Münster, Germany).

4.6. Fluorescence Microscopy

For immunofluorescence microscopy, semi-thin sections (1 μm) obtained on a Leica EM UC7 ultratome were incubated in ABB blocking buffer (5% BSA, 0.1% cold water fish skin gelatin (CWFS), 5–10% normal goat serum, 15 mM NaN_3 in PBS, pH 7.4) for 1 h at room temperature. Then nodule sections were incubated with primary antibodies diluted 1:20 in 3% BSA in PBS at 37 °C for 1 h. The samples were washed again in 3% BSA in PBS (pH 7.2) two times for 20 min each. The incubation with the secondary antibodies to the corresponding gamma globulin conjugated to AlexaFluor 488 (Molecular Probes, Thermo Fisher Scientific, Waltham, MA, USA) in 3% BSA in PBS (diluted 1:100) was conducted for 1 h at 37 °C. Then, samples were washed in a 3% BSA solution in PBS two times for 20 min. After staining complete drying, sections were covered with a drop of ProLong Gold Antifade reagent (Molecular Probes, Thermo Fisher Scientific).

The fluorescent dye SCRI Renaissance Stain 2200 (Renaissance Chemicals, North Duffield, UK) was used to detect cellulose [123]. Semi-thin sections (1 μm) were stained with a dye (diluted 1:1000) for 20 min, washed with distilled water, and mounted in the ProLong Gold® antifade reagent (Molecular Probes, Thermo Fisher Scientific). Sections were examined under a fluorescence microscope using a DAPI filter.

The following MAbs were used as primary antibodies: LM20 for high methylester homogalacturonan [124], 2F4 for a dimeric association of homogalacturonan chains through Ca^{2+} [125], CCRC-M1 fucosylated xyloglucan [126]. Sections were analyzed using an Axio Imager.Z1 microscope (Carl Zeiss). Photos were taken using a digital camera Axiocam 506 (Carl Zeiss).

Image analysis was performed using the program ImageJ [127] to detect fluorescence intensities. During the mathematical processing of fluorescent images, the areas of the presence of a signal and its absence were selected, the average fluorescence intensity was identified, and the average fluorescence intensity of the signal was normalized to the average intensity of the area without a signal.

4.7. Transcriptomic Analysis

For the RNA-seq analysis, nodules were harvested on ice and frozen in liquid nitrogen. RNA extraction was performed using the RNAeasy Plant Mini Kit (Qiagen, Hilden, Germany). The concentration of the extracted RNA was measured using Qubit 2.0 (Invitrogen, Waltham MA, USA).

The extracted RNA was used for the preparation of libraries using the RapidMACE kit (GenXPro GmbH, Frankfurt, Germany) according to the manufacturer's recommendations. Libraries were sequenced with Illumina HiSeq 2500 by the Macrogen company (Seoul, Republic of Korea). Quality assessment of the obtained reads, filtering, mapping to the reference genome, and subsequent bioinformatics analysis was performed as previously described [128].

5. Conclusions

Thus, symbiotic pea nodules are highly sensitive to the phytotoxic action of triazole fungicides Titul Duo and Vintage. All studied concentrations of fungicides, including those recommended by the manufacturer, caused structural changes in cell walls, infection structures (infection threads and symbiosomes), as well as increased autophagy processes. The data obtained by transcriptomic analysis indicate the modification of cell walls and coincide well with the data of light and electron microscopic studies. In the present study, no differences in the nodule ultrastructure between pea cultivars when treated with fungicides were revealed. Agricultural companies strive to reduce the cost of pesticides; therefore, more preparations of complex and combined action will appear in the market. A deeper understanding of the negative effects of agricultural pesticides is needed to create new and optimized strategies for growing legumes.

Supplementary Materials: The following supporting information can be downloaded at: https://www.mdpi.com/article/10.3390/ijms24108646/s1.

Author Contributions: Conceptualization, V.E.T.; investigation, A.P.G. and P.G.K.; writing—original draft preparation, A.P.G. and P.G.K.; resources, Y.G.B.; writing—review and editing, A.V.T. and V.E.T.; project administration, V.E.T. All authors have read and agreed to the published version of the manuscript.

Funding: The article was made with the support of the Ministry of Science and Higher Education of the Russian Federation in accordance with agreement No 075-15-2022-320, dated 20 April 2022, on providing a grant in the form of subsidies from the Federal budget of the Russian Federation. The grant was provided for state support for the creation and development of a World-class Scientific Centre "Agrotechnologies for the Future".

Institutional Review Board Statement: Not applicable.

Informed Consent Statement: Not applicable.

Data Availability Statement: The data presented in this study are openly available at NCBI SRA under the accession number PRJNA953817.

Acknowledgments: This work was carried out using the equipment of the Core Centrum Genomic Technologies, Proteomics and Cell Biology at the All-Russia Research Institute for Agricultural Microbiology and the Molecular and Cell Technologies Research Resource Centre at Saint Petersburg State University. We thank Alexey Afonin for his assistance with transcriptomic data preparation.

Conflicts of Interest: The authors declare no conflict of interest.

References

1. Gullino, M.L.; Leroux, P.; Smith, C.M. Uses and challenges of novel compounds for plant disease control. *Crop Prot.* **2000**, *19*, 1–11. [CrossRef]
2. Rubiales, D.; Fondevilla, S.; Chen, W.; Gentzbittel, L.; Higgins, T.J.; Castillejo, M.A.; Singh, K.B.; Rispail, N. Achievements and challenges in legume breeding for pest and disease resistance. *Crit. Rev. Plant Sci.* **2015**, *34*, 195–236. [CrossRef]
3. Desai, S.; Prasad, R.; Kumar, G.P. Fusarium wilts of chickpea, pigeon pea and lentil and their management. In *Microbial Interventions in Agriculture and Environment*; Singh, D.P., Prabha, R., Eds.; Volume 3: Soil and Crop Health Management; Springer: Singapore, 2019; pp. 49–68. [CrossRef]
4. Upadhyay, V.; Medhi, K.; Pandey, P.; Thengal, P.; Paul, S.K.; Kushwah, K. Rust disease of pea: A review. *Int. J. Curr. Microbiol. Appl. Sci.* **2019**, *8*, 416–434. [CrossRef]
5. Wille, L.; Messmer, M.M.; Studer, B.; Hohmann, P. Insights to plant–microbe interactions provide opportunities to improve resistance breeding against root diseases in grain legumes. *Plant Cell Environ.* **2019**, *42*, 20–40. [CrossRef]
6. Williamson-Benavides, B.A.; Dhingra, A. Understanding root rot disease in agricultural crops. *Horticulturae* **2021**, *7*, 33. [CrossRef]
7. Zotikov, V.; Budarina, G. Diseases of peas and the main methods of crop protection in conditions of central Russia. *Plant Prot. Quar.* **2015**, *5*, 11–15.
8. Xia, X.J.; Huang, Y.Y.; Wang, L.; Huang, L.F.; Yu, Y.L.; Zhou, Y.H.; Yu, J.Q. Pesticides-induced depression of photosynthesis was alleviated by 24-epibrassinolide pretreatment in *Cucumis sativus* L. *Pestic. Biochem. Physiol.* **2006**, *86*, 42–48. [CrossRef]
9. Schelkovo Agrohim. Available online: https://eng.betaren.ru/catalog/pesticides/fungicides/ (accessed on 23 March 2023).
10. Villani, S.M.; Hulvey, J.; Hily, J.-M.; Cox, K.D. Overexpression of the *CYP51A1* gene and repeated elements are associated with differential sensitivity to DMI fungicides in *Venturia inaequalis*. *Phytopathology* **2016**, *106*, 562–571. [CrossRef]
11. Rodrigues, M.L. The multifunctional fungal ergosterol. *MBio* **2018**, *9*, e01755-18. [CrossRef]
12. Sant, D.; Tupe, S.; Ramana, C.V.; Deshpande, M. Fungal cell membrane—Promising drug target for antifungal therapy. *J. Appl. Microbiol.* **2016**, *121*, 1498–1510. [CrossRef]
13. Remya, V.R.; George, J.S.; Thomas, S. Manufacturing of slow- and controlled-release pesticides. In *Controlled Release of Pesticides for Sustainable Agriculture*; Rakhimol, K.R., Thomas, S., Volova, T., Jayachandran, K., Eds.; Springer International Publishing: Cham, Switzerland, 2020; pp. 127–139. [CrossRef]
14. Fletcher, R.A.; Gilley, A.; Sankhla, N.; Davis, T.D. Triazoles as plant growth regulators and stress protectants. *Hortic. Rev.* **2000**, *24*, 55–138.
15. Manivannan, P.; Jaleel, C.A.; Kishorekumar, A.; Sankar, B.; Somasundaram, R.; Sridharan, R.; Panneerselvam, R. Changes in antioxidant metabolism of *Vigna unguiculata* (L.) Walp. by propiconazole under water deficit stress. *Colloids Surf. B Biointerfaces* **2007**, *57*, 69–74. [CrossRef] [PubMed]
16. Gomathinayagam, M.; Jaleel, C.A.; Lakshmanan, G.A.; Panneerselvam, R. Changes in carbohydrate metabolism by triazole growth regulators in cassava (*Manihot esculenta* Crantz); effects on tuber production and quality. *Comptes Rendus Biol.* **2007**, *330*, 644–655. [CrossRef] [PubMed]

17. Junqueira, V.B.; Müller, C.; Rodrigues, A.A.; Amaral, T.S.; Batista, P.F.; Silva, A.A.; Costa, A.C. Do fungicides affect the physiology, reproductive development and productivity of healthy soybean plants? *Pestic. Biochem. Physiol.* **2021**, *172*, 104754. [CrossRef] [PubMed]
18. Gopi, R.; Jaleel, C.A.; Sairam, R.; Lakshmanan, G.; Gomathinayagam, M.; Panneerselvam, R. Differential effects of hexaconazole and paclobutrazol on biomass, electrolyte leakage, lipid peroxidation and antioxidant potential of *Daucus carota* L. *Colloids Surf. B Biointerfaces* **2007**, *60*, 180–186. [CrossRef]
19. Hajihashemi, S. Physiological, biochemical, antioxidant and growth characterizations of gibberellin and paclobutrazol-treated sweet leaf (*Stevia rebaudiana* B.) herb. *J. Plant Biochem. Biotechnol.* **2018**, *27*, 237–240. [CrossRef]
20. Kamran, M.; Ahmad, S.; Ahmad, I.; Hussain, I.; Meng, X.; Zhang, X.; Javed, T.; Ullah, M.; Ding, R.; Xu, P. Paclobutrazol application favors yield improvement of maize under semiarid regions by delaying leaf senescence and regulating photosynthetic capacity and antioxidant system during grain-filling stage. *Agronomy* **2020**, *10*, 187. [CrossRef]
21. Jaleel, C.A.; Gopi, R.; Manivannan, P.; Panneerselvam, R. Responses of antioxidant defense system of *Catharanthus roseus* (L.) G. Don. to paclobutrazol treatment under salinity. *Acta Physiol. Plant.* **2007**, *29*, 205–209. [CrossRef]
22. Manivannan, P.; Jaleel, C.A.; Kishorekumar, A.; Sankar, B.; Somasundaram, R.; Panneerselvam, R. Protection of *Vigna unguiculata* (L.) Walp. plants from salt stress by paclobutrazol. *Colloids Surf. B Biointerfaces* **2008**, *61*, 315–318. [CrossRef]
23. Rezayian, M.; Niknam, V.; Ebrahimzadeh, H. Improving tolerance against drought in canola by penconazole and calcium. *Pestic. Biochem. Physiol.* **2018**, *149*, 123–136. [CrossRef]
24. Babarashi, E.; Rokhzadi, A.; Pasari, B.; Mohammadi, K. Ameliorating effects of exogenous paclobutrazol and putrescine on mung bean [*Vigna radiata* (L.) Wilczek] under water deficit stress. *Plant Soil Environ.* **2021**, *67*, 40–45. [CrossRef]
25. Jiang, Y.; Sun, Y.; Zheng, D.; Han, C.; Cao, K.; Xu, L.; Liu, S.; Cao, Y.; Feng, N. Physiological and transcriptome analyses for assessing the effects of exogenous uniconazole on drought tolerance in hemp (*Cannabis sativa* L.). *Sci. Rep.* **2021**, *11*, 1–15. [CrossRef] [PubMed]
26. Maheshwari, C.; Garg, N.K.; Hasan, M.; Prathap, V.; Meena, N.L.; Singh, A.; Tyagi, A. Insight of PBZ mediated drought amelioration in crop plants. *Front. Plant Sci.* **2022**, *13*, 1008993. [CrossRef] [PubMed]
27. Stagnari, F.; Maggio, A.; Galieni, A.; Pisante, M. Multiple benefits of legumes for agriculture sustainability: An overview. *Chem. Biol. Technol.* **2017**, *4*, 2. [CrossRef]
28. Ahemad, M.; Khan, M.S. Alleviation of fungicide-induced phytotoxicity in greengram [*Vigna radiata* (L.) Wilczek] using fungicide-tolerant and plant growth promoting Pseudomonas strain. *Saudi J. Biol. Sci.* **2012**, *19*, 451–459. [CrossRef]
29. Shahid, M.; Khan, M.S.; Kumar, M. Kitazin-pea interaction: Understanding the fungicide induced nodule alteration, cytotoxicity, oxidative damage and toxicity alleviation by *Rhizobium leguminosarum*. *RSC Adv.* **2019**, *9*, 16929–16947. [CrossRef]
30. Ghosh, D.; Sen, S.; Mohapatra, S. Drought-mitigating *Pseudomonas putida* GAP-P45 modulates proline turnover and oxidative status in *Arabidopsis thaliana* under water stress. *Ann. Microbiol.* **2018**, *68*, 579–594. [CrossRef]
31. Shahid, M.; Khan, M.S. Fungicide tolerant *Bradyrhizobium japonicum* mitigate toxicity and enhance greengram production under hexaconazole stress. *J. Environ. Sci.* **2019**, *78*, 92–108. [CrossRef]
32. Zaidi, A.; Khan, M.S.; Ahmad, E.; Saif, S.; Rizvi, A.; Shahid, M. Growth stimulation and management of diseases of ornamental plants using phosphate solubilizing microorganisms: Current perspective. *Acta Physiol. Plant.* **2016**, *38*, 1–21. [CrossRef]
33. Shahid, M.; Ahmed, B.; Zaidi, A.; Khan, M.S. Toxicity of fungicides to *Pisum sativum*: A study of oxidative damage, growth suppression, cellular death and morpho-anatomical changes. *RSC Adv.* **2018**, *8*, 38483–38498. [CrossRef]
34. Shahid, M.; Khan, M.S.; Zaidi, A. Fungicide toxicity to legumes and its microbial remediation. In *Pesticides in Crop Production*; Srivastava, P.K., Singh, V.P., Singh, A., Tripathi, D.K., Singh, S., Prasad, S.M., Chauhan, D.K., Eds.; John Wiley & Sons Ltd.: New York, NY, USA, 2020; pp. 15–33. [CrossRef]
35. Uthman, T.O.; Awosanya, A.O.; Ayanwale, A.O.S.; Ayinde, T.G. The effect of fungicides on the nodulation of cowpea (*Vigna unguiculata* L. Walp) in a tropical rainforest environment. *J. Sustain. Dev.* **2017**, *7*, 56–61.
36. Erohin, A.I. Influence of joint application of preparations biostim maslichnyj, fungicide Titul Duo, KKR and insecticide Kinfos, CE on the productivity of peas with foliar treatment of plants. *Legumes Groat Crops* **2018**, *1*, 36. [CrossRef]
37. Erokhin, A.I.; Tsukanova, Z.R.; Latyntseva, E.V. Prospects for the use of liquid peat potassium humate and fungicide Titul Duo, KKR for foliar treatment of pea plants. *Legumes Groat Crops* **2020**, *3*, 32–36. [CrossRef]
38. Rezvyakova, S.B.; Eryomin, L.P. Increasing soybean yield based on protection against fungal diseases. *Bull. Agric. Sci.* **2021**, *3*, 77–83. [CrossRef]
39. Chakrabarty, M.; Hossen, F.; Begum, A.; Akhter, H. Plant growth promoting (PGP) activities of Rhizobial isolates from *Sesbania bispinosa* in response to pesticides. *Bangladesh J. Med. Microbiol.* **2021**, *38*, 31–37. [CrossRef]
40. Erokhin, A.I. Effectiveness of foliar (leaf) treatment of pea plants with the preparation Humate+7. *Legumes and Groat Crops* **2022**, *1*, 55–60. [CrossRef]
41. Ahemad, M. A comparative analysis of Tebuconazole mediated phytotoxicity to legumes. *J. Environ. Sci. Technol.* **2011**, *4*, 630–637. [CrossRef]
42. Ahemad, M.; Khan, M.S. Effect of tebuconazole-tolerant and plant growth promoting *Rhizobium* isolate MRP1 on pea–*Rhizobium* symbiosis. *Sci. Hortic.* **2011**, *129*, 266–272. [CrossRef]
43. Mamenko, T.P.; Kots, S.Y.; Khomenko, Y.O. The intensity of ethylene release by soybean plants under the influence of fungicides in the early stages of legume-rhizobial symbiosis. *Regul. Mech. Biosyst.* **2020**, *11*, 98–104. [CrossRef]

44. You, M.P.; Lamichhane, J.R.; Aubertot, J.-N.; Barbetti, M.J. Understanding why effective fungicides against individual soilborne pathogens are ineffective with soilborne pathogen complexes. *Plant Dis.* **2020**, *104*, 904–920. [CrossRef]
45. Roman, D.L.; Voiculescu, D.I.; Filip, M.; Ostafe, V.; Isvoran, A. Effects of triazole fungicides on soil microbiota and on the activities of enzymes found in soil: A review. *Agriculture* **2021**, *11*, 893. [CrossRef]
46. Ahemad, M.; Khan, M.S. Productivity of greengram in tebuconazole-stressed soil, by using a tolerant and plant growth-promoting *Bradyrhizobium* sp. MRM6 strain. *Acta Physiol. Plant.* **2012**, *34*, 245–254. [CrossRef]
47. Tsyganov, V.E.; Morzhina, E.V.; Stefanov, S.Y.; Borisov, A.Y.; Lebsky, V.K.; Tikhonovich, I.A. The pea (*Pisum sativum* L.) genes *sym33* and *sym40* control infection thread formation and root nodule function. *Mol. Gen. Genet.* **1998**, *259*, 491–503. [CrossRef]
48. Gorshkov, A.P.; Tsyganova, A.V.; Vorobiev, M.G.; Tsyganov, V.E. The fungicide tetramethylthiuram disulfide negatively affects plant cell walls, infection thread walls, and symbiosomes in pea (*Pisum sativum* L.) symbiotic nodules. *Plants* **2020**, *9*, 1488. [CrossRef] [PubMed]
49. Jeyakumar, P.; Boominathan, P. Combined effect of trifloxystrobin and tebuconazole on growth and yield attributes of groundnut. *Legume Res.* **2016**, *39*, 605–609. [CrossRef]
50. Machado, F.J.; Santana, F.M.; Lau, D.; Del Ponte, E.M. Quantitative review of the effects of triazole and benzimidazole fungicides on Fusarium head blight and wheat yield in Brazil. *Plant Dis.* **2017**, *101*, 1633–1641. [CrossRef]
51. Tak, P.S.; Pannu, P.P.S.; Kaur, J. Effect of foliar application of fungicides, Neem extract and Butter Milk on yellow rust, plant growth and yield parameters of wheat. *Agric. Res. J.* **2021**, *58*, 232–239. [CrossRef]
52. Kang, Z.; Huang, L.; Krieg, U.; Mauler-Machnik, A.; Buchenauer, H. Effects of tebuconazole on morphology, structure, cell wall components and trichothecene production of *Fusarium culmorum in vitro*. *Pest Manag. Sci.* **2001**, *57*, 491–500. [CrossRef]
53. Li, N.; Zhao, J.; Zhang, R.; Deng, L.; Li, J.; Gao, Y.; Liu, C. Effect of tebuconazole enantiomers and environmental factors on fumonisin accumulation and FUM gene expression in *Fusarium verticillioides*. *J. Agric. Food Chem.* **2018**, *66*, 13107–13115. [CrossRef]
54. Li, C.; Fan, S.; Wen, Y.; Tan, Z.; Liu, C. Enantioselective effect of flutriafol on growth, deoxynivalenol production, and TRI gene transcript levels in *Fusarium graminearum*. *J. Agric. Food Chem.* **2021**, *69*, 1684–1692. [CrossRef]
55. Rafique, T.; Naseem, S.; Usmani, T.H.; Bashir, E.; Khan, F.A.; Bhanger, M.I. Geochemical factors controlling the occurrence of high fluoride groundwater in the Nagar Parkar area, Sindh, Pakistan. *J. Hazard. Mater.* **2009**, *171*, 424–430. [CrossRef] [PubMed]
56. Johnson, R.C.; Cantonwine, E.G. Post-infection activities of fungicides against *Cercospora arachidicola* of peanut (*Arachis hypogaea*). *Pest Manag. Sci.* **2014**, *70*, 1202–1206. [CrossRef] [PubMed]
57. Melloy, P.; Del Ponte, E.M.; Sparks, A.H. Timing of triazole-based spray schedules for managing mungbean powdery mildew in Australia: A meta-analysis. *Plant Dis.* **2022**, *106*, 918–924. [CrossRef]
58. Vasić, T.; Vojinović, U.; Žujović, S.; Krnjaja, V.; Živković, S.; Marković, J.; Stević, M. *In vitro* toxicity of fungicides with different modes of action to alfalfa anthracnose fungus, *Colletotrichum destructivum*. *J. Environ. Sci. Health B* **2019**, *54*, 964–971. [CrossRef] [PubMed]
59. Fahad, S.; Hussain, S.; Bano, A.; Saud, S.; Hassan, S.; Shan, D.; Khan, F.A.; Khan, F.; Chen, Y.; Wu, C. Potential role of phytohormones and plant growth-promoting rhizobacteria in abiotic stresses: Consequences for changing environment. *Environ. Sci. Pollut. Res.* **2015**, *22*, 4907–4921. [CrossRef]
60. Liu, C.; Feng, N.; Zheng, D.; Cui, H.; Sun, F.; Gong, X. Uniconazole and diethyl aminoethyl hexanoate increase soybean pod setting and yield by regulating sucrose and starch content. *J. Sci. Food Agric.* **2019**, *99*, 748–758. [CrossRef]
61. Karikalan, L.; Rajan, S.; Gopi, R.; Sujatha, B.; Panneerselvam, R. Induction of salt tolerants by triadimefon in pigeon pea (*Cajanus cajan* L.) Millsp. *Indian J Exp. Biol.* **1999**, *37*, 825–829.
62. Hu, H.; Feng, N.; Shen, X.; Zhao, L.; Zheng, D. Transcriptomic analysis of *Vigna radiata* in response to chilling stress and uniconazole application. *BMC Genom.* **2022**, *23*, 205. [CrossRef]
63. Petit, A.-N.; Fontaine, F.; Vatsa, P.; Clément, C.; Vaillant-Gaveau, N. Fungicide impacts on photosynthesis in crop plants. *Photosynth. Res.* **2012**, *111*, 315–326. [CrossRef]
64. Zhu, L.-H.; van de Peppel, A.; Li, X.-Y.; Welander, M. Changes of leaf water potential and endogenous cytokinins in young apple trees treated with or without paclobutrazol under drought conditions. *Sci. Hortic.* **2004**, *99*, 133–141. [CrossRef]
65. Zhang, M.; Duan, L.; Tian, X.; He, Z.; Li, J.; Wang, B.; Li, Z. Uniconazole-induced tolerance of soybean to water deficit stress in relation to changes in photosynthesis, hormones and antioxidant system. *J. Plant Physiol.* **2007**, *164*, 709–717. [CrossRef] [PubMed]
66. Fletcher, R.; Hofstra, G. Triazoles as potential plant protectants. In *Sterol Biosynthesis Inhibitors in Plant Protection*; Berg, D., Plempel, M., Eds.; Ellis Horwood Ltd.: Cambridge, UK, 1988; pp. 31–331.
67. Pal, S.; Zhao, J.; Khan, A.; Yadav, N.S.; Batushansky, A.; Barak, S.; Rewald, B.; Fait, A.; Lazarovitch, N.; Rachmilevitch, S. Paclobutrazol induces tolerance in tomato to deficit irrigation through diversified effects on plant morphology, physiology and metabolism. *Sci. Rep.* **2016**, *6*, 39321. [CrossRef] [PubMed]
68. Wang, L.-H.; Lin, C.-H. The effect of paclobutrazol on physiological and biochemical changes in the primary roots of pea. *J. Exp. Bot.* **1992**, *43*, 1367–1372. [CrossRef]
69. Shishatskaya, E.; Menzyanova, N.; Zhila, N.; Prudnikova, S.; Volova, T.; Thomas, S. Toxic effects of the fungicide tebuconazole on the root system of fusarium-infected wheat plants. *Plant Physiol. Biochem.* **2018**, *132*, 400–407. [CrossRef]
70. Svobodníková, L.; Kummerová, M.; Zezulka, Š.; Babula, P.; Sendecká, K. Root response in *Pisum sativum* under naproxen stress: Morphoanatomical, cytological, and biochemical traits. *Chemosphere* **2020**, *258*, 127411. [CrossRef] [PubMed]

71. Kummerová, M.; Zezulka, Š.; Babula, P.; Váňová, L. Root response in *Pisum sativum* and *Zea mays* under fluoranthene stress: Morphological and anatomical traits. *Chemosphere* **2013**, *90*, 665–673. [CrossRef]
72. Zahran, H.H. *Rhizobium*-legume symbiosis and nitrogen fixation under severe conditions and in an arid climate. *Microbiol. Mol. Biol. R.* **1999**, *63*, 968–989. [CrossRef]
73. Gaind, S.; Rathi, M.S.; Kaushik, B.D.; Nain, L.; Verma, O.P. Survival of bio-inoculants on fungicides-treated seeds of wheat, pea and chickpea and subsequent effect on chickpea yield. *J. Environ. Sci. Health B* **2007**, *42*, 663–668. [CrossRef]
74. Yang, C.; Hamel, C.; Vujanovic, V.; Gan, Y. Nontarget effects of foliar fungicide application on the rhizosphere: Diversity of *nifH* gene and nodulation in chickpea field. *J. Appl. Microbiol.* **2012**, *112*, 966–974. [CrossRef]
75. Dhungana, S.K.; Kim, I.-D.; Kwak, H.-S.; Shin, D.-H. Unraveling the effect of structurally different classes of insecticide on germination and early plant growth of soybean [*Glycine max* (L.) Merr.]. *Pestic. Biochem. Physiol.* **2016**, *130*, 39–43. [CrossRef]
76. Schneider, M.; Keiblinger, K.M.; Paumann, M.; Soja, G.; Mentler, A.; Golestani-Fard, A.; Retzmann, A.; Prohaska, T.; Zechmeister-Boltenstern, S.; Wenzel, W. Fungicide application increased copper-bioavailability and impaired nitrogen fixation through reduced root nodule formation on alfalfa. *Ecotoxicology* **2019**, *28*, 599–611. [CrossRef] [PubMed]
77. Mishra, V.; Lal, A.A.; Simon, S. Efficacy of botanicals and bio-agents against powdery mildew disease of garden pea (*Pisum sativum* L.). *J. Pharmacogn. Phytochem.* **2017**, *6*, 1125–1126.
78. Ju, C.; Xu, J.; Wu, X.; Dong, F.; Liu, X.; Tian, C.; Zheng, Y. Effects of hexaconazole application on soil microbes community and nitrogen transformations in paddy soils. *Sci. Total Environ.* **2017**, *609*, 655–663. [CrossRef]
79. Kengar, Y.D.; Patil, B.J. Ureide content of guar under influence of hexaconazole and triazophos. *J. Fertil. Pestic* **2017**, *8*, 176. [CrossRef]
80. Li, Y.Y.; Yang, H. Bioaccumulation and degradation of pentachloronitrobenzene in *Medicago sativa*. *J. Environ. Manag.* **2013**, *119*, 143–150. [CrossRef]
81. Yakubu, H.; Buji, I.; Sandabe, M. Effects of seed-dressing fungicides on germination, nodulation, N2-fixation and yields of two groundnut varieties in semi-arid region of Nigeria. *Int. J. Appl. Agric. Res.* **2011**, *6*, 121–129.
82. Zemenu, E.A.; Gizachew, H.B.; Ahmed, M.Y. Evaluations of different seed-dressing fungicides with chickpea varieties for the management of fusarium wilt in Eastern Amhara, Ethiopia. *Adv. Agric.* **2022**, *2022*, 9324559. [CrossRef]
83. Mourad, B.; Baha-Eddine, B.; Mokhtar, B. The responses of the antioxidant defence system of a legume green bean *Phaseolus Vulgaris* cv. Djedida, exposed to a xenobiotic hexaconazole. *Int. J. Adv. Eng. Manag.* **2017**, *2*, 270–278. [CrossRef]
84. Sankar, B.; Jaleel, C.A.; Manivannan, P.; Kishorekumar, A.; Somasundaram, R.; Panneerselvam, R. Effect of paclobutrazol on water stress amelioration through antioxidants and free radical scavenging enzymes in *Arachis hypogaea* L. *Colloids Surf. B Biointerfaces* **2007**, *60*, 229–235. [CrossRef]
85. Liu, R.; Li, J.; Zhang, L.; Feng, T.; Zhang, Z.; Zhang, B. Fungicide difenoconazole induced biochemical and developmental toxicity in wheat (*Triticum aestivum* L.). *Plants* **2021**, *10*, 2304. [CrossRef]
86. Matamoros, M.A.; Baird, L.M.; Escuredo, P.R.; Dalton, D.A.; Minchin, F.R.; Iturbe-Ormaetxe, I.; Rubio, M.C.; Moran, J.F.; Gordon, A.J.; Becana, M. Stress-induced legume root nodule senescence. Physiological, biochemical, and structural alterations. *Plant Physiol.* **1999**, *121*, 97–112. [CrossRef] [PubMed]
87. Tsyganova, A.V.; Seliverstova, E.V.; Tsyganov, V.E. Influence of mutation in pea (*Pisum sativum* L.) *cdt* (*cadmium tolerance*) gene on histological and ultrastructural nodule organization. *Ecol. Genet.* **2019**, *17*, 71–80. [CrossRef]
88. Tsyganov, V.E.; Tsyganova, A.V.; Gorshkov, A.P.; Seliverstova, E.V.; Kim, V.E.; Chizhevskaya, E.P.; Belimov, A.A.; Serova, T.A.; Ivanova, K.A.; Kulaeva, O.A. Efficacy of a plant-microbe system: *Pisum sativum* (L.) cadmium-tolerant mutant and *Rhizobium leguminosarum* strains, expressing pea metallothionein genes *PsMT1* and *PsMT2*, for cadmium phytoremediation. *Front. Microbiol.* **2020**, *11*, 15. [CrossRef] [PubMed]
89. Dhanushkodi, R.; Matthew, C.; McManus, M.T.; Dijkwel, P.P. Drought-induced senescence of *Medicago truncatula* nodules involves serpin and ferritin to control proteolytic activity and iron levels. *New Phytol.* **2018**, *220*, 196–208. [CrossRef]
90. Tenhaken, R. Cell wall remodeling under abiotic stress. *Front. Plant Sci.* **2015**, *5*, 771. [CrossRef]
91. Sánchez-Pardo, B.; Fernández-Pascual, M.; Zornoza, P. Copper microlocalisation, ultrastructural alterations and antioxidant responses in the nodules of white lupin and soybean plants grown under conditions of copper excess. *Environ. Exp. Bot.* **2012**, *84*, 52–60. [CrossRef]
92. de María, N.; de Felipe, M.R.; Fernández-Pascual, M. Alterations induced by glyphosate on lupin photosynthetic apparatus and nodule ultrastructure and some oxygen diffusion related proteins. *Plant Physiol. Biochem.* **2005**, *43*, 985–996. [CrossRef]
93. Bolaños, L.; El-Hamdaoui, A.; Bonilla, I. Recovery of development and functionality of nodules and plant growth in salt-stressed *Pisum sativum-Rhizobium leguminosarum* symbiosis by boron and calcium. *J. Plant Physiol.* **2003**, *160*, 1493–1497. [CrossRef]
94. Wilmowicz, E.; Kućko, A.; Alché, J.D.D.; Czeszewska-Rosiak, G.; Florkiewicz, A.B.; Kapusta, M.; Karwaszewski, J. Remodeling of cell wall components in root nodules and flower abscission zone under drought in yellow lupine. *Int. J. Mol. Sci.* **2022**, *23*, 1680. [CrossRef]
95. Sujkowska-Rybkowska, M.; Kasowska, D.; Gediga, K.; Banasiewicz, J.; Stępkowski, T. *Lotus corniculatus*-rhizobia symbiosis under Ni, Co and Cr stress on ultramafic soil. *Plant Soil* **2020**, *451*, 459–484. [CrossRef]
96. Díaz-Cacho, P.; Moral, R.; Encina, A.; Luis Acebes, J.; Alvarez, J. Cell wall modifications in bean (*Phaseolus vulgaris*) callus cultures tolerant to isoxaben. *Physiol. Plant.* **1999**, *107*, 54–59. [CrossRef]

97. Manfield, I.W.; Orfila, C.; McCartney, L.; Harholt, J.; Bernal, A.J.; Scheller, H.V.; Gilmartin, P.M.; Mikkelsen, J.D.; Paul Knox, J.; Willats, W.G. Novel cell wall architecture of isoxaben-habituated *Arabidopsis* suspension-cultured cells: Global transcript profiling and cellular analysis. *Plant J.* **2004**, *40*, 260–275. [CrossRef] [PubMed]
98. Paniagua, C.; Bilkova, A.; Jackson, P.; Dabravolski, S.; Riber, W.; Didi, V.; Houser, J.; Gigli-Bisceglia, N.; Wimmerova, M.; Budínská, E. Dirigent proteins in plants: Modulating cell wall metabolism during abiotic and biotic stress exposure. *J. Exp. Bot.* **2017**, *68*, 3287–3301. [CrossRef] [PubMed]
99. Coculo, D.; Lionetti, V. The plant invertase/pectin methylesterase inhibitor superfamily. *Front. Plant Sci.* **2022**, *13*, 863892. [CrossRef]
100. Redondo-Nieto, M.; Pulido, L.; Reguera, M.; Bonilla, I.; Bolaños, L. Developmentally regulated membrane glycoproteins sharing antigenicity with rhamnogalacturonan II are not detected in nodulated boron deficient *Pisum sativum*. *Plant Cell Environ.* **2007**, *30*, 1436–1443. [CrossRef]
101. Sujkowska-Rybkowska, M.; Borucki, W. Pectins esterification in the apoplast of aluminum-treated pea root nodules. *J. Plant Physiol.* **2015**, *184*, 1–7. [CrossRef]
102. Ivanova, K.A.; Tsyganova, A.V.; Brewin, N.J.; Tikhonovich, I.A.; Tsyganov, V.E. Induction of host defences by *Rhizobium* during ineffective nodulation of pea (*Pisum sativum* L.) carrying symbiotically defective mutations *sym40* (*PsEFD*), *sym33* (*PsIPD3/PsCYCLOPS*) and *sym42*. *Protoplasma* **2015**, *252*, 1505–1517. [CrossRef]
103. Tsyganova, A.V.; Tsyganov, V.E. Plant cell wall in symbiotic interactions. Pectins. *Agric. Biol.* **2019**, 446. [CrossRef]
104. Sujkowska-Rybkowska, M.; Borucki, W.; Znojek, E. Structural changes in *Medicago truncatula* root nodules caused by short-term aluminum stress. *Symbiosis* **2012**, *58*, 161–170. [CrossRef]
105. Sujkowska-Rybkowska, M.; Borucki, W. Localization of hydrogen peroxide accumulation and diamine oxidase activity in pea root nodules under aluminum stress. *Micron* **2014**, *57*, 13–22. [CrossRef]
106. Sujkowska-Rybkowska, M.; Ważny, R. Metal resistant rhizobia and ultrastructure of *Anthyllis vulneraria* nodules from zinc and lead contaminated tailing in Poland. *Int. J. Phytoremediat.* **2018**, *20*, 709–720. [CrossRef] [PubMed]
107. Tsyganova, A.V.; Seliverstova, E.V.; Brewin, N.J.; Tsyganov, V.E. Bacterial release is accompanied by ectopic accumulation of cell wall material around the vacuole in nodules of *Pisum sativum sym33-3* allele encoding transcription factor PsCYCLOPS/PsIPD3. *Protoplasma* **2019**, *256*, 1449–1453. [CrossRef] [PubMed]
108. Ruiz-Díez, B.; Quiñones, M.A.; Fajardo, S.; Morcillo, C.; Fernández-Pascual, M. Possible reasons for tolerance to mercury of *Lupinus albus* cv. G1 inoculated with Hg-resistant and sensitive *Bradyrhizobium canariense* strains. *Symbiosis* **2015**, *67*, 91–102. [CrossRef]
109. Trainer, M.A.; Charles, T.C. The role of PHB metabolism in the symbiosis of rhizobia with legumes. *Appl. Microbiol. Biotechnol.* **2006**, *71*, 377–386. [CrossRef]
110. Borucki, W.; Sujkowska, M. The effects of sodium chloride-salinity upon growth, nodulation, and root nodule structure of pea (*Pisum sativum* L.) plants. *Acta Physiol. Plant.* **2008**, *30*, 293–301. [CrossRef]
111. Tsyganova, A.V.; Kitaeva, A.B.; Gorshkov, A.P.; Kusakin, P.G.; Sadovskaya, A.R.; Borisov, Y.G.; Tsyganov, V.E. *Glycyrrhiza uralensis* nodules: Histological and ultrastructural organization and tubulin cytoskeleton dynamics. *Agronomy* **2021**, *11*, 2508. [CrossRef]
112. Pankhurst, C.E.; Craig, A.S.; Jones, W.T. Effectiveness of Lotus root nodules: I. Morphology and flavolan content of nodules formed on *Lotus pedunculatus* by fast-growing Lotus rhizobia. *J. Exp. Bot.* **1979**, *30*, 1085–1093. [CrossRef]
113. Li, X.; Bao, H.; Wang, Z.; Wang, M.; Fan, B.; Zhu, C.; Chen, Z. Biogenesis and function of multivesicular bodies in plant immunity. *Front. Plant Sci.* **2018**, *9*, 979. [CrossRef]
114. Wang, M.; Li, X.; Luo, S.; Fan, B.; Zhu, C.; Chen, Z. Coordination and crosstalk between autophagosome and multivesicular body pathways in plant stress responses. *Cells* **2020**, *9*, 119. [CrossRef]
115. Krishnan, H.B.; Kim, W.-S.; Sun-Hyung, J.; Kim, K.Y.; Jiang, G. Citrate synthase mutants of *Sinorhizobium fredii* USDA257 form ineffective nodules with aberrant ultrastructure. *Appl. Environ. Microbiol.* **2003**, *69*, 3561–3568. [CrossRef]
116. Engvild, K.C. Nodulation and nitrogen fixation mutants of pea, *Pisum sativum*. *Theor. Appl. Genet.* **1987**, *74*, 711–713. [CrossRef] [PubMed]
117. Jensen, E.S. Variation in nitrate tolerance of nitrogen fixation in the pea/*Rhizobium* symbiosis. *Plant Breed.* **1987**, *98*, 130–135. [CrossRef]
118. Glenn, A.; Poole, P.; Hudman, J. Succinate uptake by free-living and bacteroid forms of *Rhizobium leguminosarum*. *Microbiology* **1980**, *119*, 267–271. [CrossRef]
119. Beringer, J.E. R factor transfer in *Rhizobium leguminosarum*. *Microbiology* **1974**, *84*, 188–198. [CrossRef] [PubMed]
120. Fåhraeus, G. The infection of clover root hairs by nodule bacteria studied by a simple glass slide technique. *J. Gen. Microbiol.* **1957**, *16*, 374–381. [CrossRef] [PubMed]
121. Serova, T.A.; Tsyganova, A.V.; Tsyganov, V.E. Early nodule senescence is activated in symbiotic mutants of pea (*Pisum sativum* L.) forming ineffective nodules blocked at different nodule developmental stages. *Protoplasma* **2018**, *255*, 1443–1459. [CrossRef] [PubMed]
122. Humphrey, C.D.; Pittman, F.E. A simple methylene blue-azure II-basic fuchsin stain for epoxy-embedded tissue sections. *Stain. Technol.* **1974**, *49*, 9–14. [CrossRef]
123. Musielak, T.J.; Schenkel, L.; Kolb, M.; Henschen, A.; Bayer, M. A simple and versatile cell wall staining protocol to study plant reproduction. *Plant Reprod.* **2015**, *28*, 161–169. [CrossRef] [PubMed]

124. Verhertbruggen, Y.; Marcus, S.E.; Haeger, A.; Ordaz-Ortiz, J.J.; Knox, J.P. An extended set of monoclonal antibodies to pectic homogalacturonan. *Carbohydr. Res.* **2009**, *344*, 1858–1862. [CrossRef]
125. Liners, F.; Letesson, J.-J.; Didembourg, C.; Van Cutsem, P. Monoclonal antibodies against pectin: Recognition of a conformation induced by calcium. *Plant Physiol.* **1989**, *91*, 1419–1424. [CrossRef]
126. Pattathil, S.; Avci, U.; Baldwin, D.; Swennes, A.G.; McGill, J.A.; Popper, Z.; Bootten, T.; Albert, A.; Davis, R.H.; Chennareddy, C. A comprehensive toolkit of plant cell wall glycan-directed monoclonal antibodies. *Plant Physiol.* **2010**, *153*, 514–525. [CrossRef] [PubMed]
127. Schneider, C.A.; Rasband, W.S.; Eliceiri, K.W. NIH Image to ImageJ: 25 years of image analysis. *Nat. Methods* **2012**, *9*, 671–675. [CrossRef] [PubMed]
128. Kusakin, P.G.; Serova, T.A.; Gogoleva, N.E.; Gogolev, Y.V.; Tsyganov, V.E. Laser microdissection of *Pisum sativum* L. nodules followed by RNA-Seq analysis revealed crucial transcriptomic changes during infected cell differentiation. *Agronomy* **2021**, *11*, 2504. [CrossRef]

International Journal of
Molecular Sciences

MDPI

Article

Molecular Characterizations of the *er1* Alleles Conferring Resistance to *Erysiphe pisi* in Three Chinese Pea (*Pisum sativum* L.) Landraces

Suli Sun [1], Dong Deng [1], Wenqi Wu [1], Yuhua He [2], Gaoling Luo [3], Chengzhang Du [4], Canxing Duan [1] and Zhendong Zhu [1,*]

1 Institute of Crop Sciences, Chinese Academy of Agricultural Sciences, Beijing 100081, China
2 Yunnan Academy of Agricultural Sciences, Kunming 650205, China
3 Rice Research Institute, Guangxi Academy of Agricultural Sciences, Nanning 530007, China
4 Institute of Specialty Crop, Chongqing Academy of Agricultural Sciences, Chongqing 402160, China
* Correspondence: zhuzhendong@caas.cn; Tel.: +86-10-82109609; Fax: +86-10-82109608

Abstract: Powdery mildew caused by *Erysiphe pisi* DC. is a major disease affecting pea worldwide. This study aimed to confirm the resistance genes contained in three powdery mildew-resistant Chinese pea landraces (Suoshadabaiwan, Dabaiwandou, and Guiwan 1) and to develop the functional markers of the novel resistance genes. The resistance genes were identified by genetic mapping and *PsMLO1* gene sequence identification. To confirm the inheritance of powdery mildew resistance in the three Landraces, the susceptible cultivars Bawan 6, Longwan 1, and Chengwan 8 were crossed with Suoshadabaiwan, Dabaiwandou, and Guiwan 1 to produce F_1, F_2, and $F_{2:3}$ populations, respectively. All F_1 plants were susceptible to *E. pisi*, and phenotypic segregation patterns in all the F_2 and $F_{2:3}$ populations fit the 3:1 (susceptible: resistant) and 1:2:1 (susceptible homozygotes: heterozygotes: resistant homozygotes) ratios, respectively, indicating powdery mildew resistance in the three Landraces were controlled by a single recessive gene, respectively. The analysis of *er1*-linked markers and genetic mapping in the F_2 populations suggested that the recessive resistance genes in three landraces could be *er1* alleles. The cDNA sequences of 10 homologous *PsMLO1* cDNA clones from the contrasting parents were obtained. A known *er1* allele, *er1*-4, was identified in Suoshadabaiwan. Two novel *er1* alleles were identified in Dabaiwandou and Guiwan 1, which were designated as *er1*-13 and *er1*-14, respectively. Both novel alleles were characterized with a 1-bp deletion (T) in positions 32 (exon 1) and 277 (exon 3), respectively, which caused a frame-shift mutation to result in premature termination of translation of *PsMLO1* protein. The co-dominant functional markers specific for *er1*-13 and *er1*-14, KASPar-*er1*-13, and KASPar-*er1*-14 were developed and effectively validated in populations and pea germplasms. Here, two novel *er1* alleles were characterized and their functional markers were validated. These results provide powerful tools for marker-assisted selection in pea breeding.

Keywords: *Erysiphe pisi*; *er1*-13; *er1*-14; KASPar marker; pea

1. Introduction

Pea (*Pisum sativum* L.) is an important and widely distributed cool season legume crop, which frequently suffers from abiotic and biotic stresses during the whole growth season [1,2]. Among the biotic factors, the disease is the main cause affecting pea production [2]. Powdery mildew caused by *Erysiphe pisi* DC. is a major constraint for pea yield and quality worldwide [3]. *E. pisi* infections of peas can lead to yield losses of up to 80% in regions that are suitable for disease infection [3]. To date, the use of resistant cultivars carrying the *E. pisi*-resistant gene *er1* has been considered to be the most effective and environmentally friendly way to control this disease [4,5].

E. pisi-resistance in pea has been proved to be controlled by three different genes in different germplasms, including two single recessive genes (*er1* and *er2*) and one dominant gene (*Er3*) [6–9]. The *er1*, *er2*, and *Er3* genes have been mapped on different linkage groups of peas using linked markers [10–18]. The two recessive genes *er1* and *er2* were mapped to pea linkage groups (LGs) VI and III, respectively [10,19]. The dominant gene *Er3* isolated from wild pea (*Pisum fulvum*) was located to pea LG IV recently [20].

To date, the recessive gene *er1* is the most widely used gene in pea production due to *er1* conferring high resistance or immunity to *E. pisi* in most pea germplasms [21]. In contrast, resistance conferred by *er2* is unstable and easily affected by leaf development stage and plant location [7,21–23]. *er2* is only found in a few pea germplasms [21]. *Er3* was known from wild pea (*P. fulvum*), and there have not been any extensive studies conducted to date [8,24].

Gene *er1* inhibits the incursion of *E. pisi* into pea epidermal cells, which confers stable and durable resistance to *E. pisi* [23]. Recent studies have shown that the *er1*-resistant phenotype is caused by loss-of-function mutations in the pea *MLO* (Mildew Resistance Locus O) homolog (*PsMLO1*). The *MLO* gene family has been identified in both dicots (e.g., *Arabidopsis thaliana*, and tomato—*Solanum lycopersicum*) and monocots (e.g., barley—*Hordeum vulgare*) [9,25–29].

To date, total of 12 *er1* alleles have been identified conferring resistance to *E. pisi* by natural mutation or obtained by mutagenesis in pea germplasms: *er1*-1 (also known as *er1mut1*) [9,13,16,30,31], *er1*-2 [9,15,16], *er1*-3 [9], *er1*-4 [9], *er1*-5 [28], *er1*-6 [18], *er1*-7 [17], *er1*-8, *er1*-9 [32], *er1*-10 (also known as *er1mut2*) [13,30,33], and *er1*-11 [33,34]; *er1*-12 was more recently identified in pea germplasm JI2019 from India [35]. Each *er1* allele corresponds to a different *PsMLO1* mutation site and pattern. Among the 12 *er1* alleles identified, only *er1*-1 and *er1*-2 are commonly applied in pea breeding programs [9,28]. Previous studies revealed that the functional markers of the known *er1* alleles have been developed and applied for the rapid selection of pea germplasms resistant to *E. pisi* in pea breeding [15,17,18,28,33,34,36].

E. pisi severely affects the yield and quality of pea crops in China [2]. The disease infects up to 100% of pea plants in some regions of planting susceptible cultivars. In our previous studies, we have focused on the identification of pea germplasms resistant to *E. pisi* [31,37]. A novel *er1* allele *er1*-6 had been identified in a Chinese pea germplasm [17] and new alleles *er1*-7, *er1*-8, and *er1*-9 were identified in pea germplasms from India, Afghanistan, and Australia, respectively [17,32]. *er1*-6 was also identified in some pea landraces from Yunnan Province of China [18]. Thus, a natural mutation of the *er1* gene conferring *E. pisi*-resistance has been observed in some Chinese pea landraces, which provides rich resistant sources that can be used to improve the *E. pisi* resistance of Chinese pea cultivars. The allelic diversity of this locus in the cultivated pea has been well characterized; however, relatively few studies have investigated and characterized the *E. pisi*-resistant gene in Chinese pea landraces. Thus, this study aimed to identify and characterize the *E. pisi*-resistant gene in three *E. pisi*-resistant Chinese pea landraces by genetic mapping and homologous *PsMLO1* gene sequence cloning. Additionally, any novel *er1* alleles were performed to develop their functional markers to improve marker-assisted selection in *E. pisi*-resistant pea breeding programs.

2. Results

2.1. Phenotypic Evaluation and Inheritence Analysis for Resistance

Six parental cultivars and contrasting controls were evaluated for their resistance to the *E. pisi* isolate EPYN. At 10 days post-inoculation, the *E. pisi* disease severity of the susceptible control was rated as score 4, indicating susceptibility to *E. pisi*. As expected, the three resistant pea parents, Suoshadabaiwan, Dabaiwandou, and Guiwan 1, and resistant control (Xucai 1) were immune to *E. pisi* infection (disease severity 0), while the three susceptible parents (Bawan 6, Longwan 1, and Chengwan 8) were susceptible to *E. pisi* (disease severity 4) (Figure 1). The segregation patterns of *E. pisi* resistance in the F_1, F_2, and $F_{2:3}$ populations derived from the crosses Bawan 6 × Suoshadabaiwan, Longwan 1 × Dabaiwandou, and Chengwan 8 × Guiwan 1 are presented in Table 1.

Figure 1. Phenotypic evaluation of the three *Erysiphe pisi*-resistant pea cultivars Suoshadabaiwan, Dabaiwandou, and Guiwan 1, as well as the three *E. pisi*-susceptible cultivars Bawan 6, Longwan 1, and Chengwan 8, after inoculation with *E. pisi* isolate EPYN. (**A**) Suoshadabaiwan and *E. pisi*-susceptible cultivar Bawan 6 (BW6). (**B**) Dabaiwandou and *E. pisi*-susceptible cultivar Longwan 1. (**C**) Guiwan 1 and *E. pisi*-susceptible cultivar Chengwan 8.

Table 1. Segregation patterns of pea resistance to powdery mildew in the F_1, F_2, and $F_{2:3}$ populations derived from three crosses, Bawan 6 × Suoshadabaiwan, Longwan 1 × Dabaiwandou, Chengwan 8 × Guiwan 1.

Parents and the Cross	Generation	Amount	No. of Plant or Families			Expected Ratio and Goodness of Fit		
			R	Rs	S	R:Rs:S	χ^2	P
Bawan 6	P1	30	-	-	30	-		
Suoshadabaiwan	P2	30	30	-	-	-		
Bawan 6 × Suoshadabaiwan	F_1	5	-	-	5	-		
	F_2	102	26	-	76	1:3	0.02	0.88
	$F_{2:3}$	102	26	51	25	1:2:1	0.03	0.99
Longwan 1	P_1	30	-	-	30	-		
Dabaiwandou	P_2	30	30	-	-	-		
	F_1	6	-	-	6	-		
Longwan 1 × Dabaiwandou	F_2	121	29	-	92	1:3	0.07	0.79
	$F_{2:3}$	121	29	56	36	1:2:1	1.41	0.49
Chengwan 8	P_1	30	-	-	30	-		
Guiwan 1	P_2	30	30	-	-	-		
	F_1	8	-	-	8	-		
Chengwan 8 × Guiwan 1	F_2	131	36	-	95	1:3	0.43	0.51
	$F_{2:3}$	131	36	61	34	1:2:1	0.67	0.71

"R", "Rs", and "S" stand for resistant, segregating, and susceptible, respectively.

Five F_1 plants produced from the cross Bawan 6 × Suoshadabaiwan were susceptible to *E. pisi* (Table 1). One of the five plants generated 102 F_2 and $F_{2:3}$ offspring through self-pollination. Of these 102 F_2 plants, 26 were resistant (R) to *E. pisi*, and 76 were susceptible (S) to *E. pisi*. This indicates that the segregation ratio (resistance: susceptibility) in the F_2 population was 1:3 ($\chi^2 = 0.02$; $p = 0.88$), indicating the inheritance of a single recessive gene. Moreover, a segregation ratio of 26 (homozygous resistant):51 (segregating):25 (homozygous susceptible) in the $F_{2:3}$ population fitted well with the genetic model of 1:2:1

ratio ($\chi^2 = 0.03$, $p = 0.99$) (Table 1), confirming that the *E. pisi* resistance in Suoshadabaiwan was controlled by a single recessive gene.

The cross of Longwan 1 × Dabaiwandou generated six F_1 plants, which showed *E. pisi*-susceptibility (Table 1). One of six F_1 plants generated 121 F_2 offspring. Of 121, 29 were resistant, and 92 of 121 were susceptible to *E. pisi*. The segregation ratio in the F_2 population of resistance to susceptibility fitted a genetic model ratio of 1:3 ($\chi^2 = 0.07$; $p = 0.79$), also indicating the inheritance of a single recessive gene. Moreover, a segregation ratio of 29 (homozygous resistant):56 (segregating):36 (homozygous susceptible) in the $F_{2:3}$ population (121 families) fitted well with the genetic model of 1:2:1 ratio ($\chi^2 = 1.41$; $p = 0.49$), indicating that *E. pisi* resistance in Dabaiwandou was also controlled by a single recessive gene (Table 1).

The cross of Chengwan 8 × Guiwan 1 generated eight F_1 plants which showed *E. pisi*-susceptibility (Table 1). One of eight F_1 plants generated 131 F_2 offspring. Of 131, 36 were resistant, and 95 of 131 were susceptible to *E. pisi*. The segregation ratio in the F_2 population of resistance to susceptibility fitted a genetic model ratio of 1:3 ($\chi^2 = 0.43$; $p = 0.51$), also indicating the recessive inheritance of a single gene. Moreover, a segregation ratio of 36 (homozygous resistant):61 (segregating):34 (homozygous susceptible) in the $F_{2:3}$ population (131 families) fitted well with the genetic model of 1:2:1 ratio ($\chi^2 = 0.67$; $p = 0.71$), indicating that *E. pisi* resistance in Guiwan 1 was also controlled by a single recessive gene (Table 1).

2.2. Mapping of Resistance Genes

Of the molecular markers tested, six (c5DNAmet, AD160, AC74, AD51, AD59, and AD60) were polymorphic between contrasting parents Bawan 6 and Suoshadabaiwan, and three (c5DNAmet, AA220, and AD51) were polymorphic between Longwan 1 and Dabaiwandou, Unfortunately, no polymorphic marker appeared between Longwan 1 and Dabaiwandou among the above markers tested. Thus, the additional eight SSR markers (16410, 28516, 26140, 23309, 29872, 26514, 23949, and 22903) developed recently were used to test the polymorphism between Longwan 1 and Dabaiwandou [38]. Two (26514 and 22903) were polymorphic between the contrasting parents, Longwan 1 and Dabaiwandou. All polymorphic markers between the parents were likely linked to the *E. pisi* resistance gene, respectively. Thus, the six, three, and the two parental polymorphic markers were used to confirm the genotypes of each F_2 plant derived from Bawan 6 × Suoshadabaiwan, Longwan 1 × Dabaiwandou, and Chengwan 8 × Guiwan 1, respectively. This genetic linkage analysis suggested that six markers (c5DNAmet, AD160, AC74, AD51, AD59, and AD60), three markers (c5DNAmet, AA220, and AD51), and two markers (26514 and 22903) were linked to the resistance gene *er1* in Suoshadabaiwan, Dabaiwandou, and Guiwan 1, respectively (Figure 2). Our results also indicated that the resistance genes in the three resistant cultivars were located in the *er1* region. In Suoshadabaiwan, the linkage map indicated that the markers (AD59 and AD60) were mapped on both sides of the target gene with 3.4 cM and 8.3 cM genetic distances, respectively (Figure 2A). In Dabaiwandou, two other markers (c5DNAmet and AA220) were located on both sides of the target gene with 2.6 cM and 11.6 cM genetic distances, respectively (Figure 2B). In Guiwan 1, two markers (26514 and 22903) were located on both sides of the target gene with 12.8 cM and 19.3 cM genetic distances, respectively (Figure 2C). Our linkage and genetic map analyses confirmed that an *er1* allele controlled *E. pisi* resistance in Suoshadabaiwan, Dabaiwandou, and Guiwan 1, respectively (Figure 2).

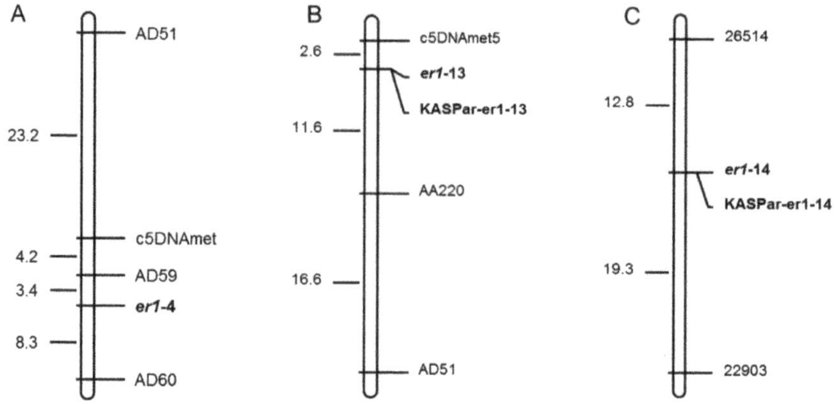

Figure 2. Genetic linkage maps constructed using the *er1*-linked markers and the functional markers for *er1*-13 and *er1*-14, based on the F_2 populations derived from (**A**) Bawan 6 × Suoshadabaiwan, (**B**) Longwan 1 × Dabaiwandou, and (**C**) Chengwan 8 × Guiwan 1. Map distances and loci order were determined with MAPMAKER v3.0. Estimated genetic distances between loci are shown to the left of the maps in centiMorgans (cM).

2.3. PsMLO1 Sequence Analysis

The *PsMLO1* cDNA sequences of Bawan 6, Longwan 1, Chengwan 8, and the susceptible parents, were consistent with that of the wild-type *PsMLO1* cDNA.

In landrace Suoshadabaiwan, a 1-bp deletion (A) was identified in a previously reported position 91 in exon 1 of the *PsMLO1* cDNA sequence. This result is consistent with the mutation in the *er1* gene carried by germplasm YI (JI1591), named *er1*-4. In landrace Dabaiwandou, a novel mutation pattern was found in the Dabaiwandou cDNA fragment homologous to *PsMLO1*: a 1-bp deletion (T) corresponding to positions 32 in exon 1 (the first exon) of the *PsMLO1* cDNA sequence. This deletion caused a substitution of the amino acid leucine with tryptophan at position 11 of the PsMLO1 protein sequence (Figure 3A). This change caused the early termination of protein translation, probably also resulting in a functional change of PsMLO1 (Figure 3A). In Guiwan 1, a 1-bp deletion (T) was also identified in a previously unreported position 277 in exon 3 of the *PsMLO1* cDNA sequence. This deletion caused a substitution of the amino acid tryptophan with glycine at position 93 of the PsMLO1 protein sequence (Figure 3B). This change caused the early termination of protein translation, probably also resulting in a functional change of PsMLO1 (Figure 3B). The two natural mutations differed from all known *er1* alleles, indicating that the *E. pisi* resistance of Dabaiwandou and Guiwan 1 was controlled by the novel alleles of *er1*. These novel alleles were designated *er1*-13 and *er1*-14, respectively, following the accepted nomenclature [9,17,18,32,33,35,36]. Thus, a known and two novel *er1* alleles were discovered in the three resistant cultivars, Suoshadabaiwan (from Chongqing), Dabaiwandou (from Yunnan), and Guiwan 1 (from Guangxi), respectively.

2.4. Development of Functional Markers for er1-13 and er1-14

The KASPar markers flanking the 1-bp (T) deletion sequences from Dabaiwandou and Guiwan 1 were designed as functional markers specific for KASPar-*er1*-13 and KASPar-*er1*-14, respectively.

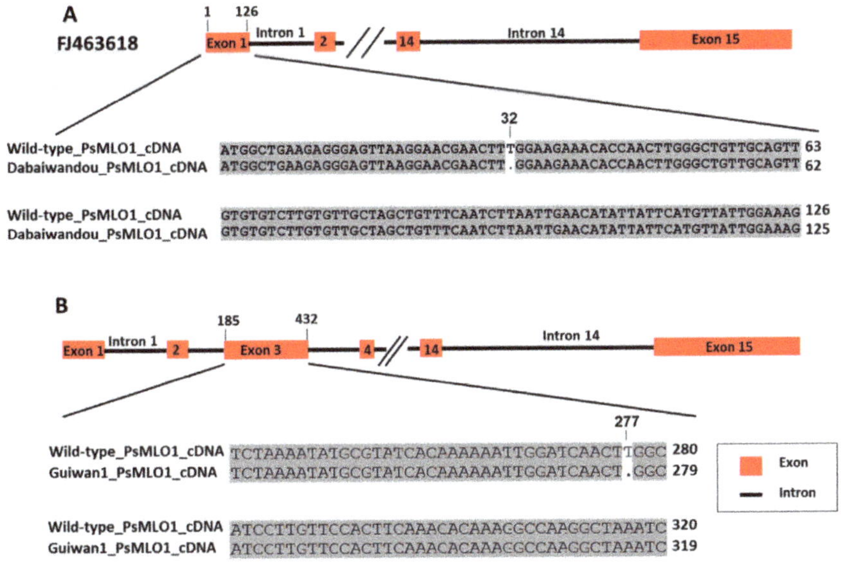

Figure 3. *PsMLO1* cDNA sequence comparisons of those from the powdery mildew-resistant pea landrace Dabaiwandou and Guiwan 1 with the wild-type pea cultivar Sprinter (GenBank accession number: FJ463618.1). (**A**) There is a single base deletion (T) in the *PsMLO1* cDNA of Dabaiwandou at positions 32 of exon 1. (**B**) There is a single base deletion (T) in the *PsMLO1* cDNA sequence of Guiwan 1 at position277 in exon 3. The figure shows the difference of nucleotide sequence from Dabaiwandou, Guiwan 1, and the wild-type pea cultivar Sprinter. The two mutation sites are indicated in the respective cDNA sequences.

As expected, KASPar-*er1*-13 and KASPar-*er1*-14 successfully distinguished the contrasting parents (Longwan 1 and Dabaiwandou, Chengwan 8 and Guiwan 1) into two different clusters corresponding to the FAM-labeled and HEX-labeled groups in the Kompetitive allele-specific PCR (KASPar) assay, respectively. When KASPar-*er1*-13 and KASPar-*er1*-14 were used to analyze the 121 and 131 F_2 progeny derived from Longwan 1 × Dabaiwandou and Chengwan 8 × Guiwan 1, the KASPar markers clearly separated the F_2 progeny into three clusters corresponding to three genotypes: homozygous resistant, homozygous susceptible, and heterozygous (Figure 4). In the F_2 population derived from Longwan 1 × Dabaiwandou, 29 plants were identified as homozygous resistant, 56 were heterozygous, and 36 were homozygous susceptible. In the F_2 population derived from Chengwan 8 × Guiwan 1, 36 plants were homozygous resistant, 61 were heterozygous, and 34 were homozygous susceptible. These results were completely consistent with the phenotypes of both $F_{2:3}$ populations, suggesting that KASPar-*er1*-13 and KASPar-*er1*-14 co-segregated with *er1*-13 and *er1*-14, respectively. A chi-squared (χ^2) test showed that both segregation ratios of KASPar-*er1*-13 and KASPar-*er1*-14 in respective F_2 populations fit 1:2:1 (KASPar-*er1*-13: $\chi^2 = 1.41$, $p = 0.49$; KASPar-*er1*-14: $\chi^2 = 0.67$; $p = 0.71$), indicating co-dominant markers.

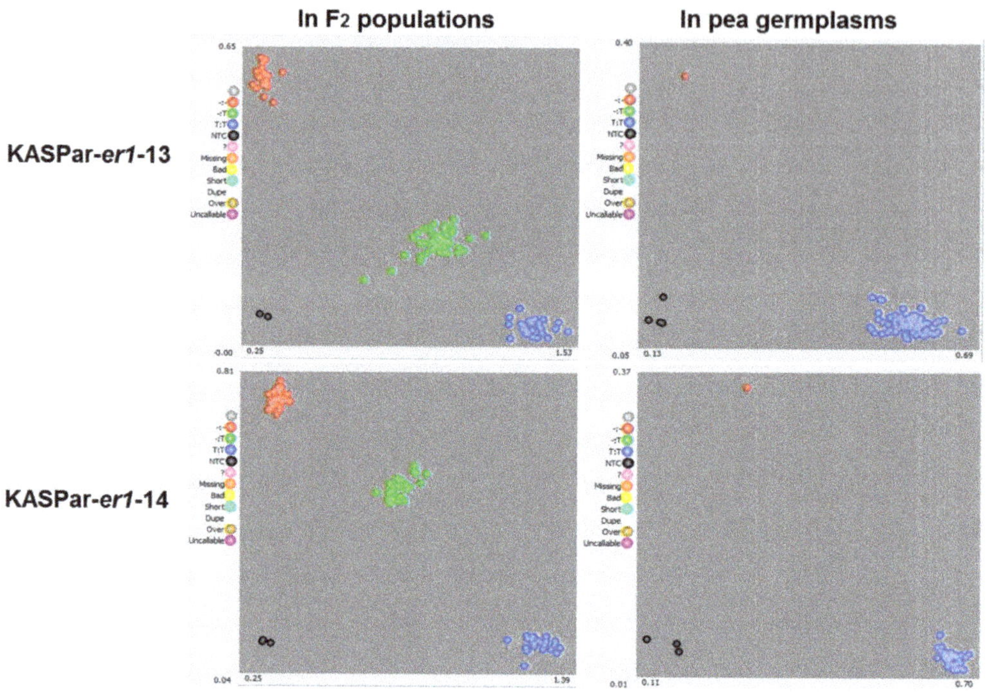

Figure 4. KASPar genotyping detection with markers KASPar-er1-13 and KASPar-er1-14 in the F_2 populations derived from Longwan 1 × Dabaiwandou and Chengwan 8 × Guiwan 1, as well as in other pea germplasms. Red dots indicate er1-13/er1-14 homozygous *Erysiphe pisi*-resistant lines/germplasms, blue dots indicate Er1-13/Er1-14 homozygous *E. pisi*-susceptible lines/germplasms, and green dots indicate er1-13/Er1-13 or er1-14/Er1-14 heterozygotes. Grey dots are blank samples used as controls. In our KASPar assay, the co-dominant markers KASPar-er1-13 and KASPar-er1-14 correctly categorized all F_2 individuals into three clusters corresponding to homozygous resistant (red dots), homozygous susceptible (blue dots), and heterozygous (green dots) lines, and categorized tested pea germplasms into two clusters corresponding to homozygous resistant (red dots) and homozygous susceptible (blue dots).

2.5. Validation and Application of Functional Markers

The 56 germplasms with the known resistance phenotype to *E. pisi* isolate EPYN that carrying the known *er1* alleles (*er1*-1, *er1*-2, *er1*-4, *er1*-6, *er1*-7, *er1*-8, and *er1*-9) were selected to genotyping by KASPar-*er1*-13 and KASPar-*er1*-13 (Table S1). It included 49 that were phenotypically immune to *E. pisi* and contained known *er1* alleles; seven were resistant. The three resistant and three susceptible parents were also tested at the same time (Table S1).

When the 56 germplasms were genotyped with KASPar-*er1*-13, two distinct clusters were recovered, with one gene (*er1*-13) corresponding to Dabaiwandou and the other (non-*er1*-13) to the other germplasms, respectively. Similarly, when the germplasms were genotyped with KASPar-*er1*-14, two distinct clusters were recovered, corresponding to Guiwan 1 and all of the other germplasms, respectively (Figure 4; Table S1). Thus, markers KASPar-*er1*-13 and KASPar-*er1*-14 effectively identified pea germplasms carrying the *er1*-13 and *er1*-14 alleles, respectively. Our results also showed that KASPar-*er1*-13 and KASPar-*er1*-14 could distinguish the know *er1* alleles and *Er1* from *er1*-13 or *er1*-14, respectively.

3. Discussion

Pea powdery mildew caused by *E. pisi* DC. is an important disease and reduces considerable yield in pea production worldwide. The deployment of resistant cultivars containing the *er1* gene is the most effective way to control this disease The *E. pisi* resistance gene *er1* is recessive in pea cultivars, which is the most widely deployed gene for controlling powdery mildew worldwide.

To date, there were 12 *er1* alleles identified in resistant pea germplasms. Among the 12 known *er1* alleles, *er1*-1 and *er1*-2 are most commonly used in pea breeding programs because they confer stable resistance to *E. pisi* [9,16,28,39,40]. Previously, *er1*-1 has been identified in four *E. pisi*-resistant pea cultivars (JI1559, Tara, and Cooper from Canada; and Yunwan 8 from China), while *er1*-2 has been identified in seven *E. pisi*-resistant pea cultivars (Stratagem, Franklin, Dorian, Nadir, X9002, Xucai 1, and G0005576) [9,15,16,18,28]. Recently, the *er1* gene for *E. pisi* resistance was confirmed in 53 pea germplasms from a worldwide collection [32]. Here, more *E. pisi*-resistant germplasms carrying the *er1*-1, *er1*-2, *er1*-6, and *er1*-7 alleles were identified. To date, a dominant gene *Er3* had been found in wild pea (*Pisum fulvum*) and mapped to pea LG IV [20]. It is possible that a rich diversity of *E. pisi*-resistant genes were contained in wild pea. Thus, searching for novel *E. pisi*-resistant genes from wild pea germplasms should be a good strategy [8].

To date, more than 40 *MLO* mutant alleles have been described in the monocotyledonous plant barley [41], and *PsMLO1* allelic diversity has been widely studied in pea [9,13,16–18,28,30–36,40]. Wild-type *PsMLO1* of pea consists of 15 exons and 14 introns (NCBI accession number: KC466597). To date, a total of 12 *er1* alleles associated with the *er1*-resistance phenotype have been identified and 11 of 12 *er1* alleles *PsMLO1* mutations were caused by alterations of the coding sequence. There was 1, 1, 1, 1, 3, 1, and 2 allele mutations occurred in exons 1, 5, 6, 8, 10, 11, and 15 of wild-type *PsMLO1*, respectively. Eight alleles (*er1*-1/*er1mut1*, *er1*-3, *er1*-4, *er1*-5, *er1*-6, *er1*-9, *er1*-10/*er1mut2*, and *er1*-12) are the result of point mutations in the exons of wild-type *PsMLO1*. Four alleles result from single base substitutions in wild-type *PsMLO1* cDNA: in *er1*-1, a C→G at position 680 (exon 6); in *er1*-5, a G→A at position 570 (exon 5); in *er1*-6, a T→C at position 1121 (exon 11); and in *er1*-10, a G→A at position 939 (exon 10) [9,18,28,30]. Three alleles result from single base deletions in wild-type *PsMLO1* cDNA, including ΔG at position 862 (exon 8) in *er1*-3; ΔA at position 91 (exon 1) in *er1*-4; and ΔT at position 928 (exon 10) in *er1*-9 identified in this study [9]. Recently, *er1*-12 allele was found resulting from single base insertion (A) in front of the last exon (exon 15) in wild-type *PsMLO1* cDNA [35]. Two alleles result from small fragment deletions in wild-type *PsMLO1* cDNA, including a 10-bp deletion of positions 111–120 (exon 1) in *er1*-7 [17]; and a 3-bp deletion of positions 1339–1341 (exon 15) in *er1*-8 [32]. Only the *er1*-11 mutation is known to have resulted from an intron mutation in *PsMLO1* (a 2-bp insertion in intron 14) [33,34], and only *er1*-2 results from a large insert of unknown size in wild-type *PsMLO1* cDNA [9,15,18].

Several functional markers specific to the previously recognized *er1* alleles have already been developed to facilitate the marker-assisted breeding of pea cultivars resistant to *E. pisi* [9,15,17,18,30,32–34,36]. Pavan et al. [28] developed a functional cleaved amplified polymorphic sequence (CAPS) marker for *er1*-5, while Pavan et al. [36] developed functional markers for the five *er1* alleles, *er1*-1 through *er1*-5. Santo et al. [30] developed functional markers for *er1mut1* and *er1mut2*, and Wang et al. [15] developed a dominant marker for *er1*-2. Sudheesh et al. [34] developed a functional marker for *er1*-11, while Sun et al. [17,18] developed co-dominant functional markers for *er1*-6 and *er1*-7. Ma et al. [33] developed eight KASPar markers for eight known *er1* alleles, excluding *er1*-2, and renamed the *er1*-10 and *er1*-11. More recently, Sun et al. [32] identified the two novel *er1* alleles, *er1*-8 and *er1*-9, and developed KASPar-*er1*-8 and KASPar-*er1*-9. In this study, the developed markers, KASPar-*er1*-13 and KASPar-*er1*-14, could accurately and effectively identify pea germplasms carrying the *er1*-13 and *er1*-14 alleles and distinguish them from the know *er1* alleles or *Er1*, respectively.

This study discovered a known and two novel *er1* alleles, resulting from new mutations of wild-type *PsMLO1* cDNA: *er1*-13 and *er1*-14 was generated by a 1-bp deletion in exon 1 and 3, respectively. The co-dominant functional markers specific to *er1*-13 (KASPar-*er1*-13) and to *er1*-14 (KASPar-*er1*-14) were developed. These markers were validated in genetic populations and pea germplasms. These results will support future studies to reveal the powdery mildew resistance mechanisms. The two novel *er1* alleles and the developed co-dominant functional markers could be powerful tools for the breeding of pea cultivars resistance to *E. pisi*.

4. Materials and Methods

4.1. Plant Material and *E. pisi* Inoculum

Previously, many Chinese pea germplasms had been screened for *E. pisi* and some were found to be *E. pisi*-resistant [31,37,39]. In this study, the three *E. pisi*-resistant Chinese pea landraces, Suoshadabaiwan, Dabaiwandou, and Guiwan 1, respectively, from the Chongqing, Yunnan, and Guangxi provinces of China were conducted to reveal their *E. pisi*-resistant genes. The three *E. pisi*-susceptible Chinese pea cultivars, Bawan 6, Longwan 1, and Chengwan8, were used as susceptible controls or cross susceptible parents for genetic analysis [15,40]. The Chinese pea cultivar Xucai 1 containing *er1*-2 was used as *E. pisi*-resistant control [16].

The *E. pisi* isolate EPYN from Yunnan Province of China was used as the inoculum, which is highly virulent on pea [17,18]. The EPYN isolate was maintained through the continuous re-inoculation of healthy seedlings of Longwan 1 under controlled conditions. The inoculated plants were incubated in a growth chamber with controlled conditions [16].

4.2. Powdery Mildew Resistance Evaluation

Thirty-five seeds were planted from each of the three *E. pisi*-resistant germplasms (Suoshadabaiwan, Dabaiwandou, and Guiwan 1), three *E. pisi*-susceptible pea cultivars (Bawan 6, Longwan 1, and Chengwan 8), and from the resistant and susceptible controls (Bawan 6, Longwan 1 and Chengwan 8, and Xucai 1) [18]. The healthy seedlings were thinned to 30 per pot before the phenotypic evaluation. Three replications were planted. Seeded pots were placed in a greenhouse maintained at 18 to 26 °C. At the same time, the *E. pisi* inoculum was prepared by inoculating the 10-day-old seedlings of Longwan 1, which were incubated in a growth chamber at 20 $\pm$ 1 °C with a 12-h photoperiod. Two weeks later, all seedlings of the germplasms and controls were inoculated by gently shaking off the conidia of the Longwan 1 plants. Inoculated plants were incubated in a growth chamber at 20 $\pm$ 1 °C with a 12-h photoperiod. Ten days later, disease severity was rated based on a scale (0–4 scale) [17,18]. Plants with a score of 0 were considered *E. pisi*-immune, while those with scores of 1, 2 and 3, 4 were considered E. pisi-resistant and E. pisi-susceptible, respectively. For those identified as immune or resistant to *E. pisi*, repeated identification was performed.

4.3. Inheritance Analysis of Resistant Pea Cultivars

To reveal the inheritance controlled by *E. pisi* resistance genes in the three *E. pisi*-resistant Chinese pea landrace, Suoshadabaiwan, Dabaiwandou, and Guiwan 1, they were crossed with the *E. pisi*-susceptible cultivars Bawan 6, Longwan 1, and Chengwan 8, respectively, to generate genetic populations. The derived F_1, F_2, and $F_{2:3}$ populations from three crosses (Bawan 6 $\times$ Suoshadabaiwan, Longwan 1 $\times$ Dabaiwandou, and Chengwan 8 $\times$ Guiwan 1), which were used for the powdery mildew resistance genetic analysis of Suoshadabaiwan, Dabaiwandou, and Guiwan 1. The six parents and the derived F_1 and F_2 populations were planted in a propagation greenhouse to generate F_2 and $F_{2:3}$ family seeds, respectively.

The plants of the F_1 and F_2 populations at the fourth or fifth leaf stage were inoculated with the *E. pisi* isolate EPYN using the detached leaf method [16–18]. After inoculation, the treated leaves were placed in a growth chamber at 20 °C with a 14-h photoperiod. The six parents were also inoculated as controls. Ten days after inoculation, disease severity was rated based on a scale of 0–4 as described above. Plants with scores of 0–2 and 3–4 were classified as resistant and susceptible, respectively [16–18]. Those plants identified as *E. pisi*-resistant were tested again to confirm their resistance.

Twenty-five seeds were selected randomly from each of the 102, 121, and 131 $F_{2:3}$ families derived from Bawan 6 × Suoshadabaiwan, Longwan 1 × Dabaiwandou, and Chengwan 8 × Guiwan 1, respectively. These seeds were planted and cultivated together with their parents, following previously published protocols [25–27]. Disease severity was scored 10 days after inoculation using the 0–4 scale, as described above for the phenotypic identification of the pea germplasms. The $F_{2:3}$ families with scores of 0–2 and 3–4 were classified as homozygous resistant and homozygous susceptible, respectively. Families with scores of 0–2 and 3–4 were considered segregated to *E. pisi* resistance. The families identified as homozygous-resistant or resistance segregated were subjected to repeated testing.

A chi-squared (χ^2) analysis was used to evaluate the goodness-of-fit to Mendelian segregation ratio of the F_2 and $F_{2:3}$ phenotypes derived from Bawan 6 × Suoshadabaiwan, Longwan 1 × Dabaiwandou, and Chengwan 8 × Guiwan 1, respectively.

4.4. Genetic Mapping of the Resistance Gene

The Genomic DNA was isolated from the leaves of the F_2 populations and of their parents using a slightly modified cetyltrimethylammonium bromide (CTAB) extraction method [42]. The DNA solution was diluted and stored at −20 °C until use.

Previous studies suggested that *E. pisi* resistance was controlled by *er1* in most of all pea germplasms except for lines SVP952 and JI 2480 [7,21]. Thus, this study was performed to map the *E. pisi*-resistance gene using the known *er1*-linked markers on the pea LG VI, including five simple sequence repeat (SSR) markers (PSMPSAD51, PSMPSA5, PSMPSAD60, i.e., AD60, PSMPSAA374e, and PSMPSAA369); a gene marker (Cytosine-5, DNA-methyltransferase (c5DNAmet)) [12,15–18,37,43]; and 10 molecular markers on the pea LG VI (AD160, AC74, AC10_1, AA224, AA200, AD159, AD59, AB71, AA335, and AB86). Firstly, these markers were used to screen for polymorphisms between the crossed parents, Bawan 6 × Suoshadabaiwan, Longwan 1 × Dabaiwandou, and Chengwan 8 × Guiwan 1 [44]. The parental polymorphic markers were then used to confirm the genotypes of each F_2 plant for genetic linkage analysis. PCR amplification was conducted in a total volume of 20 μL [16–18]. PCR reactions were performed in a thermal cycler (Biometra, Göttingen, Germany). The PCR products were separated on 6%–8% polyacrylamide gels.

The segregation data of the polymorphic markers in the F_2 populations were evaluated for goodness-of-fit to Mendelian segregation patterns with a chi-squared (χ^2) test. Genetic linkage analyses were completed using MAPMAKER/EXP version 3.0b. A logarithm of odds (LOD) score > 3.0 and a distance < 50 cM were used as the thresholds to determine the linkage groups [45]. Genetic distances were determined using the Kosambi mapping function [46]. The genetic linkage map was constructed using the Microsoft Excel macro MapDraw [47].

4.5. RNA Extraction and PsMLO1 Sequence Analysis

The extraction of total RNA and synthesis of cDNA from Suoshadabaiwan, Dabaiwandou, and Guiwan 1 and controls were completed according to our previous studies [16–18].

To identify the resistance gene *er1* alleles, the full-length cDNAs of the *PsMLO1* homologs were amplified using the primers specific to *PsMLO1* [9]. The PCR cycling conditions were as follows: 95 °C for 5 min; then 35 cycles of denaturation at 94 °C for 30 s, annealing at 58 °C for 45 s, and extension at 72 °C for 1 min; and a final extension at 72 °C for 10 min. The purified amplicons were cloned with a pEasy-T5 vector (TransGen Biotech, Beijing, China). The sequencing reactions of 10 clones per parental cultivars and controls were performed by the Shanghai Shenggong Biological Engineering Co., Ltd. (Shanghai, China). The resulting sequences were aligned with the wild-type *PsMLO1* of pea (NCBI accession number: FJ463618.1) using DNAMAN v6.0 (Lynnon Biosoft, Vaudreuil, QC, Canada).

4.6. Development of Functional Markers for the Novel er1 Alleles

Primers flanking the mutation site were designed based on the *PsMLO1* gene sequence (GenBank accession number KC466597), using Primer Premier v5.0, to develop a functional marker specific to allele *er1*-13 and *er1*-14 (Table 2).

Table 2. Sequence information for the indel and Kompetitive allele-specific PCR (KASPar) markers specific to *er1*-13 and for the KASPar marker specific to *er1*-14.

Markers	Primers	Sequence Information (5′–3′)	Annealing Tm
KASPar-*er1-13*	Forward-T	GAAGAGGGAGTTAAGGAACGAACTTT	65–57 °C touchdown
	Forward	AAGAGGGAGTTAAGGAACGAACTTG	
	Common reverse	TGCAACAGCCCAAGTTGGTGTTTCT	
KASPar-*er1-14*	Forward-T	ATATGCGTATCACAAAAAATTGGATCAACTT	65–57 °C touchdown
	Forward	GCGTATCACAAAAAATTGGATCAACTG	
	Common reverse	CCTTGGCCTTTGTGTTTGAAGTGGAA	

The Kompetitive allele-specific PCR (KASPar) markers (KASPar-*er1*-13 and KASPar-*er1*-14) specific to the two novel *er1* alleles were developed based on allele *er1*-13 SNPs (1-bp deletion) and *er1*-14 SNPs (1-bp deletion) in *PsMLO1*. The forward primers and the common reverse primers specific to *er1*-13 and *er1*-14 were designed for KASPar markers by LGC KBioscience (KBioscience, Hoddesdon, UK), respectively. In brief, two KASPar markers (KASPar-*er1*-13 and KASPar-*er1*-14) were used to detect parental polymorphisms (Longwan 1 × Dabaiwandou, and Chengwan 8 × Guiwan 1), and then used to analyze the genotypes of each F_2 offspring (Longwan 1 × Dabaiwandou: 121 F_2 individuals; Chengwan 8 × Guiwan: 131 F_2 individuals).

KASPar markers were amplified with a Douglas Scientific Array Tape Platform (China Golden Marker Biotech Co., Ltd., (Beijing, China)) in a 0.8 μL Array Tape reaction volume with 10 ng dry DNA, 0.8 μL 2 × KASP master mix, and 0.011 μL primer mix (KBioscience, Hoddesdon, UK). A Nexar Liquid handling instrument was used to add the PCR solution to the Array Tape (Douglas Scientific). PCRs were performed on a Soellex PCR Thermal Cycler with the following conditions: initial denaturation at 94 °C for 15 min; followed by 10 cycles of denaturation at 94 °C for 20 s and 65 °C for 60 s at an annealing temperature that decreased by 0.8 °C per cycle; and then 26 cycles of denaturation at 94 °C for 20 s and 57 °C for 60 s; and a final cooling to 4 °C. A fluorescent end-point reading was completed with the Araya fluorescence detection system (part of the Douglas Scientific Array Tape Platform). Genotypes and clusters were visualized with Kraken (http://ccb.jhu.edu/software/kraken/MANUAL.html (accessed on 5 August 2022)).

4.7. Validation and Application of Functional Markers

To test the efficacy of the novel functional markers specific to *er1*-13 (KASPar-*er1-13*) and *er1*-14 (KASPar-*er1*-14), 56 pea germplasms with known phenotypic resistance to *E. pisi* isolate EPYN and carrying known *er1* alleles, and six parents were tested for whether they carried the *er1* alleles *er1*-13 or *er1*-14 (Table S1) [9,15,17,18,32,40]. The six parents (Suoshadabaiwan, Dabaiwandou and Guiwan 1, Bawan 6, Longwan 1, and Chengwan 8) were used as contrasting controls.

DNA was extracted from the 56 selected pea germplasms (resistant cultivars with known *er1* alleles) and the six parents using the CTAB method [42]. The PCR amplification of the KASPar markers were performed as described above (in the section "Development of Functional Markers for the novel *er1* alleles").

Supplementary Materials: The following supporting information can be downloaded at: https://www.mdpi.com/article/10.3390/ijms231912016/s1.

Author Contributions: Z.Z. conceived and designed the experiments. S.S., D.D., W.W., Y.H., G.L., C.D. (Chengzhang Du) and C.D. (Canxing Duan), performed the experiments. S.S. analyzed the data and wrote the manuscript. Z.Z. revised the manuscript. All authors have read and agreed to the published version of the manuscript.

Funding: This study was supported by the National Key R&D Program of China (2019YFD1001300, 2019YFD1001301), the China Agriculture Research System of MOF and MARA (CARS-08), National Crop Germplasm Resources Center (NCGRC-2022-09), Species and Variety Resource Protection from the Ministry of Agriculture and Rural of China, and the Scientific Innovation Program of the Chinese Academy of Agricultural Sciences from the Institute of Crop Sciences, Chinese Academy of Agricultural Sciences.

Institutional Review Board Statement: Not applicable.

Informed Consent Statement: Not applicable.

Data Availability Statement: The data presented in this study are available on request from the corresponding author.

Acknowledgments: We sincerely thank Xuxiao Zong, Xiaoming Yang, Dongxu Xu, and Dongmei Yu for providing the pea cultivars and germplasms.

Conflicts of Interest: The authors declare no conflict of interest.

Abbreviations

SSR	Simple sequence repeat
SNP	Single nucleotide polymorphism
KASPar	Kompetitive allele-specific PCR

References

1. Ali, S.M.; Sharma, B.; Ambrose, M.J. Current status and future strategy in breeding pea to improve resistance to biotic and abiotic stresses. *Euphytica* **1993**, *1*, 115–126. [CrossRef]
2. Wang, X.; Zhu, Z.; Duan, C.; Zong, X. *Identification and Control Technology of Disease and Pest on Faba Bean and Pea*; Chinese Agricultural Science and Technology Press: Beijing, China, 2007.
3. Smith, P.H.; Foster, E.M.; Boyd, L.A.; Brown, J.K.M. The early development of *Erysiphe pisi* on *Pisum sativum* L. *Plant Pathol.* **1996**, *45*, 302–309. [CrossRef]
4. Ghafoor, A.; McPhee, K. Marker assisted selection (MAS) for developing powdery mildew resistant pea cultivars. *Euphytica* **2012**, *186*, 593–607. [CrossRef]
5. Fondevilla, S.; Rubiales, D. Powdery mildew control in pea, a review. *Agron. Sustain. Dev.* **2012**, *32*, 401–409. [CrossRef]
6. Harland, S.C. Inheritance of immunity to mildew in Peruvian forms of *Pisum sativum*. *Heredity* **1948**, *2*, 263–269. [CrossRef]
7. Heringa, R.J.; Van Norel, A.; Tazelaar, M.F. Resistance to powdery mildew (*Erysiphe polygoni* D.C.) in peas (*Pisum sativum* L.). *Euphytica* **1969**, *18*, 163–169. [CrossRef]
8. Fondevilla, S.; Torres, A.M.; Moreno, M.T.; Rubiales, D. Identification of a new gene for resistance to powdery mildew in *Pisum fulvum*, a wild relative of pea. *Breed. Sci.* **2007**, *57*, 181–184. [CrossRef]
9. Humphry, M.; Reinstädler, A.; Ivanov, S.; Bisseling, T.; Panstruga, R. Durable broad-spectrum powdery mildew resistance in pea *er1* plants is conferred by natural loss-of-function mutations in *PsMLO1*. *Mol. Plant Pathol.* **2011**, *12*, 866–878. [CrossRef]
10. Timmerman, G.M.; Frew, T.J.; Weeden, N.F. Linkage analysis of *er1*, a recessive *Pisum sativum* gene for resistance to powdery mildew fungus (*Erysiphe pisi* DC). *Theor. Appl. Genet.* **1994**, *88*, 1050–1055. [CrossRef]
11. Tiwari, K.R.; Penner, G.A.; Warkentin, T.D. Identification of coupling and repulsion phase RAPD markers for powdery mildew resistance gene *er1* in pea. *Genome* **1998**, *41*, 440–444. [CrossRef]
12. Ek, M.; Eklund, M.; von Post, R.; Dayteg, C.; Henriksson, T.; Weibull, P.; Ceplitis, A.; Isaac, P.; Tuvesson, S. Microsatellite markers for powdery mildew resistance in pea (*Pisum sativum* L.). *Hereditas* **2005**, *142*, 86–91. [CrossRef]
13. Pereira, G.; Marques, C.; Ribeiro, R.; Formiga, S.; Dâmaso, M.; Sousa, T.; Farinhó, M.; Leitão, J.M. Identification of DNA markers linked to an induced mutated gene conferring resistance to powdery mildew in pea (*Pisum sativum* L.). *Euphytica* **2010**, *171*, 327–335. [CrossRef]
14. Srivastava, R.K.; Mishra, S.K.; Singh, K.; Mohapatra, T. Development of a coupling-phase SCAR marker linked to the powdery mildew resistance gene *er1* in pea (*Pisum sativum* L.). *Euphytica* **2012**, *86*, 855–866. [CrossRef]

15. Wang, Z.; Fu, H.; Sun, S.; Duan, C.; Wu, X.; Yang, X.; Zhu, Z. Identification of powdery mildew resistance gene in pea line X9002. *Acta Agron. Sin.* **2015**, *41*, 515–523. (In Chinese with English abstract) [CrossRef]
16. Sun, S.; Wang, Z.; Fu, H.; Duan, C.; Wang, X.; Zhu, Z. Resistance to powdery mildew in the pea cultivar Xucai 1 is conferred by the gene *er1*. *Crop. J.* **2015**, *3*, 489–499. [CrossRef]
17. Sun, S.; Deng, D.; Wang, Z.; Duan, C.; Wu, X.; Wang, X.; Zong, X.; Zhu, Z. A novel *er1* allele and the development and validation of its functional marker for breeding pea (*Pisum sativum* L.) resistance to powdery mildew. *Appl. Genet.* **2016**, *129*, 909–919.
18. Sun, S.; Fu, H.; Wang, Z.; Duan, C.; Zong, X.; Zhu, Z. Discovery of a novel *er1* allele conferring powdery mildew resistance in Chinese pea (*Pisum sativum* L.) landraces. *PLoS ONE* **2016**, *11*, e0147624. [CrossRef]
19. Katoch, V.; Sharma, S.; Pathania, S.; Banayal, D.K.; Sharma, S.K.; Rathour, R. Molecular mapping of pea powdery mildew resistance gene *er2* to pea linkage group III. *Mol. Breed.* **2010**, *25*, 229–237. [CrossRef]
20. Cobos, M.J.; Satovic, Z.; Rubiales, D.; Fondevilla, S. *Er3* gene, conferring resistance to *Erysiphe pisi*, is located in pea LGIV. *Euphytica* **2018**, *214*, 203. [CrossRef]
21. Tiwari, K.R.; Penner, G.A.; Warkentin, T.D. Inheritance of powdery mildew resistance in pea. *Can. J. Plant Sci.* **1997**, *77*, 307–310. [CrossRef]
22. Vaid, A.; Tyagi, P.D. Genetics of powdery mildew resistance in pea. *Euphytica* **1997**, *96*, 203–206. [CrossRef]
23. Fondevilla, S.; Carver, T.L.W.; Moreno, M.T.; Rubiales, D. Macroscopic and histological characterisation of genes *er1* and *er2* for powdery mildew resistance in pea. *Eur. J. Plant Pathol.* **2006**, *115*, 309–321. [CrossRef]
24. Fondevilla, S.; Cubero, J.I.; Rubiales, D. Confirmation that the *Er3* gene, conferring resistance to *Erysiphe pisi* in pea, is a different gene from *er1* and *er2* genes. *Plant Breed.* **2011**, *130*, 281–282. [CrossRef]
25. Bai, Y.; Pavan, S.; Zheng, Z.; Zappel, N.F.; Reinstadler, A.; Lotti, C.; DeGiovanni, C.; Ricciardi, L.; Lindhout, P.; Visser, R.; et al. Naturally occurring broad-spectrum powdery mildew resistance in a central American tomato accession is caused by loss of MLO1 function. *Mol. Plant Microbe Interact.* **2008**, *21*, 30–39. [CrossRef] [PubMed]
26. Büschges, R.; Hollricher, K.; Panstruga, R.; Simons, G.; Wolter, M.; Frijters, A.; van Daelen, R.; van der Lee, T.; Diergaarde, P.; Groenendijk, J. The barley MLO gene, a novel control element of plant pathogen resistance. *Cell* **1997**, *88*, 695–705. [CrossRef]
27. Devoto, A.; Hartmann, H.A.; Piffanelli, P.; Elliott, C.; Simmons, C.; Taramino, G.; Goh, C.S.; Cohen, F.E.; Emerson, B.C.; Schulze-Lefert, P.; et al. Molecular phylogeny and evolution of the plant-specific seven-transmembrane MLO family. *J. Mol. Evol.* **2003**, *56*, 77–88. [CrossRef] [PubMed]
28. Pavan, S.; Schiavulli, A.; Appiano, M.; Marcotrigiano, A.R.; Cillo, F.; Visser, R.G.F.; Bai, Y.; Lotti, C.; Luigi Ricciardi, L. Pea powdery mildew *er1* resistance is associated to loss-of-function mutations at a MLO homologous locus. *Appl. Genet.* **2011**, *123*, 1425–1431. [CrossRef]
29. Rispail, N.; Rubiales, D. Genome-wide identification and comparison of legume *MLO* gene family. *Sci. Rep.* **2016**, *6*, 32673. [CrossRef]
30. Santo, T.; Rashkova, M.; Alabaca, C.; Leitao, J. The ENU–induced powdery mildew resistant mutant pea (*Pisum sativum* L.) lines S (*er1mut1*) and F (*er1mut2*) harbour early stop codons in the *PsMLO1* gene. *Mol. Breed.* **2013**, *32*, 723–727. [CrossRef]
31. Fu, H.; Sun, S.; Zhu, Z.; Duan, C.; Yang, X. Phenotypic and genotypic identification of powdery mildew resistance in pea cultivars or lines from Canada. *J. Plant. Genet. Resour.* **2014**, *15*, 1028–1033. (In Chinese with English abstract)
32. Sun, S.; Deng, D.; Duan, C.; Zong, X.; Xu, D.; He, Y.; Zhu, Z. Two novel er1 alleles conferring powdery mildew (*Erysiphe pisi*) resistance identified in a worldwide collection of pea (*Pisum sativum* L.) germplasms. *Int. J. Mol. Sci.* **2019**, *20*, 5071. [CrossRef]
33. Ma, Y.; Coyne, C.J.; Main, D.; Pavan, S.; Sun, S.; Zhu, Z.; Zong, X.; Leitão, J.; McGee, R.J. Development and validation of breeder-friendly KASPar markers for *er1*, a powdery mildew resistance gene in pea (*Pisum sativum* L.). *Mol. Breed.* **2017**, *37*, 151. [CrossRef]
34. Sudheesh, S.; Lombardi, M.; Leonforte, A.; Cogan, N.O.I.; Materne, M.; Forster, J.W.; Kaur, S. Consensus Genetic Map Construction for Field Pea (*Pisum sativum* L.), Trait dissection of biotic and abiotic stress tolerance and development of a diagnostic marker for the *er1* powdery mildew resistance gene. *Plant. Mol. Biol. Rep.* **2015**, *33*, 1391–1403. [CrossRef]
35. Sulima, A.S.; Zhukov, V.A. War and Peas: Molecular bases of resistance to powdery mildew in pea (*Pisum sativum* L.) and other legumes. *Plants* **2022**, *11*, 339. [CrossRef]
36. Pavan, S.; Schiavulli, A.; Appiano, M.; Miacola, C.; Visser, R.G.F.; Bai, Y.L.; Lotti, C.; Ricciardi, L. Identification of a complete set of functional markers for the selection of *er1* powdery mildew resistance in *Pisum sativum* L. *Mol. Breed.* **2013**, *31*, 247–253. [CrossRef]
37. Wang, Z.; Bao, S.; Duan, C.; Zong, X.; Zhu, Z. Screening and molecular identification of resistance to powdery mildew in pea germplasm. *Acta Agron. Sin.* **2013**, *39*, 1030–1038. (In Chinese with English abstract) [CrossRef]
38. Liu, R.; Wang, F.; Fang, L.; Yang, T.; Zhang, H.; Huang, Y.; Wang, D.; Ji, Y.; Xu, D.; Li, G.; et al. An integrated high-density SSR genetic linkage map from two F2 population in Chinese pea. *Acta Agron. Sin.* **2020**, *46*, 1496–1506. (In Chinese with English abstract)
39. Fu, H. Phenotyping and Genotyping Powdery Mildew Resistance in Pea. Ph.D. Thesis, Gansu Agricultural University, Gansu, China, 2014. (In Chinese with English abstract)
40. Sun, S.; He, Y.; Dai, C.; Duan, C.; Zhu, Z. Two major *er1* alleles confer powdery mildew resistance in three pea cultivars bred in Yunnan Province, China. *Crop. J.* **2016**, *4*, 353–359. [CrossRef]

41. Kusch, S.; Panstruga, R. Mlo–based resistance, an apparently universal "Weapon" to defeat powdery mildew disease. *MPMI* **2017**, *30*, 179–189. [CrossRef]
42. Shure, M.; Wessler, S.; Fedoroff, N. Molecular-identification and isolation of the waxy locus in maize. *Cell* **1983**, *35*, 225–233. [CrossRef]
43. Bordat, A.; Savois, V.; Nicolas, M.; Salse, J.; Chauveau, A.; Bourgeois, M.; Potier, J.; Houtin, H.; Rond, C.; Murat, F.; et al. Translational genomics in legumes allowed placing in silico 5460 unigenes on the pea functional map and identified candidate genes in *Pisum sativum* L. *Genes Genome Genet.* **2011**, *1*, 93–103. [CrossRef]
44. Loridon, K.; McPhee, K.; Morin, J.; Dubreuil, P.; Pilet-Nayel, M.L.; Aubert, G.; Rameau, C.; Baranger, A.; Coyne, C.; Lejeune-Hénaut, I.; et al. Microsatellite marker polymorphism and mapping in pea (*Pisum sativum* L.). *Theor. Appl. Genet.* **2005**, *111*, 1022–1031. [CrossRef]
45. Lander, E.S.; Daly, M.J.; Lincoln, S.E. Constructing genetic linkage maps with MAPMAKER/EXP Version 3.0, a tutorial and reference manual. In *Institute for Biomedical Research Technical Report*, 3rd ed.; Whitehead, A., Ed.; Whitehead Institute for Biomedical Research: Cambridge, MA, USA, 1993.
46. Kosambi, D.D. The estimation of map distances from recombination values. *Ann. Eugen.* **1944**, *12*, 172–175. [CrossRef]
47. Liu, R.H.; Meng, J.L. MapDraw, A Microsoft excel macro for drawing genetic linkage maps based on given genetic linkage data. *Hereditas* **2003**, *25*, 317–321. (In Chinese with English abstract)

International Journal of
Molecular Sciences

Article

Uncovering a Phenomenon of Active Hormone Transcriptional Regulation during Early Somatic Embryogenesis in *Medicago sativa*

Jianbo Yuan, Yuehui Chao * and Liebao Han *

School of Grassland Science, Beijing Forestry University, Beijing 100083, China; yuanjingan113@163.com
* Correspondence: chaoyuehui@bjfu.edu.cn (Y.C.); hanliebao@bjfu.edu.cn (L.H.)

Abstract: Somatic embryogenesis (SE) is a developmental process in which somatic cells undergo dedifferentiation to become plant stem cells, and redifferentiation to become a whole embryo. SE is a prerequisite for molecular breeding and is an excellent platform to study cell development in the majority of plant species. However, the molecular mechanism involved in *M. sativa* somatic embryonic induction, embryonic and maturation is unclear. This study was designed to examine the differentially expressed genes (DEGs) and miRNA roles during somatic embryonic induction, embryonic and maturation. The cut cotyledon (ICE), non-embryogenic callus (NEC), embryogenic callus (EC) and cotyledon embryo (CE) were selected for transcriptome and small RNA sequencing. The results showed that 17,251 DEGs, and 177 known and 110 novel miRNAs families were involved in embryonic induction (ICE to NEC), embryonic (NEC to EC), and maturation (EC to CE). Expression patterns and functional classification analysis showed several novel genes and miRNAs involved in SE. Moreover, embryonic induction is an active process of molecular regulation, and hormonal signal transduction related to pathways involved in the whole SE. Finally, a miRNA–target interaction network was proposed during *M. sativa* SE. This study provides novel perspectives to comprehend the molecular mechanisms in *M. sativa* SE.

Keywords: somatic embryogenesis; *M. sativa*; embryonic induction; embryonic; maturation; DEGs

1. Introduction

The process of somatic embryogenesis (SE) consists mainly of dedifferentiation, in which differentiated cells reverse their developmental program during in vitro culture and again during whole-plant development [1,2]. Most plant cells have developmental plasticity, which plays an important role in their reprogramming. The stem cell condition affects the developmental plasticity of plant cells because stem cells are capable of renewing themselves and converting into new somatic embryos that can development into organs or tissues [3]. Chromatin structure is continuously reconstructed throughout plant development, and previous studies showed that chromatin structure plays a crucial role in the pluripotency of plant stem cells [4,5]. In addition, chromatin structure plays important roles in the process of early SE. It is essential, through despiralization of the super-coiled chromatin structure, for the dedifferentiation of somatic cells to produce embryos and induce callus before embryogenesis [4]. On the whole, SE is a powerful tool for research into the processes of plant development and plant stem cell culture conditions [6]. Further investigation of SE will provide opportunities for improving large-scale production of mature somatic embryos and to promote the production of artificial seeds.

Plant growth regulators (PGRs), such as auxin and cytokinins (CKs), are important trigger factors for the culture of plant stem cells for SE [6,7]. For example, 2,4-dichlorophenoxyacetic acid (2,4-D) is an active factor for dedifferentiation in vitro culture, which serves as a "stressor" for the explant [8]. Induction of SE in soybean and potato by 2,4-D is related to increased oxidative stress and expression of defense genes [9,10].

The external PGR supply changes the internal auxin concentration of the explant, which contributes to callus induction [4,9,11].

The process of SE involves various signaling pathways and differential gene expression. Transcription factors (TFs) are also important factors in SE induction. For instance, in Arabidopsis, a marked upregulation of TFs is associated with the embryo induction stage [12]. TFs encoded by genes such as *BABY BOOM* (*BBM*) [13], *PGA6* [14] and *LEAFY COTYLEDON* (*LEC*) [15] also regulate the totipotency of the plant cell, which is critical for SE. *BBM*, which belongs to the AP2/ERF family, is expressed in immature pollen grains of Brassica napus and is expressed preferentially in developing embryos [13]. The *PGA6* gene encodes a homeodomain protein and plays a key role during SE by promoting the vegetative-to-embryogenic transition and maintaining the activity of embryonic stem cells [14]. The *LEC1*, *LEC2* and *LEC3* genes are essential for SE induction in Arabidopsis [15].

Small RNAs guide regulatory processes at the DNA or RNA level in plants. Many small RNAs mediate transcriptional silencing of genes to regulate plant development, whereas other small RNAs mediate post-transcriptional silencing of genes to regulate embryonic development [16]. There are few reports of small RNAs involved in SE, but microRNAs (miRNAs) and other small noncoding RNAs regulate gene expression epigenetically, which plays a crucial role in SE [17]. miR159 regulates *LaMYB33* during the process of SE in *Larix kaempferi* (Lamb) Carr [18]. Several miRNAs, including miR397 and miR156, show positive patterns of expression during the process of dedifferentiation to redifferentiation in rice [19]. *DCL1* regulates miRNA biogenesis during early SE development in Arabidopsis, and the single mutant of dcl1 causes a loss in miRNA156 expression, which results from derepression of *SPL10* and *SPL11* genes [20].

Genome-wide profiling has made it possible to understand molecular regulatory mechanisms of SE. Recently, high-throughput sequencing technology has allowed multiple advances in genome-wide screening of quantitative gene expression in plants [21]. Gene chip technology has been used to determine mRNA abundance and to identify characteristic changes during dedifferentiation in soybeans [22]. The results of the studies mentioned imply that new cells of dedifferentiation generation that develop organized structures may rely on gene regulation to balance cell proliferation and cell death [9]. Proteomic analysis suggests the involvement of mechanisms in the transition from morphologically mature to physiologically mature somatic embryos during the partial desiccation treatment process in *Picea asperata* [23]. These studies help to identify the molecular regulatory mechanisms that are active during SE.

M. sativa is a tetraploid perennial species that is the most important cultivated forage crop due to its high regeneration capacity in vitro. Thus, *M. sativa* has been used in molecular studies and breeding. The first reported *M. sativa* regeneration was accomplished via SE [24]. In the past, the study of somatic embryo formation in *M. sativa* focused mainly on the morphological and physiological levels; the molecular mechanisms related to SE in *M. sativa* remain unclear. For example, in proteomic analysis of SE in two varieties of M. truncatula, 6-benzylaminopurine (BAP) and 1-naphthaleneacetic were added to the explant culture medium. The results suggested that more than 60% of differentially expressed protein spots exhibited different patterns of gene expression between the two varieties during 8 weeks of culture [25]. This study aimed to identify molecular mechanisms during somatic embryonic induction, embryonic and maturation in *M. sativa*.

In this study, we used Illumina high-throughput sequencing technology to analyze DEGs and miRNAs expression at the ICE, NEC, EC and CE phases in *M. sativa* SE. A total of 17,251 DEGs 177 known and 110 novel miRNAs families were obtained. Among these, several novel DEGs and miRNAs were detected during somatic embryonic induction, embryonic and maturation. For example, novel_247 targets the *LTP8* gene, which may play an important role in maturation. In addition, the results suggest a potential miRNA–target gene interaction network involved in *M. sativa* SE, which further complements the molecular mechanisms in SE.

2. Results

2.1. Morphological Comparison of SE at Different Developmental Stages

SE involves three phases: embryonic induction, embryonic and maturation [26]. To understand the detailed phases of SE, 14 developmental stages were observed using a light microscope (Figure S1). We selected four stages (ICE, NEC, EC and CE) for RNA sequencing (RNA-seq), as the morphological changes observed in these four stages were the most notable. The stages from ICE to NEC represent the embryonic induction process (Figures 1a,b and 2a,b,e,f), NEC to EC represents the embryonic stage (Figures 1b,c and 2b,c,f,g), and EC to CE represents the maturation process (Figures 1c,d and 2c,d,g,h). The cell size during the ICE was larger than during the NEC, EC and CE, whereas the number of cells was larger in the EC than ICE, NEC and CE. The cell shape was more regular in the EC than the ICE, NEC and CE. From ICE through EC, the cell number increased gradually, while the cell size decreased. However, from the EC to the CE phase, the cell size increased gradually, while the cell number decreased gradually. These findings suggest that the changes in tissue morphology differed significantly among the various development stages.

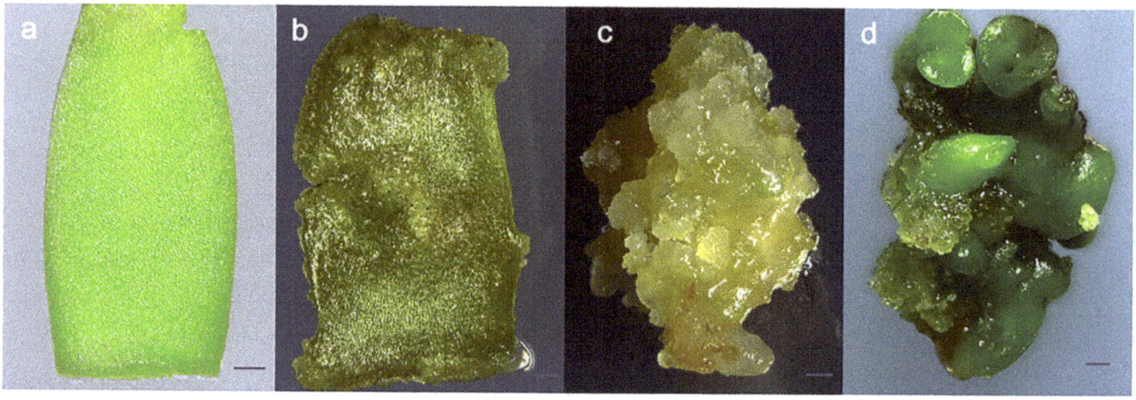

Figure 1. Somatic embryogenesis in alfalfa at four developmental phases used for RNA-Seq analysis; (**a**) Cotyledon cutting; (**b**) Non-embryogenic callus; (**c**) Embryogenic callus; (**d**) Cotyledon embryo. Bar = 1 mm.

2.2. Transcriptome Sequencing and Assembly

ICE, NEC, EC and CE cells were used as sources for RNA-seq. All data were generated using three biological replicates. The RNA-seq generated an average of 8.15 GB of data for the ICE, NEC, EC and CE databases (Table 1). The false discovery rate was less than or equal to 0.02%. All databases produced a total of 668,180,574 raw reads, including 97.60% Q20 bases with a 41.60% GC content. After 97.53% of the raw reads were selected for filtration, 207, 276, 776 clean reads were selected for further analysis using Trinity. All clean reads were assembled into 267, 977 unigenes. The mean length of the genes was 986 bp, and the maximum length was 16,765 bp. The N50 fragment length was 1392 bp, and the N90 fragment length was 466 bp. The size distributions of the unigenes and transcripts are shown in Figure 3a. Unigenes of 501–1000 bp were the most common in all sample data, accounting for 32.6% of the data. Transcripts of 301 bp or less were the most frequent, accounting for 35.85% of all data.

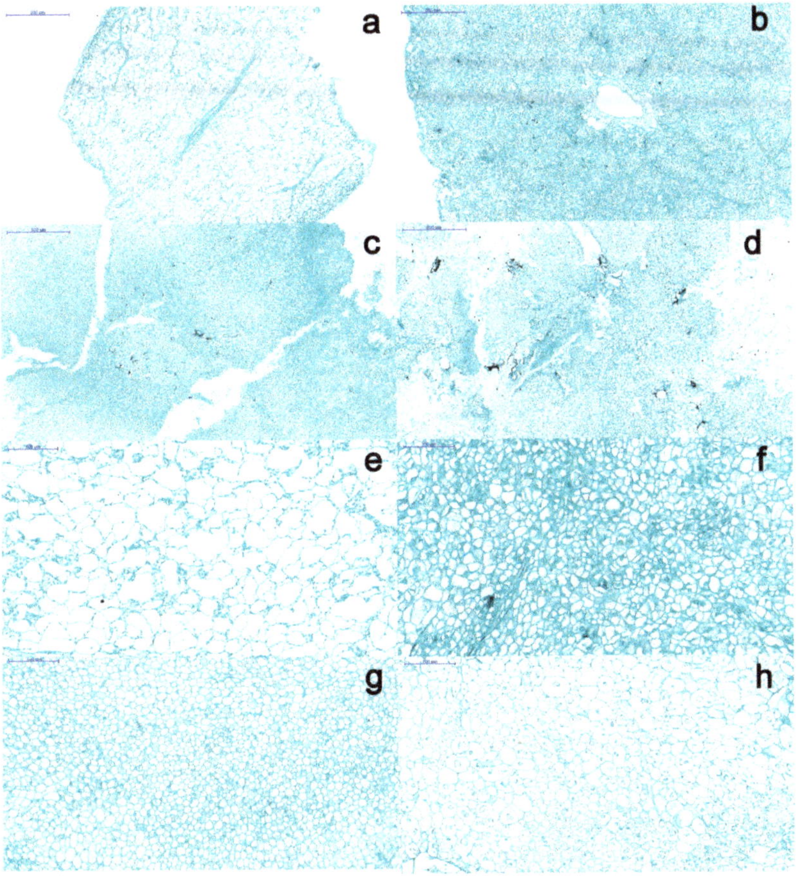

Figure 2. Morphology of somatic embryogenesis callus. (**a**) cut cotyledon; (**b**) non-embryogenic callus; (**c**) embryogenic callus; (**d**) cotyledon embryo; (**e**) Cut cotyledon; (**f**) non-embryogenic callus; (**g**) embryogenic callus, bar = 100 μm; (**h**) cotyledon embryo.

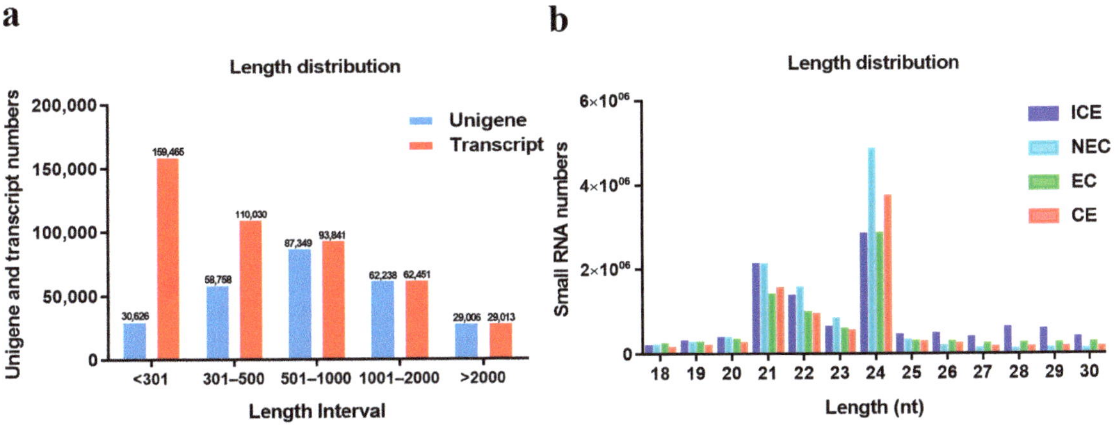

Figure 3. Unigene and small RNA length distributions. (**a**) Unigenes and transcriots; (**b**) Small RNAs.

Table 1. Transcriptome data quality profile.

Sample	Raw Reads	Clean Reads	Clean Bases	Error (%)	Q20 (%)	Q30 (%)	GC (%)
ICE_1	52,384,258	51,607,148	7.74G	0.01	97.75	94.00	41.28
ICE_2	58,306,722	56,696,602	8.5G	0.02	97.20	92.69	41.30
ICE_3	50,810,168	50,103,162	7.52G	0.01	97.86	94.21	41.34
NEC_1	57,200,884	55,844,904	8.38G	0.01	97.89	94.30	41.60
NEC_2	60,219,472	56,035,448	8.41G	0.02	96.97	92.05	41.86
NEC_3	53,763,200	52,838,680	7.93G	0.01	97.97	94.50	41.60
EC_1	52,977,216	51,478,324	7.72G	0.01	97.97	94.51	41.45
EC_2	55,361,562	54,341,900	8.15G	0.01	98.02	94.63	41.39
EC_3	57,268,222	55,969,446	8.4G	0.01	97.64	93.76	41.79
CE_1	63,997,186	62,745,116	9.41G	0.02	97.47	93.47	41.93
CE_2	50,846,076	49,993,036	7.5G	0.02	97.49	93.49	41.95
CE_3	55,045,608	54,018,626	8.1G	0.02	97.44	93.39	41.65

ICE, non-embryogenic callus; NEC, non-embryogenic callus; EC embryogenic callus; CE, cotyledon embryo; Q20 , phred percentage of base greater than 20, pecentage of base population; Q30, phred percentage of base greater than 30, percentage of base population; GC, GC content.

2.3. Gene Annotation and Functional Classification in M. sativa

After matching the sequences with the KOG/COG database, a total of 43,572 unigenes were classified into 26 categories (Figure S2). The majority of unigenes (5207) were predicted to be associated with post-translational modification, protein turnover and chaperones (11.95%), followed by general functions (11.90%), translation, ribosomal structure and biogenesis (8.18%), signal transduction mechanisms (7.73%) and RNA processing and modification (6.85%). Five unigenes (0.01%) had unknown functions.

A total of 101,969 unigenes were selected for GO classification using Blast2GO v2.5 (BioBam, Valencia, Spain), which classified into biological process, cellular component and molecular function groups (Table S1). The highest category was nucleoside binding (14.03%) of molecular function, followed by intracellular membrane-bounded organelle (12.84%), small molecule metabolic process (10.77%), cytoplasm (8.78%) and cell communication (7.55%). In the cellular component, many of unigenes conducted various functions, which contained nuclear chromatin (0.01%), cell wall (0.04%), plasma membrane (1.07%) and transporter complex (0.35%). In the distribution of molecular function, the nucleoside binding (14.03%), substrate-specific transmembrane transporter activity (4.29%) and hydrolase activity, acting on ester bonds (3.65%) were mainly representation. Interestingly, 83 unigenes were predicted to be involved in embryonic development and 237 in post-embryonic development. The unigenes associated with embryogenesis are summarized in Table S2, which includes 39 GO functional terms. There were embryo development (83), reproductive structure development (34), regulation of cell differentiation (49), post-embryonic morphogenesis (201) and root morphogenesis (3), these GO terms were important developmental processes during SE.

All unigenes were analyzed using KEGG classification to identify the biological functions in *M. sativa*. After mapping against the KEGG database, 47,092 unigenes were classified into four main categories and 128 pathways (Table S3). The largest category of unigenes was metabolism, accounting for 62.7%, followed by genetic information processing (26.89%), cellular processes (6.41%) and environmental information processing (4%). Interestingly, carbon metabolism was the largest representative in the metabolism category, accounting for 4.66% of unigenes. This finding was consistent with that described in *Lilium* SE [27], suggesting that carbon metabolism is active in *M. sativa*. In addition, zeatin biosynthesis pathway (182) components of metabolism and plant hormone signal transduction genes involved in environmental information processing (1046) were predicted to be active in *M. sativa* SE, suggesting that these pathways play important roles in SE.

2.4. Identification and Analysis of DEGs in SE

After four unigene libraries were compared with each other to identify DEGs (p-value < 0.05 and $|\log2\ FC| \geq 1$), three DEG groups were then produced (ICE vs. NEC, NEC vs. EC, EC vs. CE) that included 17,251 DEGs (Table S4). A comparative analysis of the 17,251 DEGs at different SE phases is shown in Figure 4a. The number of DEGs was significantly higher in the ICE compared to the NEC, and higher in the NEC compared to the EC, and in the EC compared to the CE. In addition, 9206 genes were upregulated and 5522 genes were downregulated in the ICE compared to the NEC. Only 668 genes were upregulated and 600 genes were down-regulated in the NEC compared to the EC, and 619 genes were upregulated and 636 genes were downregulated in the EC compared to the CE. These results indicate that the early SE is active during the complicated development process.

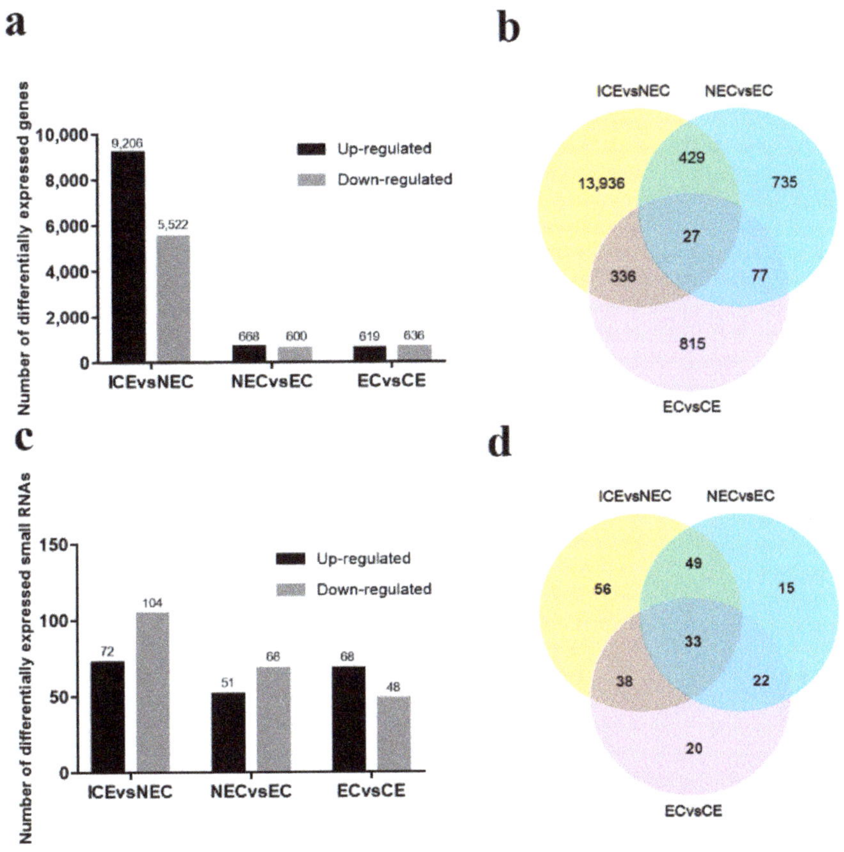

Figure 4. Histograms and Venn diagrams of DEGs for three phrases in SE: embryonic induction (ICE vs. NEC), embryonic induction (NEC vs. EC), and maturation (EC vs. CE). (**a**) The numbers of DEGs up- or down-regulated during SE; (**b**) Venn diagram showing similarly or differently regulated genes over the three phreases; (**c**) The numbers of small RNAs up- or down-regulated during SE; (**d**) Venn diagram showing similarly or differentially regulated small RNAs during SE.

All DEGs were selected for Venn diagram analysis (Figure 4b). In total, 27 genes were expressed in all the phases of SE, 429 genes were expressed in the ICE to the NEC and the NEC to the EC two development phrases; 77 genes were expressed in the ICE to NEC, and the EC to the CE two development groups; 336 genes were expressed in the ICE to

NEC and NEC to EC two development phrases. There were few DEG involved in multiple SE stages.

2.5. Some DEGs Involved in Embryonic Induction, Embryonic and Maturation

To further investigate the function of DEGs, we identified 61 DEGs in 18 families involved in SE (Table S5), including *BABY BOOM* (*BBM*), *WUSCHEL related homeobox* (*WOX*), *PKL* (*PICKLE*) and somatic embryogenesis receptor-like kinase (*SERK*). Most of the DEGs (80.32%) were involved in embryonic induction, followed by embryonic (14.75%) and maturation (4.91%). In addition, *LATERAL ORGAN BOUNDARIES DOMAIN* (*LBD*), *argonaute* (*AGO*), *glycogen debranching enzyme* (*AGL*) and authentic response regulator (*ARR*) were observed in embryonic induction and embryonic, and *WOX* and *Pyoluteorin* (*PLT*) were involved in embryonic induction and maturation. Interestingly, no single gene family was observed throughout SE, indicating that stage-specific studies are necessary in the future. Among all DEGs, 14 were upregulated and 35 downregulated in embryonic induction, particularly *LBD41*, *LBD16*, *flavin monooxygenase5* (*YUC5*) and *YUC9*, with | log2 FC | values > 9. 5 genes were upregulated and 3 genes were downregulated in embryonic. Such as *PKL* and *LBD4* were upregulated more than 3-fold. *WOX6*, *PLT1* and *PLT2* was upregulated more than 3-folds in maturation.

2.6. Expression of Plant Hormone Signal Transduction Genes in SE

We detected ~118 DEGs involved in plant hormone signal transduction pathways (Table S6), including IAA, zeatin, ethylene (ET), ABA and GA pathways; 106 DEGs were involved in the transition from embryonic induction (Table S6), 8 in maturation, and 5 in embryonic (Table S6). Several DEGs exhibited higher expression levels during embryonic induction. *IAA*, *ARR*, and *Auxin Response Factor* (*ARF*) family genes were upregulated more than 2-fold (| log2 FC | > 2). However, Auxin-induced protein (SAUR), *SAUR32* and *SAUR36* was downregulated (Figure 5a). KEGG analysis indicated that *IAA*, *ARF* and *SAUR* are involved in auxin pathways during embryonic induction. Endogenous auxin expression increased nearly 2-fold from ICE to NEC (Figure 6a). Elhiti et al. found that expression of *LEC* gene directly induce AGAMOUS expression during early embryogenesis, which in turn upregulates GA_{2OX} and decreases GA synthesis [26]. In our data, which are GA_{2OX} homologs as DEGs. During the embryonic induction, LEC2 gene expression was upregulated more than 5-fold, and GA_{2OX1} was downregulated more than 2-fold. However, GA_{2OX2} was downregulated more than 3-fold. The expression of these genes did not change significantly during other phases. GA3 expression decreased sharply at four sampling sites (Figure 6b). Li et al. found that several MYBs are positive regulators of ABA responses [28]. We also detected genes homologous to MYB among the DEGs, including *MYB4* and *MYB48*. In maturation, *MYB4* and *MYB48* were downregulated more than 2.03-fold, with no significant changes during the other phases. We detected changes in ABA among the phases (Figure 6c). From ICE to NEC and NEC to EC, the level of ABA decreased sharply. However, from EC to CE, the ABA content increased slightly. We also detected changes in ZR levels among the phases (Figure 6d). From ICE to NEC, the ZR content was slightly upregulated, and from NEC to EC, it was decreased by nearly half. However, from EC to CE, the ZR content more than doubled. In addition, *Pathogenesis-related protein 1* (*PRB1*) and *Abscisic acid receptor* (*PYL4*) were upregulated, and *IAA27*, *ARF3* and *ARR9* were downregulated in embryonic (Figure 5b). *Serine/threonine-protein kinase* (*SAPK3*) was upregulated, and *PYL1*, *IAA30* and *ABA response element-binding factor* (*ABF2*) were downregulated in maturation (Figure 5c).

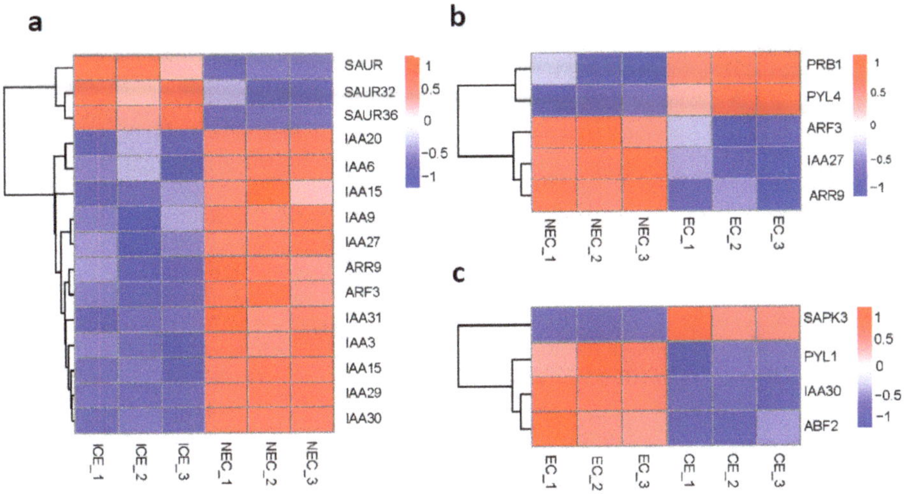

Figure 5. Plant hormone signal transduction genes in SE. (**a**) Expression of gene in ICE to NEC; (**b**) Expression of gene in NEC to EC; (**c**) Expression of gene in EC to CE. |log2FC| > 0.5. ICE_1, ICE_2 and ICE_3 represent three biological replicates.

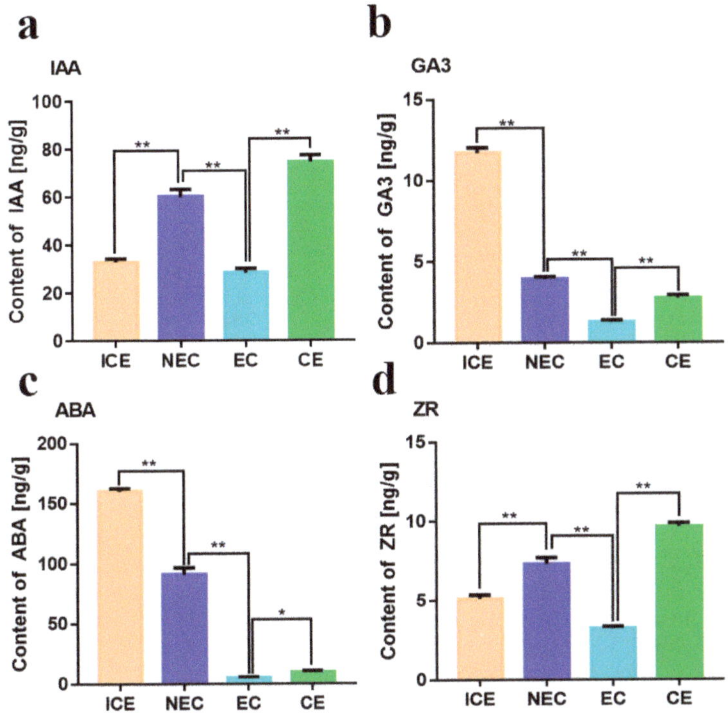

Figure 6. Hormone changes during different phases of SE. (**a**) Auxin changes at cut cotyledon (ICE). non-embyogenic callus (NEC), embyogenic callus (EC), and cotyledon embryo (CE) phases; (**b**) Gibberellin changes at ICE, NEC, EC and CE phases; (**c**) Abscisic acid changes At ICE, NEC, EC and CE; (**d**) Cytokinin changes at ICE, NEC, EC and CE. Indirect competitive enzyme-linked immunoassay (icELISA) for hormone detection. * $p < 0.05$, ** $p < 0.01$ (Student' s t-test).

2.7. Annotation of Small RNA-Seq Data

To investigate SE in *M. sativa*, four small RNA libraries (ICE, NEC, EC and CE) were sequenced. A total of 51,465,356 raw reads were obtained from each library. After filtering, 46,453,585 clean reads were obtained from four libraries (Table 2). The most common small RNAs were 21–24 nucleotides (nt) in length (Figure 3b), with the majority being 24 nt. All unique sequences were annotated and mapped in the Rfam database using BLAST. A total of 19,765,615 small RNAs were annotated in four libraries. ICE was the most abundant, accounting for 30.19% (Table 3). The annotated small RNAs included known miRNAs, novel miRNAs, ribosomal RNAs (rRNAs), tRNAs, snRNAs, small nucleolar RNAs (snoRNAs) and trans-acting small interfering RNAs (TASs) in ICE. The known miRNAs were the most abundant (12.04%), followed by novel rRNAs (7.24%), novel miRNAs (2.23%), TASs (0.32%), snoRNAs (0.16%), snRNAs (0.07%) and tRNAs (0.00%). However, 77.58% of the other small RNAs involved in ICE are unknown. These results indicated that small RNAs are more active in embryonic induction than in other development stages.

Table 2. Small RNA filter profile.

Sample	ICE	NEC	EC	CE
total reads	14,669,757 (100.00%)	14,696,331 (100.00%)	11,817,247 (100.00%)	10,282,021 (100.00%)
N% > 10%	18 (0.00%)	25 (0.00%)	0 (0.00%)	0 (0.00%)
low quality	753,071 (5.13%)	1,954,044 (13.30%)	488,075 (4.13%)	225,087 (2.19%)
5 adapter contamine	23,386 (0.16%)	16,804 (0.11%)	9703 (0.08%)	11,016 (0.11%)
3 adapter null or insert null	649,842 (4.43%)	316,352 (2.15%)	212,613 (1.80%)	189,171 (1.84%)
With ployA/T/G/C	39,873 (0.27%)	55,440 (0.38%)	35,208 (0.30%)	32,043 (0.31%)
clean reads	13,203,567 (90.01%)	12,353,666 (84.06%)	11,071,648 (93.69%)	9,824,704 (95.55%)

Table 3. Annotation of small RNA.

Types	ICE	NEC	EC	CE
known miRNA	740,026 (12.40%)	302,053 (5.92%)	299,136 (6.74%)	479,121 (11.25%)
novel miRNA	132,783 (2.23%)	63,563 (1.25%)	34,189 (0.77%)	46,934 (1.10%)
rRNA	432,250 (7.24%)	121,355 (2.38%)	252,580 (5.69%)	149,802 (3.52%)
tRNA	3 (0.00%)	0	1 (0.00%)	0
snRNA	3988 (0.07%)	2290 (0.04%)	5984 (0.13%)	2902 (0.07%)
snoRNA	9607 (0.16%)	23,276 (0.46%)	22,688 (0.51%)	15,722 (0.37%)
TAS	19,322 (0.32%)	31,369 (0.61%)	25,992 (0.59%)	132,783 (0.46%)
other	4,629,365 (77.58%)	4,560,608 (89.34%)	25,992 (0.59%)	19,645 (0.46%)
total	5,967,344	5,104,514	4,436,671	4,257,086

2.8. Identification and Expression Analysis of Known miRNAs in *M. sativa*

After annotating the small RNAs, ~308 known miRNAs were identified from four libraries. Table S7 shows the number of known miRNAs. All identified known miRNAs belong to 177 miRNA families. Several miRNA families contained more than one member, such as the mi156 (11), miR166 (5), miR169 (7) and miR171 (8) families. Approximately 45 families contained only one member, such as the miR5218, miR5237, miR2625 and miR2605 families. Small RNA-seq data revealed 124 known families with significantly different expression levels ranging from 0 to 1,000,000 transcripts per kilobase million (TPM) among all libraries (Table S8). These 124 families were classified into four groups based on maximum expression levels. The first group contained 13 families, including miR5213, miR159, miR166 and miR167, which expressed more than 10,000 TPM and were detected in at least one sample. Among these 13 families, miR5213 showed the highest expression, exceeding 200,000 TPM in each sample (Figure 7a). The second group contained 11 known families that ranged in expression from 1000 to 10,000 TPM, with miR1510 showing the highest expression of more than 5000 TPM in all samples (Figure 7b).

The third group contained 21 families that ranged in expression from 100 to 1000 TPM, with miR5214 exhibiting the highest expression level in this group (Figure 7c). The fourth group contained the remaining 79 families, which ranged in expression from 0 to 100 TPM. As shown in Figure 7d, miR2603 showed the highest level of expression.

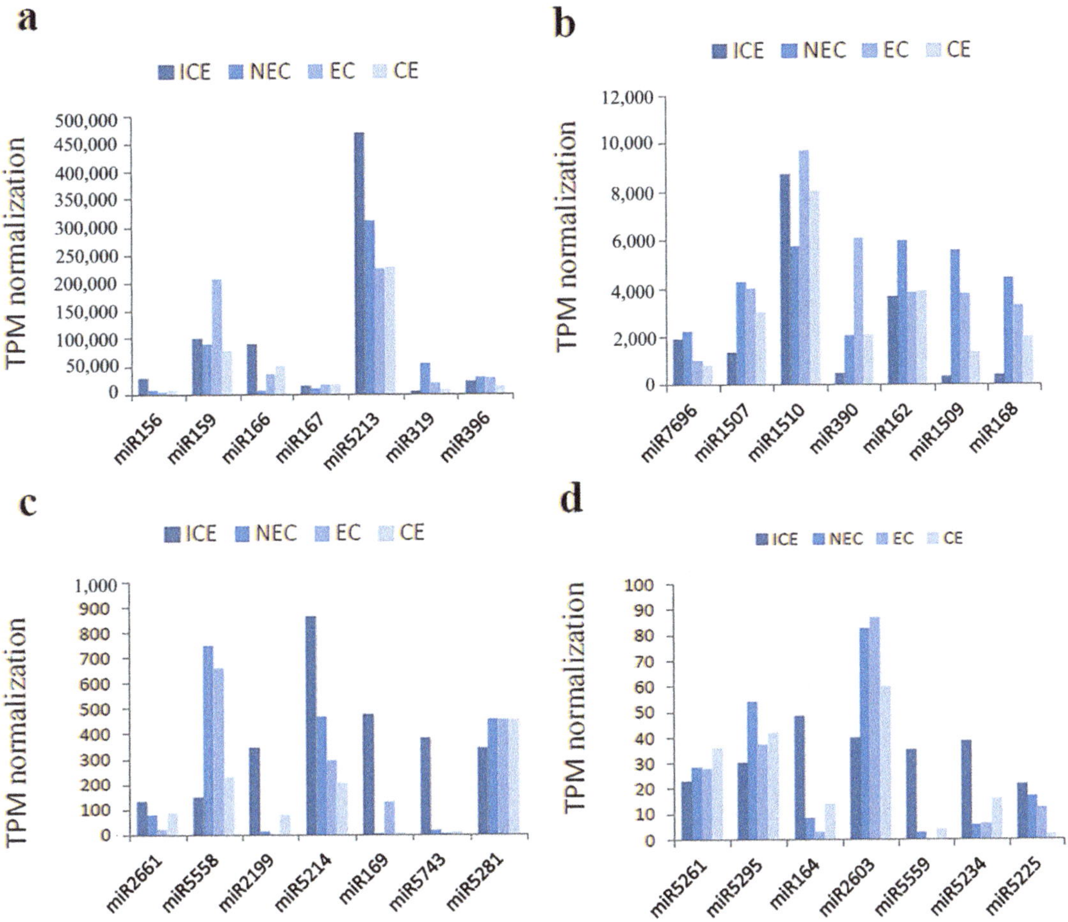

Figure 7. Transcripts reads per million reads known miRNAs in *Medicago sativa*. (**a**) miRNAs families with TPM value over 10,000; (**b**) miRNA families TPM value between 1000 and 10,000; (**c**) miRNA families TPM value between 100 and 1000; (**d**) miRNA families TPM value between 0 and 100.

2.9. Identification of Potential Novel miRNAs in M. sativa

The miREvo and miRdeep2 software (Berlin, Germany) were used to predict several potential miRNAs due to their unique hairpin structure. This mature sequence is a common trait in potential novel miRNAs in *M. sativa*, and these novel miRNA candidates might be regarded as a new miRNA family if they originated from different loci. We identified 110 novel families distributed in four samples (Table S9). Interestingly, 10 families were involved throughout SE (novel_322, novel_263, novel_249, novel_70, novel_67, novel_186, novel_94, novel_46, novel_231 and novel_287), which accounting for 9.09%. We detected 21 families that were expressed exclusively during embryonic induction, 12 during maturation and 9 during embryonic. The length of the novel family members expressed during the maturation phase ranged from 19 to 24 nt. Gene expression profiles revealed that 28 families were upregulated and 46 families downregulated in embryonic induction, and 27 families were upregulated and 30 families downregulated in embryonic formation. More families were downregulated than upregulated during the embryonic induction and embryonic formation stages. However, more families were upregulated (32) than downregulated (24) during the maturation stage. These results suggest that several new novel miRNAs play roles in maturation in a manner that differs from those involved in embryonic induction and embryonic.

2.10. Expression of Known and Novel miRNAs during SE in *M. sativa*

The numbers of miRNAs involved in the three developmental processes are shown in Figure 4c. The miRNA changes from the ICE to the NEC phase were significantly higher than in the NEC to the EC phase and the EC to the CE phase. From the EC to CE stage, more miRNAs were upregulated than downregulated. Venn diagram analysis of 418 miRNAs showed different expression levels among the three developmental processes (Figure 4d). A total of 33 miRNAs were expressed throughout SE; 49 miRNAs overlapped between the ICE to the NEC phase and the NEC to the EC phase, 22 miRNAs overlapped between the ICE to the NEC phase and the EC to the CE phase, and 38 miRNAs overlapped between the ICE to the NEC phase and the EC to the CE phase.

As mentioned previously, SE in *M. sativa* was classified into three major stages: embryonic induction (ICE to NEC), embryonic (NEC to EC) and maturation (EC to CE). The differentially expressed miRNA families were defined as having an $|\log2\ FC|$ value ≥ 2. A total of 39 known miRNA families exhibited differential expression during embryonic induction. Among these families, seven were selected for further comparison analysis in embryonic induction; five families showed downregulated expression of varying degrees, and two families were upregulated, including miR156 and miR2111 (Figure 8a). A total of 18 known miRNA families showed differential expression during the embryonic phase. We selected seven of these families for analysis (Figure 8b), of which five were upregulated and two were downregulated (miR5256 and miR169). A total of 17 known miRNA families were differentially expressed during maturation. Five families were upregulated and two downregulated (Figure 8c). In general, these results indicate that each developmental stage relies on major clusters of known miRNAs to regulate SE in *M. sativa*.

To identify the major miRNA clusters involved in the various stages of SE, we analyzed the differential expression of 110 novel miRNA families. A total of 51 families showed differential expression in embryonic induction, 33 in embryonic formation and 32 in maturation. We selected seven families for analysis in each stage. There was no significant difference in the number of miRNAs that were up- or downregulated (Figure 8d–f). Interestingly, the novel miRNA families showed differential expression in each stage, a phenomenon similar to that observed with known miRNAs. These results indicated that several major novel miRNA clusters showed stage-specific expression.

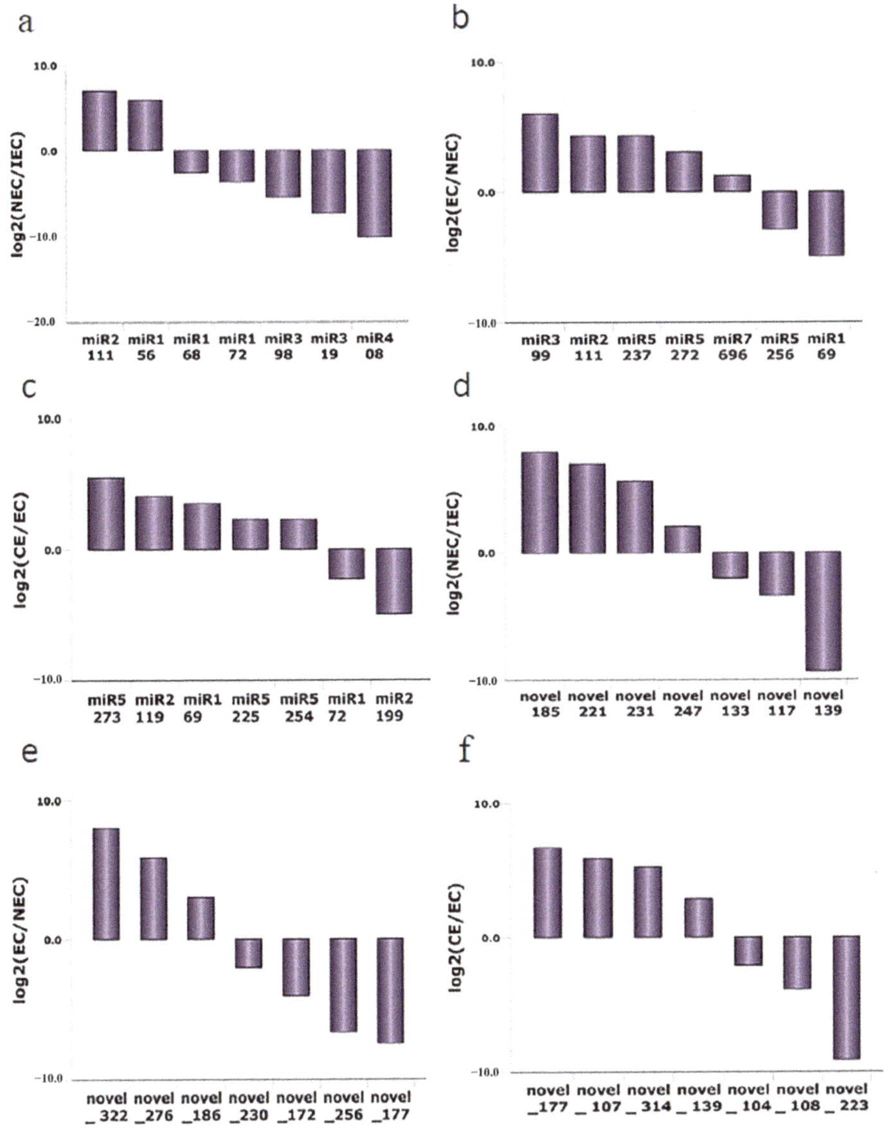

Figure 8. Differentially expressed known and novel miRNAs in *M. sativa* SE. (**a**–**c**) comparison of diferentially known miRNA families among ICE to NEC, NEC to EC and EC to CE; (**d**–**f**) comparison of diferentially novel miRNA families among ICE to NEC, NEC to EC and EC to CE. $|\log2FC| \geq 2$.

2.11. Analysis of Target Genes of miRNAs

To investigate the miRNA-mediated pathways during different stages of SE, we analyzed 17,253 unigenes with annotations from transcriptome databases and used the psRNATarget website to predict target genes. We identified 5785 potential target genes potentially related to 408 miRNAs (Table S11). More than 92.89% of miRNAs had multiple target unigenes, and only 29 miRNAs (7.11%) had a single unigene or no unigene from all miRNA–target gene pairs. Among the miRNA–target gene pairs, most 43.14% (176) were detected during embryonic induction (Table S10); 29.16% (119) were detected in the embryonic stage and 27.7% (113) in the maturation stage. These results suggest that the embryonic induction stage was the most active period. The prediction of miRNA–target pairs suggests that miRNAs can mediate multiple pathways during the various phases of SE in *M. sativa*. Most target genes exhibited significant differences in expression, revealing a complex biological regulation process in SE. These target genes included hormone-related genes, such as *ARF*, *LBD*, and *ARR2*, TFs and certain kinases. KEGG analysis showed the involvement of these target genes in plant hormone signal transduction pathways. As *ARF* is involved in tryptophan metabolism, it may function to promote cell growth in SE. This study showed the involvement of four members of the ARF family in *M. sativa* SE (Table S11). *ARF17*, *ARF18* and *ARF10* were targeted by miR160, whereas ARF8 was targeted by miR156, miR167 and novel_272 in embryonic induction (Table S11). *LBD11* was targeted by miR7696 in embryonic. miR2604 targeted *ARR2*, which is involved in the zeatin biosynthesis pathway that promote cell division in SE. In addition, we detected several TFs related to embryonic development, such as *SHORT-ROOT* (*SHR*) and *Polyol transporter 5* (*PLT5*). *SHR* was targeted by miR156 in embryonic induction, and *PLT5* was targeted by *novel_299* in embryonic. Several kinase genes were targeted in SE; SERK5, mitogen-activated protein kinase 8 (Mapk8) and cyclin-dependent kinase C-1 (CDKC-1) were targeted by miR5561, miR5559 and miR7701, respectively.

To further investigate the function of miRNAs in SE, 5785 target genes were selected for GO analysis. We identified 1350 GO terms involved in embryonic induction, 416 in maturation and 266 in embryonic (Table S12). The biological (GO: 0008150), metabolic (GO: 0008152), cellular (GO: 0009987), organic substance metabolic (GO: 0071704) and primary metabolic (GO: 0044238) processes were the main biological process categories enriched among the target genes. Among the cellular categories, cellular component (GO: 0005575), cell (GO: 0005623) and cell part (GO: 0044464) accounted for the highest proportion of target genes. Molecular function (GO: 0003674), binding (GO: 0005488) and catalytic activity (GO: 0003824) were the enriched molecular function categories among the target genes. As shown in Table S12, several miRNAs with target genes were involved in biological processes related to SE in *M. sativa*. For example, novel_249 play a role in post-embryonic development, plant epidermal development, and regulation of multicellular organismal development; novel_211 regulate cell growth and cell differentiation in SE; and miR2673 play an important role in cell development, cell–cell signaling, and cell activation. Interestingly, 22 miRNA families with 44 target genes were involved in cellular developmental processes, representing the most active biological processes in SE. In addition, several miRNAs were involved in programmed cell death, such as miR5207, miR2630 and miR319.

A total of 5785 target genes showed KEGG enrichment using the KOBAS, revealing the involvement of 112 pathways and 1953 target genes in SE (Table S13). The spliceosome pathway showed the highest enrichment (91 target genes), followed by protein processing in the endoplasmic reticulum (62) and ubiquitin-mediated proteolysis (61) (Figure 9). Moreover, several other pathways may play key roles in M. sativa SE, such as plant hormone signal transduction, amino sugar and nucleotide sugar metabolism, and RNA degradation (Table S13).

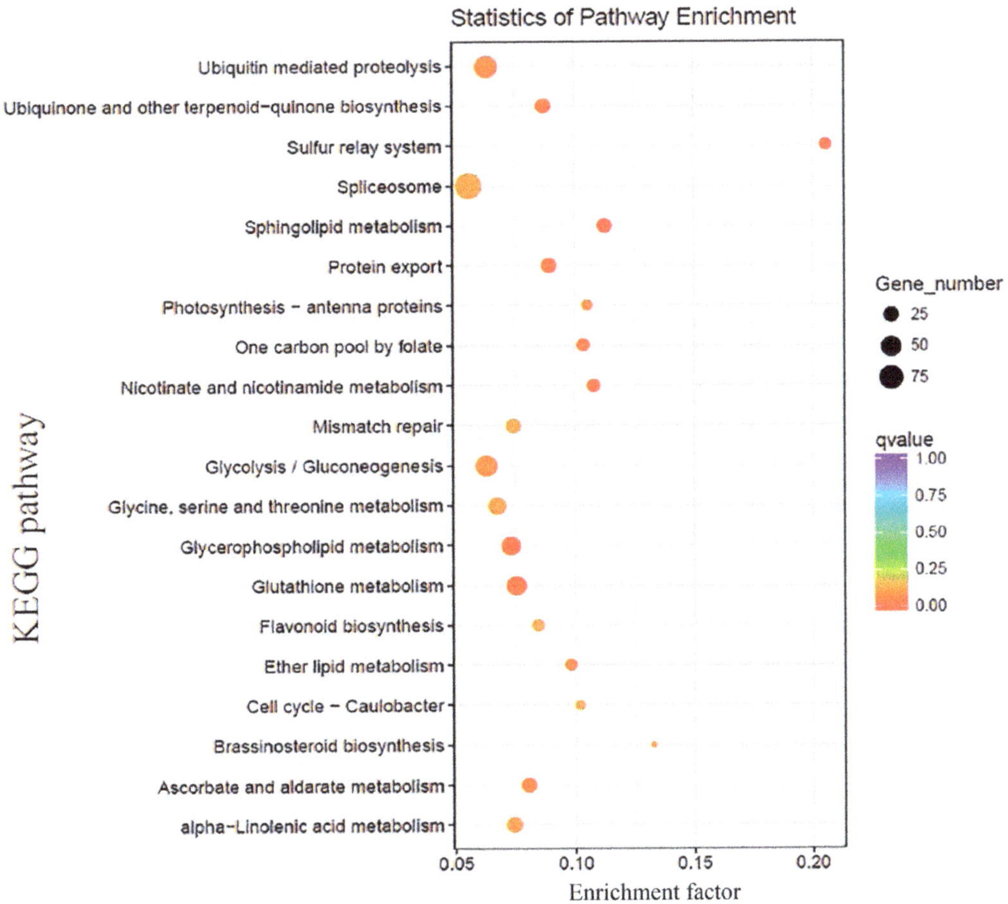

Figure 9. KEGG enrichment target genes in *M. sativa* SE.

2.12. Validation of Various Expression Patterns of DEGs and miRNAs

To validate the expression patterns of genes and miRNAs during ICE, NEC, EC and CE, we determined the expression levels of several genes and miRNAs by qRT-PCR and transcriptome analysis. qRT-PCR analysis revealed that *PNC1* expression was downregulated in SE (Figure 10a); likewise, transcriptome analysis showed that *Cationic peroxidase 1* (*PNC1*) was downregulated more than 4-fold throughout SE. Our results showed that *PNC1* was targeted by novel_244, and novel_244 was downregulated slightly from ICE to NEC. The *IPT5* expression level did not differ significantly from ICE to EC. However, *IPT5* gene expression was significantly upregulated from EC to CE (Figure 10a). Similarly, the transcriptome results showed that *IPT5* was upregulated more than 5-fold from EC to CE. However, no obvious changes in *IPT5* levels were observed in the other stages. miR156 families play crucial roles during SE. From ICE to NEC, the miR156a expression level was downregulated more than 3-fold (Figure 10b), as revealed by small RNA-seq data. *SPL6* was targeted by miR156a, which was upregulated from ICE to NEC. No obvious changes in miR156a levels were observed in the other stages. miR166a expression was downregulated from ICE to NEC and EC to CE and upregulated from NEC to EC. Small RNA-seq and qRT-PCR analysis of miR156a showed consistent results. *2-succinylbenzoate–CoA ligase*

(*AAE*) was targeted by miR166a. *AAE* was upregulated from ICE to NEC but showed no obvious changes in the other stages.

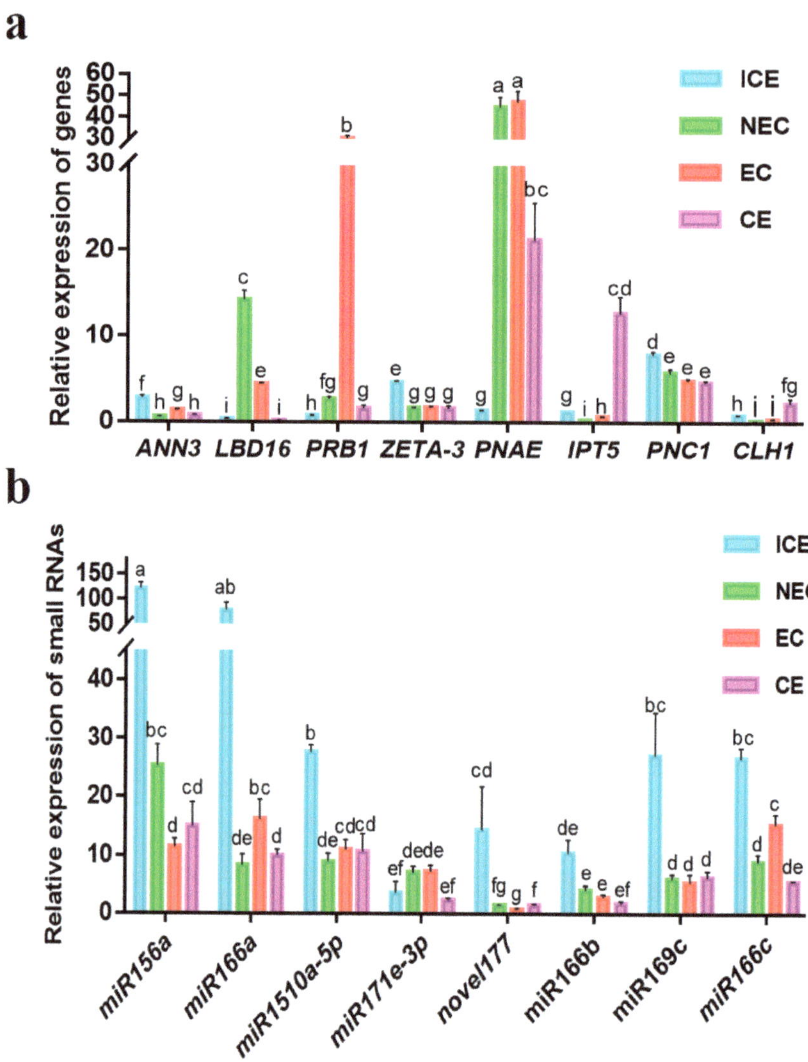

Figure 10. Relative expression of genes and miRNAs during cut cotyledon (ICE), non-embyogenic callus (NEC), embryogenic callus (EC), cotyledon embryo (CE) phases. (**a**) The relative expression of DEGs; (**b**) The relative expression of miRNAs. genes and miRNAs calculated by the equation = $2^{-\Delta\Delta Ct}$, The 18sRNA is housekeeping gene for gene expression detection, and the U6 gene is housekeeping gene for miRNA expression detection. The different letters represent significant differences ($p < 0.05$, (ANOVA)).

3. Discussion

SE, a complex developmental process involved in completing plant regeneration of somatic cells, was first described in the carrot [29,30]. Previous studies identified single or multiple genes involved in SE using mRNA differential display [31]. However, the more recent technology of RNA-seq has clear advantages, as it can be used to generate millions of clear reads for transcriptome and miRNA analyses. Xu et al. suggested that these analyses can be used to generate a comprehensive view of several gene and miRNA families involved in SE in plants [32]. However, no such studies of DEGs or miRNAs in SE in *M. sativa* have been reported. This study investigated DEGs and miRNA roles during somatic embryonic induction, embryonic and maturation in *M. sativa* SE. Four libraries from four SE stages (ICE, NEC, EC and CE) were constructed, and pairwise comparisons of the data resulted in annotation of 101,969 unigenes and 418 miRNAs and prediction of several target genes of miRNAs. Finally, we proposed a potential miRNA–target interaction network involved in *M. sativa* SE.

3.1. The Early Phase of SE Is an Active Process of Molecular Regulation

Although numerous studies have examined the mechanism of SE in the past decade [4,33], the molecular mechanisms underlying somatic embryonic induction, embryonic and maturation in *M. sativa* is unclear. The embryonic induction can be used to study early embryonic development. In the embryonic induction phase, which occurs early in SE, 9206 genes were upregulated and 5522 genes downregulated, with greater numbers of DEGs detected than during the other phases of SE. We identified several genes that play a regulatory role in early somatic embryonic development, including polycomb repressive complex (*PRC1*), *WUS*, *SERK1* and *heat shock protein17* (*HSP17*) (Table S4). *PRC1* and *PRC2* modify chromatin to repress the expression of genes not required for a specific differentiated state [34]. *WUS* is a marker of embryonic cells [26], and several studies have shown that *WUS* is important in totipotent embryogenic stem cells [14,26]. *SERK1* encodes a leucine repeat receptor protein kinase, which promotes early embryogenesis [35]. *HSP17* shows transient accumulation during embryonic maturation and germination in the oak and increases in level during dedifferentiation in SE [36]. In addition, we identified several new genes that were upregulated 10-fold from the embryonic induction phase, including *Phosphoglycolate phosphatase 1B* (*PGLP1B*), *Benzyl alcohol O-benzoyltransferase* (*HSR201*) and *Probable glucuronoxylan glucuronosyltransferase* (*IRX7*) (Table S4). Therefore, we infer that the early phase of SE in *M. sativa* involves an active process of molecular regulation.

3.2. Identification of Hormonal Signal Transduction during SE in M. sativa

Plant hormones and PGRs play critical roles during SE. Plant hormones specify the endogenous compounds produced by diverse cells, and PGRs complement synthetic compounds added exogenously [37]. Among all phytohormones, auxin plays important roles in regulating plant development, while IAA has been recognized as the most important auxin [38]. Auxin biosynthesis is regulated by *Aux/IAA*, *TIR1*, *ARF*, *CH3* and *SAUR* [39–43]. Our study suggests that the key genes *Aux/IAA*, *ARF*, *CH3* and *SAUR* are involved in auxin biosynthesis. The expression of *Aux/IAA* and CH3 were significantly upregulated in the embryonic induction and maturation stages. Auxin/IAA (Aux/IAA) proteins are transcriptional regulators of plant responses to auxin during fruit development and leaf morphogenesis. *IAA9* belongs to the Aux/IAA gene family and is downregulated in Arabidopsis mutants [44].

Cytokinins play important roles in promoting cell division [6]. Cytokinin biosynthesis is regulated by *CRE1*, *histidine phosphotransfer proteins* (*AHPs*), *A-ARRs* and *B-ARRs* [45–48]. *AHPs*, *A-ARRs* and *B-ARRs* are involved in embryonic induction. *A-ARRs* are also involved in embryogenic. The genes encoding *AHP*, *A-ARR* and *B-ARR* exhibited significant downregulation during embryonic induction, whereas *A-ARR* showed upregulation during the embryogenic formation stage. *AHP* is a phosphorelay carrier between cytokinin receptors and nuclear cytokinin responses. AHP mutants function study exhibited that AHP were positive factors in cytokinin signaling [49,50]. *B-ARRs* are mediators of cytokinin signaling, while *A-ARRs* are negative regulators of cytokinin signaling that function via feedback mechanisms to the primary cytokinin signal response [48,51].

GA regulates cell elongation during seedling development [52]. Several TFs are involved in GA biosynthesis during SE in *M. sativa*, including *PHYTOCHROME INTERACT-*

ING FACTOR (*PIF4*), *NONEXPRESSOR OF PRGENES1* (*NPR1*), *ABRE-BINDING FACTORS* (*ABF*), *SALT-RESPONSIVE ERF1* (*ERF1*), *bHLH transcription factors* (*MYC2*) and *Thermal Gravimetric Analysis* (*TGA*). *NPR1*, *ABF* and *ERF1* showed significant upregulation in embryonic induction. In addition, *ABF* showed positive upregulation in maturation. ABF is a key factor involved in the transition of embryogenesis to seed germination, playing an intermediate role in the cross-talk between ABA and GA signaling [53]. *NPR1* regulates the cross-talk between salicylic acid and jasmonic acid signaling pathways [54]. As a role of *NPR1* in GA synthesis has not been reported, further studies are required. *ERF1* is involved in not only the ET signaling pathway but also the ABA and GA biosynthetic pathways [55].

ABA supplementation promotes SE during maturation in *Podocarpus lambertii* [56]. Our study showed that ABA biosynthesis plays important roles in SE in *M. sativa*; the key molecules regulating ABA biosynthesis during embryonic induction and maturation include *PYRABACTIN RESISTANCE 1* (*PYR/PYL*), *2C protein phosphatase* (*PP2C*), *subfamily 2 Snfl-related kinases* (*SnRK2*) and *ABF*. *PYR/PYL*, encoding ABA receptors involved in ABA signal transduction [57], showed involvement throughout SE and exhibited significant upregulation. A key component of ABA biosynthesis is PP2C, which interacts with ABA receptors and SnRK2s [58]. In addition, biosyntheses of ET, brassinosteroid, jasmonic acid and salicylic acid are involved in SE in M. sativa, consistent with previous reports. Our transcriptome analysis revealed differential expression patterns of genes involved in hormonal signal transduction in SE in *M. sativa*. All hormonal signal transduction pathways are shown in Figure S3.

3.3. miRNA and Target Genes Form a Potential Molecular Regulatory Network in SE in M. sativa

The T0 generation cut cotyledon functions as an explant to induce somatic cell, and the addition of plant growth regulators to the medium promotes the formation of embryonic cells. *M. sativa* SE was divided into the embryonic induction, embryonic and maturation phases. Four types of calli (ICE, NEC, EC and CE) were selected to construct four transcriptome and small RNA libraries. The DEGs and differentially expressed miRNAs among four types of calli were identified. The expression levels of eight DEGs and eight miRNAs were detected by qRT-PCR. The target genes of the miRNAs were predicted and their functions annotated. The DEGs involved in the pathways of SE in *M. sativa* were predicted using KEGG analysis (Table S3). The miRNAs and target genes in the differentially expressed RNA libraries were compared. Previous reports suggested the existence of a potential molecular regulatory network, and the present study identified several novel genes and miRNAs involved in *M. sativa* SE (Figure 11).

The miR172 family is involved in a variety of processes including flowering time and floral organ identity [12], developmental timing [59], promotion of vegetative phase changes [60], soybean nodulation [61] and regulation of stem cells [62]. In addition, miR172 regulates *AP2* to control embryogenic and non-embryogenic callus development [63]. The miR172 target genes include *TOE1*, *TOE2*, *SMZ*, *SNZ* and SPL10 [59,64]. In our study, *AP2*, *NPK1*, *PUB21*, *ERF054* and *BHLH35* genes were found to be targeted by miR172d in embryonic induction, which might promote embryonic callus formation.

The miR156 family is involved in multiple plant developmental processes via targeting of *SPL* genes, including regulation of flowering [65], plastochrone length and organ size [66] and anthocyanin biosynthesis [20]. miR156 also regulates shoot development by targeting the SPL3 gene [67]. In addition, miR156 plays roles in SE. The *SPL10* and *SPL11* genes were repressed by miR156, which affects the precocious accumulation of maturation-phase transcripts, in Arabidopsis eight-cell embryos [20]. miR156 is also involved in early SE in the yellow poplar [18] and the regulation of CE formation [27,47]. Our results showed significantly upregulated miR156 expression in embryonic induction (Figure 8a) and identified *SPL6*, *AHL*, *ALS3*, *ARF8* and *SRG1* as target genes of miR156. However, aside from *SPL6*, the functions of the other target genes remain unclear.

miR159 regulates *LaMYB33* in the embryogenic formation and maturation stages of *larix kaempferi* SE [18]. In Arabidopsis, miR159 regulates GAMYB-like genes to promote programmed cell death [47]. In the present study, miR159 targeted *Pheophytinase* (*PPH*) in embryonic induction in *M. sativa*, suggesting that miR159 plays important roles in early SE.

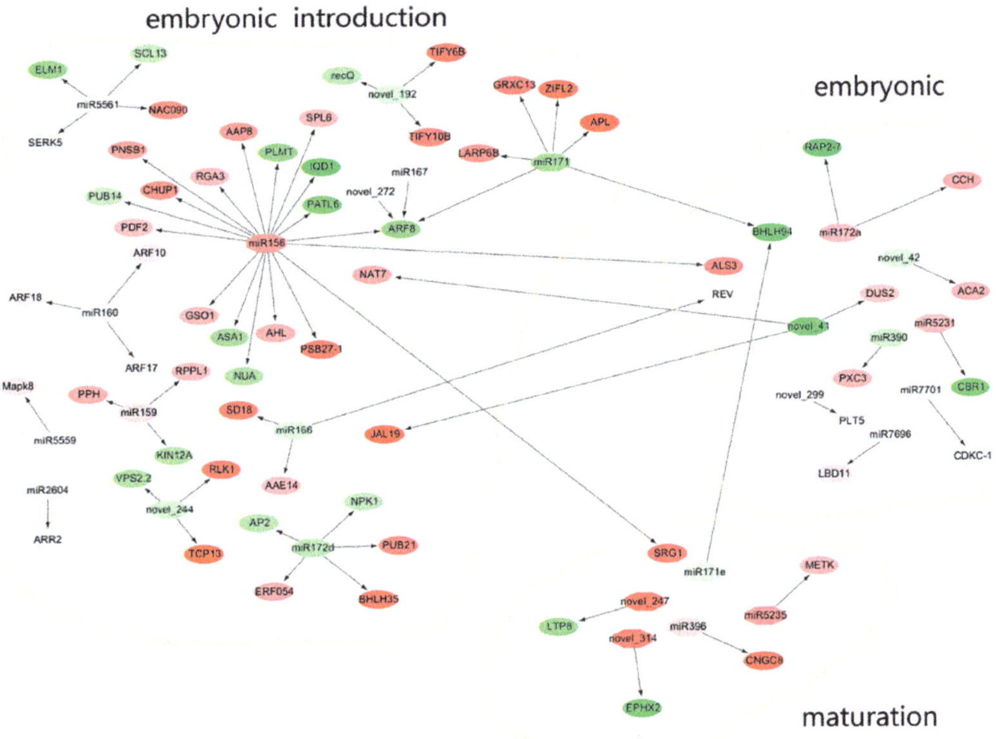

Figure 11. The miRNAs and target genes formed potential miRNA-target gene interaction network during embryonic induction, embyonic and maturation in *M.sative* SE. Red represents gene upregulation and green represents gene down regulation. The darker the color represent more obvious the difference in expression. Arrows represent the targeting of miRNAs.

In Arabidopsis, miR166 targeted *HD-ZIP III* TFs to regulate shoot apical meristem development during embryogenesis [68,69]. Moreover, miR166 regulated floral [70] and root development [71]. In *M. sativa* SE, miR166 targeted *AAE14* and *REV* in embryonic induction and maturation. In this study, miR166 was downregulated in embryonic induction; however, AAE14 was upregulated. Therefore, the miR166–AAE14 interaction may promote the transition from embryonic induction to embryonic formation. miR171 maintained embryogenic potential in *Larix kaempferi* Carr SE via regulation of its target gene *SCL6* [70]. In addition, miR171 targeted SCL6/22/27 to negatively regulate chlorophyll biosynthesis in Arabidopsis [72], and miR171 targeting of *GRAS* regulated GA and auxin homeostasis in the tomato [73]. In our study, *ARF8* was identified as a target gene of miR171 during embryonic induction. However, our results revealed that miR171 was upregulated, whereas *ARF8* was downregulated. *ARF8* is involved in auxin biosynthesis, which plays a crucial role in SE.

In addition, miR5561, miR390, miR5231, miR5559, novel_41 and novel_247 were differentially expressed during various stages of M. sativa SE (Table S10, Figure 11). According to predictions, miR5561 targeted *SERK5*, *Mitochondrial fission protein* (*ELM1*) and *NAC domain-containing protein 90* (*NAC090*) in embryonic introduction (Figure 11). miR390 was found to target ARF to regulate lateral root growth [74]. In our study, miR390 targeted

Leucine-rich repeat receptor-like tyrosine-protein kinase (*PXC3*) in embryonic. Although we identified several novel miRNAs and target genes, further studies are needed to define their functions.

4. Materials and Methods

4.1. Plant Materials

SE consists of three major phases in *M. sativa*: embryonic induction (ICE to NEC), embryonic formation (NEC to EC) and maturation (EC to CE). EC was induced in MS medium supplemented with 0.2 mg/L 6-benzylaminopurine and 4 mg/L 2,4-D. Then, the embryos were collected at 0, 10 and 40 d, representing the ICE, NEC and EC stages, respectively. CEs were obtained in MS medium containing 1 mg/L kinetin and 0.5 mg/L 6-benzylaminopurine and collected at 69 d. Calli were induced from cotyledons as explants. All samples included three biological replicates.

4.2. Morphological Analysis

We compared tissue morphology at different periods of SE using a light microscope. An explant showing significant changes was selected for further study. The organization of different morphological features was observed using a stereomicroscope (MZ FLIII; Leica, Wetzlar, Germany). The sample explants at different phases (0, 3, 6, 12, 24, 48 and 72 h, and 5, 7, 10, 15, 25, 40 and 69 d) were fixed in glutaraldehyde. Samples were stained with safranin for 1–2 h and rinsed with water. Then, a 50%, 70% and 80% alcohol gradient was applied for 1 min to decolorize the samples. Paraffin sections of the samples were made for observation under a light microscope (ECLIPSE Ci-L, NIKON, Tokyo, Japan).

4.3. RNA Extraction

Total RNA was extracted from the explants during the ICE, NEC, EC and CE phases using TRIzol reagent (Invitrogen, Burlington, ON, Canada) according to the manufacturer's protocol. RNA quality was visualized on a 1% agarose gel. RNA purity was measured using a NanoPhotometer spectrophotometer (IMPLEN, Westlake Village, CA, USA). RNA concentrations were measured using the Qubit RNA Assay Kit in a Qubit 2.0 Fluorometer (Life Technologies, Carlsbad, CA, USA). RNA integrity was assessed using the Bioanalyzer 2100 system (Agilent Technologies, Santa Clara, CA, USA).

4.4. Sample Preparation for Sequencing and Data Quality Control

RNA (6 μg per sample) was used as input material for the RNA sample preparations. All four samples had RNA integrity number (RIN) values above 8.0. Transcriptome and small RNA libraries of ICE, NEC, EC and CE were constructed and sequenced using the Illumina system [75,76]. Raw data (raw reads) in fastq format were processed using in-house Perl scripts. In this step, we obtained the clean data (clean reads), and removed reads containing the adapter, reads containing poly-N, and low-quality reads from the raw data. Simultaneously, Q20, Q30, GC content, and the sequence duplication level of the clean data were calculated. All the subsequent analyses were based on clean data.

4.5. Annotation of DEGs and miRNAs

To obtain DEGs from ICE to NEC, NEC to EC and EC to CE, the fold change (FC) in expression was assessed by taking the log2 ratio of Reads Per Kilobase per Million mapped reads (RPKM). Differential expression analysis of two conditions was performed using the DEGSeq R package (1.12.0; TNLIST, Beijing, China). The adjustment of *p*-values was performed using the Liszkay method [77]. The corrected *p*-value of 0.005 and log2 of ± 1 were set as the threshold for significant differential gene expression. DEGs were annotated using Blast2GO v2.5 (BioBam, Valencia, Spain). The genes and miRNA expression patterns of each phase (ICE, NEC, EC and CE) were clustered according to their log2 value using corset v1.05 software (Melbourne, Australia), and the database used for comparison was union_for_cluster. Heat-maps of cluster data were constructed using Java Tree View

(Stanford, CA, USA) [78]. miRNAs were annotated using the Rfam database, and novel miRNAs were predicted using miREvo and miRdeep2 software. miRNA target genes were predicted using the psRNATarget server. The miRNA–target interaction network was drawn using cityscape (San Diego, CA, USA).

4.6. Functional Annotation of DEGs and miRNAs

GO analysis was performed using GOseq [79]. and the GO database (http://www.geneontology.org/, accessed on 5 June 2022). The KOG/COG database can be found at (http://www.ncbi.nlm.nih.gov/COG/, accessed on 5 June 2022), and diamond v0.8.22 was used as the analysis software for the KOG database. KEGG classification was constructed based on the KEGG database (http://www.genome.jp/kegg/, accessed on 5 June 2022). The KOBAS (Beijing, China) was used for the KEGG analysis.

4.7. Indirect Competitive Enzyme-Linked Immunoassay (icELISA) for Detection of Indole-3-Acetic Acid (IAA), Gibberellin (GA3), Zeatin Riboside (ZR) and Abscisic Acid (ABA)

The icELISA protocol was previously described [80]. Reagents purchased from Sigma-Aldrich (St. Louis, MO, USA) included IAA, GA3, ZR, ABA, goat anti-mouse IgG conjugated to horseradish peroxidase (IgG-HRP), o-phenylenediamine, potassium periodate and citric acid monohydrate ($C_6H_8O_7{\cdot}H_2O$). SE was divided into four phases (ICE, NEC, EC and CE), and each group of *M. sativa* explants (8.0 g) was prepared for detection using icELISA. IAA, GA3, ZR and ABA were extracted as previously reported [81]. Detected samples were used for icELISA analysis.

4.8. Quantitative RT-PCR

Multiple genes and miRNAs were sorted for validation using reverse transcriptase-polymerase chain reaction (RT-PCR) (Bio-Rad C1000, Hercules, CA, USA). The genes and miRNA primers (Table S14) were designed using primer premier 5.0. The 18S RNA was used as an internal control for gene validation [82]. The small nuclear RNA (snRNA) U6 was used as an internal control for miRNA validation [83]. Quantitative RT-PCR (qRT-PCR) was performed as described by Liu et al. [84].

4.9. Statistical Analyses

Differences in the means for hormone detection were assessed by one-way Student's *t*-test at the 0.05 significance level. Differences in the means for gene expression data were assessed by one-way ANOVA at the 0.05 significance level.

5. Conclusions

Analyses of small RNAs and transcriptomes involved in the somatic embryonic induction, embryonic and maturation phases of SE provided information regarding the molecular mechanisms specific to *M. sativa*. Morphological observations revealed significant changes in tissues in SE, particularly during the stages of ICE, NEC, EC and CE. Several novel DEGs were identified in *M. sativa* SE, including *LBD*, *AGO* and *AGL*, which play important roles in promoting embryonic formation. Our analysis suggested that many DEGs playing roles in plant hormone signal transduction are involved in regulatory processes, e.g., *IAA* and *ARF* regulation of auxin biosynthesis, *LEC2* regulation of GA biosynthesis, and *MYB4* and *MYB48* regulation of ABA biosynthesis. In addition, hormonal signal transduction is regulated by the interactions of target genes with various miRNAs such as miR156, miR160 and miR167. This study predicted 110 novel miRNA families involved in embryonic induction, embryonic and maturation. Further studies are needed to determine whether these novel miRNAs play additional roles in *M. sativa*. Finally, we analyzed several miRNAs exhibiting significantly different expression patterns in SE and predicted their target genes. A potential miRNA–target gene interaction network is presented in Figure 11, which outlines the molecular mechanisms of SE in *M. sativa*.

Supplementary Materials: The following supporting information can be downloaded at: https://www.mdpi.com/article/10.3390/ijms23158633/s1.

Author Contributions: J.Y., Y.C. and L.H. wrote the main manuscript text. L.H. prepared Figures 1–11 and Tables S1–S14. All authors have read and agreed to the published version of the manuscript.

Funding: This work is supported by the Fundamental Research Funds for the Central Universities (Grant No. 2021ZY84).

Institutional Review Board Statement: Not applicable.

Informed Consent Statement: Not applicable.

Data Availability Statement: The raw assembled data has been uploaded The National Center for Biotechnology Information (NCBI). The accession number is PRJNA474427. The URL is https://www.ncbi.nlm.nih.gov/bioproject/474427, accessed on 5 June 2022.

Conflicts of Interest: The authors declare no conflict of interest.

References

1. Zimmerman, J.L. Somatic embryogenesis: A model for early development in higher plants. *Plant Cell* **1993**, *5*, 1411. [CrossRef] [PubMed]
2. Wu, X.M.; Kou, S.J.; Liu, Y.L.; Fang, Y.N.; Xu, Q.; Guo, W.W. Genomewide analysis of small RNAs in nonembryogenic and embryogenic tissues of citrus: MicroRNA- and siRNA-mediated transcript cleavage involved in somatic embryogenesis. *Plant Biotechnol. J.* **2015**, *13*, 383–394. [CrossRef]
3. Altamura, M.M.; Della Rovere, F.; Fattorini, L.; D'Angeli, S.; Falasca, G. Recent Advances on Genetic and Physiological Bases of In Vitro Somatic Embryo Formation. *Methods Mol. Biol.* **2016**, *1359*, 47–85. [PubMed]
4. Feher, A.; Pasternak, T.P.; Dudits, D. Transition of somatic plant cells to an embryogenic state. *Plant Cell Tissue Organ Cult.* **2003**, *74*, 201–228. [CrossRef]
5. Zhao, X.Y.; Su, Y.H.; Cheng, Z.J.; Zhang, X.S. Cell fate switch during in vitro plant organogenesis. *J. Integr Plant Biol.* **2008**, *50*, 816–824. [CrossRef]
6. Mantiri, F.R.; Kurdyukov, S.; Lohar, D.P.; Sharopova, N.; Saeed, N.A.; Wang, X.D.; Vandenbosch, K.A.; Rose, R.J. The transcription factor MtSERF1 of the ERF subfamily identified by transcriptional profiling is required for somatic embryogenesis induced by auxin plus cytokinin in *Medicago truncatula*. *Plant Physiol.* **2008**, *146*, 1622–1636. [CrossRef] [PubMed]
7. Zhao, Z.; Andersen, S.U.; Ljung, K.; Dolezal, K.; Miotk, A.; Schultheiss, S.J.; Lohmann, J.U. Hormonal control of the shoot stem-cell niche. *Nature* **2010**, *465*, 1089–1092. [CrossRef] [PubMed]
8. Arnholdt-Schmitt, B. Physiological aspects of genome variability in tissue culture. II. Growth phase-dependent quantitative variability of repetitive BstNI fragments of primary cultures of *Daucus carota* L. *Theor. Appl. Genet.* **1995**, *91*, 816–823. [CrossRef]
9. Thibaud-Nissen, F.; Shealy, R.T.; Khanna, A.; Vodkin, L.O. Clustering of microarray data reveals transcript patterns associated with somatic embryogenesis in soybean. *Plant Physiol.* **2003**, *132*, 118–136. [CrossRef]
10. Sharma, S.K.; Millam, S.; Hein, I.; Bryan, G.J. Cloning and molecular characterisation of a potato SERK gene transcriptionally induced during initiation of somatic embryogenesis. *Planta* **2008**, *228*, 319–330. [CrossRef] [PubMed]
11. Quiroz-Figueroa, F.R.; Rojas-Herrera, R.; Galaz-Avalos, R.M.; Loyola-Vargas, V.M. Embryo production through somatic embryogenesis can be used to study cell differentiation in plants. *Plant Cell Tissue Organ Cult.* **2006**, *86*, 285–301. [CrossRef]
12. Gliwicka, M.; Nowak, K.; Balazadeh, S.; Mueller-Roeber, B.; Gaj, M.D. Extensive modulation of the transcription factor transcriptome during somatic embryogenesis in Arabidopsis thaliana. *PLoS ONE* **2013**, *8*, e69261. [CrossRef]
13. Boutilier, K.; Offringa, R.; Sharma, V.K.; Kieft, H.; Ouellet, T.; Zhang, L.; Hattori, J.; Liu, C.M.; van Lammeren, A.A.; Miki, B.L.; et al. Ectopic expression of BABY BOOM triggers a conversion from vegetative to embryonic growth. *Plant Cell* **2002**, *4*, 1737–1749. [CrossRef]
14. Zuo, J.; Niu, Q.W.; Frugis, G.; Chua, N.H. The WUSCHEL gene promotes vegetative-to-embryonic transition in Arabidopsis. *Plant J.* **2002**, *30*, 349–359. [CrossRef] [PubMed]
15. Gaj, M.D.; Zhang, S.; Harada, J.J.; Lemaux, P.G. Leafy cotyledon genes are essential for induction of somatic embryogenesis of Arabidopsis. *Planta* **2005**, *222*, 977–988. [CrossRef] [PubMed]
16. Chen, X. Small RNAs and their roles in plant development. *Annu. Rev. Cell Dev. Biol.* **2009**, *25*, 21–44. [CrossRef]
17. Yakovlev, I.A.; Fossdal, C.G.; Johnsen, Ø. MicroRNAs, the epigenetic memory and climatic adaptation in Norway spruce. *New Phytol.* **2010**, *187*, 1154–1169. [CrossRef] [PubMed]
18. Li, T.; Chen, J.; Qiu, S.; Zhang, Y.; Wang, P.; Yang, L.; Lu, Y.; Shi, J. Deep sequencing and microarray hybridization identify conserved and species-specific microRNAs during somatic embryogenesis in hybrid yellow poplar. *PLoS ONE* **2012**, *7*, e43451. [CrossRef] [PubMed]
19. Luo, Y.C.; Zhou, H.; Li, Y.; Chen, J.Y.; Yang, J.H.; Chen, Y.Q.; Qu, L.H. Rice embryogenic calli express a unique set of microRNAs, suggesting regulatory roles of microRNAs in plant post-embryogenic development. *FEBS Lett.* **2006**, *580*, 5111–5116. [CrossRef] [PubMed]

20. Gou, J.Y.; Felippes, F.F.; Liu, C.J.; Weigel, D.; Wang, J.W. Negative regulation of anthocyanin biosynthesis in Arabidopsis by a miR156-targeted SPL transcription factor. *Plant Cell* **2011**, *23*, 1512–1522. [CrossRef]
21. Tang, W.; Newton, R.J. Genome-wide expression analysis of genes involved in somatic embryogenesis. *Somat. Embryog.* **2005**, *2*, 69–83.
22. Ye, C.Y.; Xu, H.; Shen, E.; Liu, Y.; Wang, Y.; Shen, Y.; Qiu, J.; Zhu, Q.H.; Fan, L. Genome-wide identification of non-coding RNAs interacted with microRNAs in soybean. *Front. Plant Sci.* **2014**, *23*, 743. [CrossRef]
23. Jing, D.; Zhang, J.; Xia, Y.; Kong, L.; OuYang, F.; Zhang, S.; Zhang, H.; Wang, J. Proteomic analysis of stress-related proteins and metabolic pathways in Picea asperata somatic embryos during partial desiccation. *Plant Biotechnol. J.* **2017**, *15*, 27–38. [CrossRef] [PubMed]
24. Saunders, J.W.; Bingham, E.T. Production of Alfalfa Plants from Callus Tissue 1. *Crop Sci.* **1972**, *12*, 804–808. [CrossRef]
25. Imin, N.; Nizamidin, M.; Daniher, D.; Nolan, K.E.; Rose, R.J.; Rolfe, B.G. Proteomic analysis of somatic embryogenesis in Medicago truncatula. Explant cultures grown under 6-benzylaminopurine and 1-naphthaleneacetic acid treatments. *Plant Physiol.* **2005**, *137*, 1250–1260. [CrossRef] [PubMed]
26. Elhiti, M.; Stasolla, C.; Wang, A. Molecular regulation of plant somatic embryogenesis. *Vitr. Cell. Dev. Biol.-Plant* **2013**, *49*, 631–642. [CrossRef]
27. Zhang, J.; Xue, B.; Gai, M.; Song, S.; Jia, N.; Sun, H. Small RNA and Transcriptome Sequencing Reveal a Potential miRNA-Mediated Interaction Network That Functions during Somatic Embryogenesis in Lilium pumilum DC. Fisch. *Front Plant Sci.* **2017**, *8*, 566. [CrossRef] [PubMed]
28. Li, W.F.; Zhang, S.G.; Han, S.Y.; Wu, T.; Zhang, J.H.; Qi, L.W. Regulation of LaMYB33 by miR159 during maintenance of embryogenic potential and somatic embryo maturation in Larix kaempferi (Lamb.) Carr. *Plant Cell Tissue Organ Cult. (PCTOC)* **2013**, *113*, 131–136. [CrossRef]
29. Reinert, J. Morphogenese und ihre Kontrolle an Gewebekulturen aus Carotten. *Naturwissenschaften* **1958**, *45*, 344–345. [CrossRef]
30. Steward, F.C.; Mapes, M.O.; Smith, J. Growth and organized development of cultured cells. I. Growth and division of freely suspended cells. *Am. J. Bot.* **1958**, *45*, 693–703. [CrossRef]
31. Galland, R.; Randoux, B.; Vasseur, J.; Hilbert, J.L. A glutathione S-transferase cDNA identified by mRNA differential display is upregulated during somatic embryogenesis in Cichorium. *Biochim. Biophys. Acta* **2001**, *1522*, 212–216. [CrossRef]
32. Xu, H.; Gao, Y.; Wang, J. Transcriptomic analysis of rice (Oryza sativa) developing embryos using the RNA-Seq technique. *PLoS ONE* **2012**, *7*, e30646. [CrossRef] [PubMed]
33. Goldberg, R.B.; Barker, S.J.; Perez-Grau, L. Regulation of gene expression during plant embryogenesis. *Cell* **1989**, *56*, 149–160. [CrossRef]
34. Bratzel, F.; López-Torrejón, G.; Koch, M.; Del Pozo, J.C.; Calonje, M. Keeping cell identity in Arabidopsis requires PRC1 RING-finger homologs that catalyze H2A monoubiquitination. *Curr. Biol.* **2010**, *20*, 1853–1859. [CrossRef] [PubMed]
35. Hecht, V.; Vielle-Calzada, J.P.; Hartog, M.V.; Schmidt, E.D.; Boutilier, K.; Grossniklaus, U.; de Vries, S.C. The Arabidopsis SOMATIC EMBRYOGENESIS RECEPTOR KINASE 1 gene is expressed in developing ovules and embryos and enhances embryogenic competence in culture. *Plant Physiol.* **2001**, *127*, 803–816. [CrossRef] [PubMed]
36. Puigderrajols, P.; Jofré, A.; Mir, G.; Pla, M.; Verdaguer, D.; Huguet, G.; Molinas, M. Developmentally and stress-induced small heat shock proteins in cork oak somatic embryos. *J. Exp. Bot.* **2002**, *53*, 1445–1452.
37. Jiménez, V.M. Involvement of plant hormones and plant growth regulators on in vitro somatic embryogenesis. *Plant Growth Regul.* **2005**, *47*, 91–110. [CrossRef]
38. Mashiguchi, K.; Tanaka, K.; Sakai, T.; Sugawara, S.; Kawaide, H.; Natsume, M.; Hanada, A.; Yaeno, T.; Shirasu, K.; Yao, H.; et al. The main auxin biosynthesis pathway in Arabidopsis. *Proc. Natl. Acad. Sci. USA* **2011**, *108*, 18512–18517. [CrossRef]
39. Swarup, R.; Friml, J.; Marchant, A.; Ljung, K.; Sandberg, G.; Palme, K.; Bennett, M. Localization of the auxin permease AUX1 suggests two functionally distinct hormone transport pathways operate in the Arabidopsis root apex. *Genes Dev.* **2001**, *15*, 2648–2653. [CrossRef] [PubMed]
40. Dharmasiri, N.; Dharmasiri, S.; Estelle, M. The F-box protein TIR1 is an auxin receptor. *Nature* **2005**, *435*, 441–445. [CrossRef]
41. Zenk, M.H.; Müller, G. In vivo destruction of exogenously applied indolyl-3-acetic acid as influenced by naturally occurring phenolic acids. *Nature* **1963**, *200*, 761–763. [CrossRef]
42. Hagen, G.; Guilfoyle, T. Auxin-responsive gene expression: Genes, promoters and regulatory factors. *Plant Mol. Biol.* **2002**, *49*, 373–385. [CrossRef]
43. Abe, H.; Uchiyama, M.; Sato, R. Isolation and identification of native auxins in marine algae. *Agric. Biol. Chem.* **1972**, *36*, 2259–2260. [CrossRef]
44. Wang, H.; Jones, B.; Li, Z.; Frasse, P.; Delalande, C.; Regad, F.; Chaabouni, S.; Latché, A.; Pech, J.C.; Bouzayen, M. The tomato Aux/IAA transcription factor IAA9 is involved in fruit development and leaf morphogenesis. *Plant Cell* **2005**, *17*, 2676–2692. [CrossRef] [PubMed]
45. Gonzalez-Rizzo, S.; Crespi, M.; Frugier, F. The Medicago truncatula CRE1 cytokinin receptor regulates lateral root development and early symbiotic interaction with Sinorhizobium meliloti. *Plant Cell* **2006**, *18*, 2680–2693. [CrossRef] [PubMed]
46. Tun, N.N.; Livaja, M.; Kieber, J.J.; Scherer, G.F. Zeatin-induced nitric oxide (NO) biosynthesis in Arabidopsis thaliana mutants of NO biosynthesis and of two-component signaling genes. *New Phytol.* **2008**, *178*, 515–531. [CrossRef] [PubMed]

47. Argueso, C.T.; Ferreira, F.J.; Kieber, J.J. Environmental perception avenues: The interaction of cytokinin and environmental response pathways. *Plant Cell Environ.* **2009**, *32*, 1147–1160. [CrossRef] [PubMed]
48. Hwang, I.; Sheen, J. Two-component circuitry in Arabidopsis cytokinin signal transduction. *Nature* **2001**, *413*, 383–389. [CrossRef]
49. Papon, N.; Vansiri, A.; Gantet, P.; Chénieux, J.C.; Rideau, M.; Crèche, J. Histidine-containing phosphotransfer domain extinction by RNA interference turns off a cytokinin signalling circuitry in Catharanthus roseus suspension cells. *FEBS Lett.* **2004**, *558*, 85–88. [CrossRef]
50. Ferreira, F.J.; Kieber, J.J. Cytokinin signaling. *Curr. Opin. Plant Biol.* **2005**, *8*, 518–525. [CrossRef] [PubMed]
51. To, J.P.; Haberer, G.; Ferreira, F.J.; Deruère, J.; Mason, M.G.; Schaller, G.E.; Alonso, J.M.; Ecker, J.R.; Kieber, J.J. Type-A Arabidopsis response regulators are partially redundant negative regulators of cytokinin signaling. *Plant Cell.* **2004**, *16*, 658–671. [CrossRef] [PubMed]
52. de Lucas, M.; Davière, J.M.; Rodríguez-Falcón, M.; Pontin, M.; Iglesias-Pedraz, J.M.; Lorrain, S.; Fankhauser, C.; Blázquez, M.A.; Titarenko, E.; Prat, S. A molecular framework for light and gibberellin control of cell elongation. *Nature* **2008**, *451*, 480–484. [CrossRef]
53. Schoonheim, P.J.; Costa Pereira, D.D.; De Boer, A.H. Dual role for 14-3-3 proteins and ABF transcription factors in gibberellic acid and abscisic acid signalling in barley (Hordeum vulgare) aleurone cells. *Plant Cell Environ.* **2009**, *32*, 439–447. [CrossRef] [PubMed]
54. Spoel, S.H.; Koornneef, A.; Claessens, S.M.; Korzelius, J.P.; Van Pelt, J.A.; Mueller, M.J.; Buchala, A.J.; Métraux, J.P.; Brown, R.; Kazan, K.; et al. NPR1 modulates cross-talk between salicylate- and jasmonate-dependent defense pathways through a novel function in the cytosol. *Plant Cell* **2003**, *15*, 760–770. [CrossRef]
55. Yoong, F.Y.; O'Brien, L.K.; Truco, M.J.; Huo, H.; Sideman, R.; Hayes, R.; Michelmore, R.W.; Bradford, K.J. Genetic Variation for Thermotolerance in Lettuce Seed Germination Is Associated with Temperature-Sensitive Regulation of ETHYLENE RESPONSE FACTOR1 (ERF1). *Plant Physiol.* **2016**, *170*, 472–488. [CrossRef]
56. Fraga, H.P.; Vieira, L.D.; Puttkammer, C.C.; Dos Santos, H.P.; Garighan, J.A.; Guerra, M.P. Glutathione and abscisic acid supplementation influences somatic embryo maturation and hormone endogenous levels during somatic embryogenesis in Podocarpus lambertii Klotzsch ex Endl. *Plant Sci.* **2016**, *253*, 98–106. [CrossRef]
57. Lee, S.C.; Luan, S. ABA signal transduction at the crossroad of biotic and abiotic stress responses. *Plant Cell Environ.* **2012**, *35*, 53–60. [CrossRef]
58. Soon, F.F.; Ng, L.M.; Zhou, X.E.; West, G.M.; Kovach, A.; Tan, M.H.; Suino-Powell, K.M.; He, Y.; Xu, Y.; Chalmers, M.J.; et al. Molecular mimicry regulates ABA signaling by SnRK2 kinases and PP2C phosphatases. *Science* **2012**, *335*, 85–88. [CrossRef] [PubMed]
59. Wu, G.; Park, M.Y.; Conway, S.R.; Wang, J.W.; Weigel, D.; Poethig, R.S. The sequential action of miR156 and miR172 regulates developmental timing in Arabidopsis. *Cell* **2009**, *138*, 750–759. [CrossRef] [PubMed]
60. Lauter, N.; Kampani, A.; Carlson, S.; Goebel, M.; Moose, S.P. microRNA172 down-regulates glossy15 to promote vegetative phase change in maize. *Proc. Natl. Acad. Sci. USA* **2005**, *102*, 9412–9417. [CrossRef] [PubMed]
61. Yan, Z.; Hossain, M.S.; Wang, J.; Valdés-López, O.; Liang, Y.; Libault, M.; Qiu, L.; Stacey, G. miR172 regulates soybean nodulation. *Mol. Plant Microbe Interact.* **2013**, *26*, 1371–1377. [CrossRef] [PubMed]
62. Zhao, L.; Kim, Y.; Dinh, T.T.; Chen, X. miR172 regulates stem cell fate and defines the inner boundary of APETALA3 and PISTILLATA expression domain in Arabidopsis floral meristems. *Plant J.* **2007**, *51*, 840–849. [CrossRef]
63. Zhang, S.; Zhou, J.; Han, S.; Yang, W.; Li, W.; Wei, H.; Li, X.; Qi, L. Four abiotic stress-induced miRNA families differentially regulated in the embryogenic and non-embryogenic callus tissues of Larix leptolepis. *Biochem. Biophys. Res. Commun.* **2010**, *398*, 355–360. [CrossRef]
64. Glazińska, P.; Zienkiewicz, A.; Wojciechowski, W.; Kopcewicz, J. The putative miR172 target gene InAPETALA2-like is involved in the photoperiodic flower induction of Ipomoea nil. *J. Plant Physiol.* **2009**, *166*, 1801–1813. [CrossRef]
65. Wang, J.W.; Czech, B.; Weigel, D. miR156-regulated SPL transcription factors define an endogenous flowering pathway in *Arabidopsis thaliana*. *Cell* **2009**, *138*, 738–749. [CrossRef] [PubMed]
66. Wang, J.W.; Schwab, R.; Czech, B.; Mica, E.; Weigel, D. Dual effects of miR156-targeted SPL genes and CYP78A5/KLUH on plastochron length and organ size in *Arabidopsis thaliana*. *Plant Cell* **2008**, *20*, 1231–1243. [CrossRef]
67. Wu, G.; Poethig, R.S. Temporal regulation of shoot development in *Arabidopsis thaliana* by miR156 and its target SPL3. *Development* **2006**, *133*, 3539–3547. [CrossRef] [PubMed]
68. Zhu, H.; Hu, F.; Wang, R.; Zhou, X.; Sze, S.H.; Liou, L.W.; Barefoot, A.; Dickman, M.; Zhang, X. Arabidopsis Argonaute10 specifically sequesters miR166/165 to regulate shoot apical meristem development. *Cell* **2011**, *145*, 242–256. [CrossRef]
69. Miyashima, S.; Honda, M.; Hashimoto, K.; Tatematsu, K.; Hashimoto, T.; Sato-Nara, K.; Okada, K.; Nakajima, K. A comprehensive expression analysis of the Arabidopsis MICRORNA165/6 gene family during embryogenesis reveals a conserved role in meristem specification and a non-cell-autonomous function. *Plant Cell Physiol.* **2013**, *54*, 375–384. [CrossRef]
70. Jung, J.H.; Park, C.M. MIR166/165 genes exhibit dynamic expression patterns in regulating shoot apical meristem and floral development in Arabidopsis. *Planta* **2007**, *225*, 1327–1338. [CrossRef]
71. Li, W.F.; Zhang, S.G.; Han, S.Y.; Wu, T.; Zhang, J.H.; Qi, L.W. The post-transcriptional regulation of LaSCL6 by miR171 during maintenance of embryogenic potential in Larix kaempferi (Lamb.) Carr. *Tree Genet. Genomes* **2014**, *10*, 223–229. [CrossRef]

72. Ma, Z.; Hu, X.; Cai, W.; Huang, W.; Zhou, X.; Luo, Q.; Yang, H.; Wang, J.; Huang, J. Arabidopsis miR171-targeted scarecrow-like proteins bind to GT cis-elements and mediate gibberellin-regulated chlorophyll biosynthesis under light conditions. *PLoS Genet.* **2014**, *10*, e1004519.
73. Huang, W.; Peng, S.; Xian, Z.; Lin, D.; Hu, G.; Yang, L.; Ren, M.; Li, Z. Overexpression of a tomato miR171 target gene SlGRAS24 impacts multiple agronomical traits via regulating gibberellin and auxin homeostasis. *Plant Biotechnol. J.* **2017**, *15*, 472–488. [CrossRef] [PubMed]
74. Marin, E.; Jouannet, V.; Herz, A.; Lokerse, A.S.; Weijers, D.; Vaucheret, H.; Nussaume, L.; Crespi, M.D.; Maizel, A. miR390, Arabidopsis TAS3 tasiRNAs, and their AUXIN RESPONSE FACTOR targets define an autoregulatory network quantitatively regulating lateral root growth. *Plant Cell* **2010**, *22*, 1104–1117. [PubMed]
75. Lu, C.; Tej, S.S.; Luo, S.; Haudenschild, C.D.; Meyers, B.C.; Green, P.J. Elucidation of the small RNA component of the transcriptome. *Science* **2005**, *309*, 1567–1569. [PubMed]
76. German, M.A.; Pillay, M.; Jeong, D.H.; Hetawal, A.; Luo, S.; Janardhanan, P.; Kannan, V.; Rymarquis, L.A.; Nobuta, K.; German, R.; et al. Global identification of microRNA-target RNA pairs by parallel analysis of RNA ends. *Nat. Biotechnol.* **2008**, *26*, 941–946.
77. Liszkay, A.; van der Zalm, E.; Schopfer, P. Production of reactive oxygen intermediates (O(2)(.-), $H_{(2)}O_{(2)}$, and (.)OH) by maize roots and their role in wall loosening and elongation growth. *Plant Physiol.* **2004**, *136*, 3114–3123. [CrossRef]
78. Saldanha, A.J. Java Treeview—Extensible visualization of microarray data. *Bioinformatics* **2004**, *20*, 3246–3248. [PubMed]
79. Young, M.D.; Wakefield, M.J.; Smyth, G.K.; Oshlack, A. Gene ontology analysis for RNA-seq: Accounting for selection bias. *Genome Biol.* **2010**, *11*, R14. [CrossRef] [PubMed]
80. Zhao, J.; Li, G.; Wang, B.M.; Liu, W.; Nan, T.G.; Zhai, Z.X.; Li, Z.H.; Li, Q.X. Development of a monoclonal antibody-based enzyme-linked immunosorbent assay for the analysis of glycyrrhizic acid. *Anal. Bioanal. Chem.* **2006**, *386*, 1735–1740.
81. Jin, M.; Ren, Y.; Chen, X. Determination of 6-benzyladenine in bean sprout by LC–ESI-IT-MS–MS. *Chromatographia* **2007**, *66*, 407–410. [CrossRef]
82. Goidin, D.; Mamessier, A.; Staquet, M.J.; Schmitt, D.; Berthier-Vergnes, O. Ribosomal 18S RNA prevails over glyceraldehyde-3-phosphate dehydrogenase and beta-actin genes as internal standard for quantitative comparison of mRNA levels in invasive and noninvasive human melanoma cell subpopulations. *Anal. Biochem.* **2001**, *295*, 17–21. [CrossRef]
83. Miyagishi, M.; Taira, K. U6 promoter-driven siRNAs with four uridine 3′ overhangs efficiently suppress targeted gene expression in mammalian cells. *Nat. Biotechnol.* **2002**, *20*, 497–500. [CrossRef] [PubMed]
84. Liu, Q.; Xu, J.; Liu, Y.; Zhao, X.; Deng, X.; Guo, L.; Gu, J. A novel bud mutation that confers abnormal patterns of lycopene accumulation in sweet orange fruit (*Citrus sinensis* L. Osbeck). *J. Exp. Bot.* **2007**, *58*, 4161–4171. [CrossRef] [PubMed]

International Journal of
Molecular Sciences

MDPI

Article

Transcriptome Analysis of Differentially Expressed Genes Associated with Salt Stress in Cowpea (*Vigna unguiculata* L.) during the Early Vegetative Stage

Byeong Hee Kang [1,2], Woon Ji Kim [1], Sreeparna Chowdhury [1], Chang Yeok Moon [1,2], Sehee Kang [1,2], Seong-Hoon Kim [3], Sung-Hwan Jo [4], Tae-Hwan Jun [5], Kyung Do Kim [6,*] and Bo-Keun Ha [1,2,*]

1 Department of Applied Plant Science, Chonnam National University, Gwangju 61186, Republic of Korea
2 BK21 Interdisciplinary Program in IT-Bio Convergence System, Chonnam National University, Gwangju 61186, Republic of Korea
3 National Agrobiodiversity Center, National Institute of Agricultural Sciences, RDA, Jeonju 5487, Republic of Korea
4 SEEDERS Inc., Daejeon 34912, Republic of Korea
5 Department of Plant Bioscience, Pusan National University, Miryang 50463, Republic of Korea
6 Department of Bioscience and Bioinformatics, Myongji University, Yongin 17058, Republic of Korea
* Correspondence: kyungdokim@mju.ac.kr (K.D.K.); bkha@jnu.ac.kr (B.-K.H.); Tel.:+82-31-330-6192 (K.D.K.); +82-62-530-2050 (B.-K.H.)

Abstract: Cowpea (*Vigna unguiculata* (L.), $2n = 22$) is a tropical crop grown in arid and semiarid regions that is tolerant to abiotic stresses such as heat and drought. However, in these regions, salt in the soil is generally not eluted by rainwater, leading to salt stress for a variety of plant species. This study was conducted to identify genes related to salt stress using the comparative transcriptome analysis of cowpea germplasms with contrasting salt tolerance. Using the Illumina Novaseq 6000 platform, 1.1 billion high-quality short reads, with a total length of over 98.6 billion bp, were obtained from four cowpea germplasms. Of the differentially expressed genes identified for each salt tolerance type following RNA sequencing, 27 were shown to exhibit significant expression levels. These candidate genes were subsequently narrowed down using reference-sequencing analysis, and two salt stress-related genes (*Vigun_02G076100* and *Vigun_08G125100*) with single-nucleotide polymorphism (SNP) variation were selected. Of the five SNPs identified in *Vigun_02G076100*, one that caused significant amino acid variation was identified, while all nucleotide variations in *Vigun_08G125100* was classified as missing in the salt-resistant germplasms. The candidate genes and their variation, identified in this study provide, useful information for the development of molecular markers for cowpea breeding programs.

Keywords: cowpea; salt-stress; NGS; RNA sequencing; reference sequencing

1. Introduction

Cowpea (*Vigna unguiculata* (L.) Walp.; $2n = 2x = 22$) is a tropical herbaceous crop that has adapted to various abiotic stresses, including drought and heat stress [1,2]. Globally, the estimated area of cowpea cultivation is about 15 million hectares, with more than 8.8 million tons being produced annually. The whole of Africa occupies more than 95% of this cultivated area, especially the arid and semiarid regions of West Africa, which can be identified as the main cultivation areas for cowpea [3]. However, salt compounds in the soil of arid and semi-arid regions are generally not eluted due to the low frequency of rainfall [4], and the resulting accumulation of salt in the soil can increase the salt stress for cowpea and other important crops. This problem has been exacerbated by climate change, which has increased the rate of desertification and created larger areas of arid and semiarid land in West Africa [5,6], potentially reducing the yield and quality of crops, including cowpea in the region.

Salt stress causes various types of damage at all stages of a plant's life cycle, from germination to seed production [7–9], with the proportion of cropland subject to salt damage reported to be increasing worldwide [10,11]. Therefore, understanding the effects of salt stress and the mechanisms associated with it is important from the perspective of meeting food demand in the future. Plants exposed to high soil salinity generally experience high osmotic and ionic stress, which affects a range of complex physiochemical processes [12–15]. The salinity reduces the leaf water potential and turgor pressure, leading to osmotic stress that induces abscisic acid biosynthesis, which in turn causes stomatal closure [16]. As a result, photosynthesis is reduced and oxidative stress increases [17]. In addition, excessive salinity around the roots can lead to ion toxicity, which increases the reactive oxygen species (ROS) levels, resulting in nutritional imbalances and damage to the cell structure [18]. This form of ion toxicity is commonly observed with sodium and chloride ions, which accumulate in highly saline soils [19].

One way to address high soil salinity is to create more salt-tolerant crop germplasms. However, salt stress is a complex process, and there are varying degrees of tolerance, both between and within species [20,21]. It has even been found that the response to salt stress can differ depending on the time of exposure and the stage of plant growth, with more rapid exposure resulting in more stress [22]. One study has reported that the difference in germination rates within a particular species ranged from 5.8% to 94.2% [23]. Within a particular species, individual plants with salt-sensitive genotypes tend to exhibit greater ion accumulation than salt-resistant genotypes do, leading to toxic effects [24]. These results suggest that salt tolerance is an independently evolved trait that can arise from completely unrelated mechanisms. This means that the genes associated with salt tolerance found in genetically distant species may not be effective if transplanted into germplasms of cowpea. However, genetic diversity within crops can be used to create germplasms with ideal traits, including salt tolerance [25]. Thus, further research is needed to identify genes related to salt stress in cowpea and to understand the mechanisms underlying their variation. In particular, it is important to understand plant ion homeostasis, osmotic responses, and oxidative stress in relation to increases in soil salinity.

Recently, the development of high-throughput sequencing technologies, such as next-generation sequencing (NGS), has made it possible to better understand plant genomes, which is essential for understanding complex traits, including those associated with salt stress [26]. For example, NGS-based RNA sequencing (RNA-seq) makes it possible to identify differentially expressed genes (DEGs) across the genome and analyze stress-related metabolic pathways via the functional annotation of the identified DEGs [27,28]. This approach provides an opportunity to search for candidate genes involved in the stress response of crops under salt stress, including the detection of rare transcripts, thus revealing the function and pathway of genes related to salt tolerance [29–32]. Though it is challenging to identify target genes from among the numerous DEGs generated via RNA-seq, reference genome information can be used to narrow down the range of candidate genes. As an example of this, kompetitive allele-specific PCR (KASP) genotyping assays have been widely used for SNP allele scoring and indel discrimination with various crops in a way that makes use of allele-specific primers [33]. For example, DEGs have been identified using RNA-seq in rices that differed in their salt tolerance, and SNP variation in the identified DEGs was then successfully used for KASP marker development [34]. These KASP markers can subsequently be employed for plant breeding through marker-assisted selection (MAS).

In the present study, we analyzed the expression patterns of genes associated with salt stress using cowpea germplasms with different levels of salt tolerance. We also conducted transcriptome profiling based on the fact that a plant absorbs salt from the roots, and that the damage is most severe at the seedling stage. This study thus aims to analyze the genetic network and related metabolic pathways for cowpea DEGs associated with salt tolerance.

2. Results

2.1. Physiological Responses to Salt Stress in Cowpea Germplasms

In this study, four cowpea germplasms with different levels of salt tolerance were exposed to a 250 mM NaCl solution for three weeks to simulate salt stress (Figures 1 and 2).

Figure 1. Four cowpea germplasms with different levels of salt tolerance exposed to a 250 mM NaCl solution for three weeks.

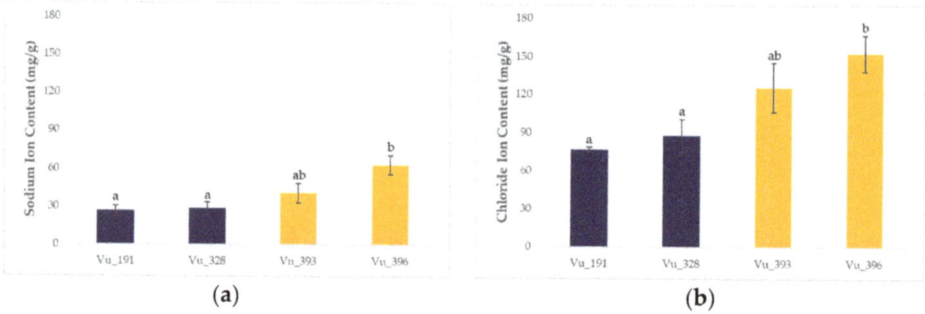

Figure 2. Ion accumulation for four cowpea germplasms exposed to 250 mM NaCl for three weeks: (**a**) sodium ions and (**b**) chloride ions. Different lowercase letters indicate statistically significant differences (LSD tests; $p < 0.05$). The error bars represent the standard deviation for five biological replicates.

Ion accumulation was generally higher in the salt-sensitive germplasms compared with the salt-resistant plants. In particular, the sodium and chloride ion levels for the salt-resistant germplasms (Vu_191 and Vu_328) were 27.65 mg/g and 82.12 mg/g, respectively, compared with 51.86 mg/g and 139.35 mg/g, respectively, for the salt-sensitive germplasms (Vu_393 and Vu_396).

2.2. Illumina Sequencing Pre-Processing, and Read Mapping

A total of 24 library samples were obtained for sequence processing, consisting of control (0 h) and NaCl treatments (24 h) for each of the four germplasms, with three replicates each. These library samples were sequenced using the Illumina Novaseq 6000 platform (Table S1). RNA-seq analysis showed that the total number of clean reads generated for each sample was 1,144,868,572 (average length 101 bp). To obtain high-quality transcriptome short reads, bases with a Phred score (Q) of less than 20 were trimmed, and those trimmed reads with a length of less than 25 bp were eliminated. The total number of filtered reads was 1,107,552,070, with an overall average of 85.31% passing through the preprocessing stage, of which 1,038,301,592 (93.81%) were uniquely mapped to the cowpea reference genome sequence (Vunguiculata_540_v1.2). Of the 31,948 standard genes used for analysis, 27,559 had expression values and 25,476 had functional descriptions.

2.3. Identification of DEGs in Cowpea Germplasms with Different Salt Tolerance Levels

DEGs were screened using DESeq2 software based on a false discovery rate (FDR) of $\leq$0.01 and absolute values for the $\log_2$fold change (FC) of >1, with up-regulation defined as $\log_2FC > 1$ and down-regulation as $\log_2FC < -1$. The gene expression profiles of the cowpea germplasms with different salt tolerance levels were compared between the salt treatment and control samples (Figure 3, Tables S2–S5). Overall, 5997 and 5532 DEGs were detected in the salt-resistant germplasms Vu_191 and Vu_328, respectively. The DEGs identified for Vu_191 typically included senescence-associated genes and genes encoding LEA proteins, while those identified for Vu_328 included genes encoding stress-induced proteins and pectin lyase. In addition, 5031 and 7444 DEGs were detected in the salt-sensitive germplasms Vu_393 and Vu_396, respectively. The DEGs identified for Vu_393 included genes encoding phosphatase family proteins and cytochrome P450 family proteins, while those identified for Vu_396 included nitrate transporters and auxin efflux carrier family proteins.

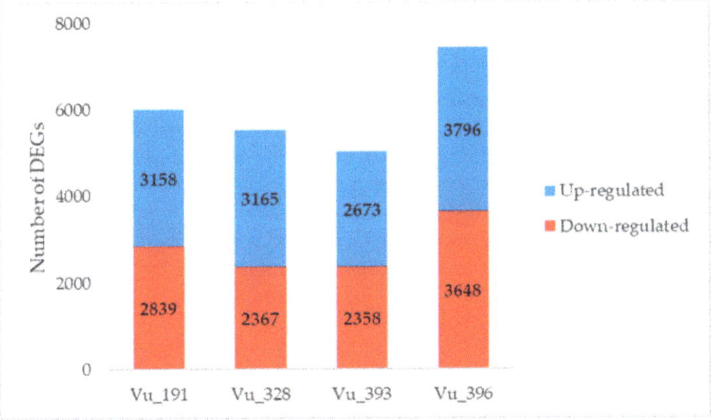

Figure 3. Number of DEGs identified in a comparison between the control and NaCl treatment for cowpea germplasms with different levels of salt tolerance.

In addition, individual DEGs induced by salt treatment in four germplasms were compared (Tables S6 and S7). Overall, a higher number of up-regulated genes were identified in the salt-resistant germplasms, while the majority of the down-regulated genes were detected in the salt-sensitive plants. In the salt-resistant germplasms, 65 common DEGs, including LEA 4–5, were up-regulated, compared with 60 in the salt-sensitive germplasms, a group which included wall-associated kinase 3 (Figure 4a). In addition, 59 common DEGs, including metallothionein 2A, were down-regulated in the salt-resistant germplasms, compared with 99 for the salt-sensitive germplasms, including the cytochrome P450 family (CyP-89-A-5) (Figure 4b).

Gene ontology (GO) and Kyoto Encyclopedia of Genes and Genomes (KEGG) enrichment analyses were conducted in order to understand and classify the functions of the common DEGs identified for the different germplasms (Tables S8 and S9). In the salt-resistant germplasms, the up-regulated genes had nine enriched GO terms in the molecular function (MF) category, with many of the DEGs associated with catalytic and transferase activity. In addition, the down-regulated genes had only one enriched GO term in the MF category, with one DEG identified associated with ADP binding, unlike the up-regulated genes. On the other hand, there were no GO terms identified as being at a significant level ($p < 0.05$) from among the common DEGs for the salt-sensitive germplasms. KEGG analysis classified the common DEGs into five major groups for the salt tolerance germplasms. Most of the DEGs, for both salt tolerance levels, were associated with metabolism at the major classification and with the global/overview maps related to pathways or metabolism at the sub-classification.

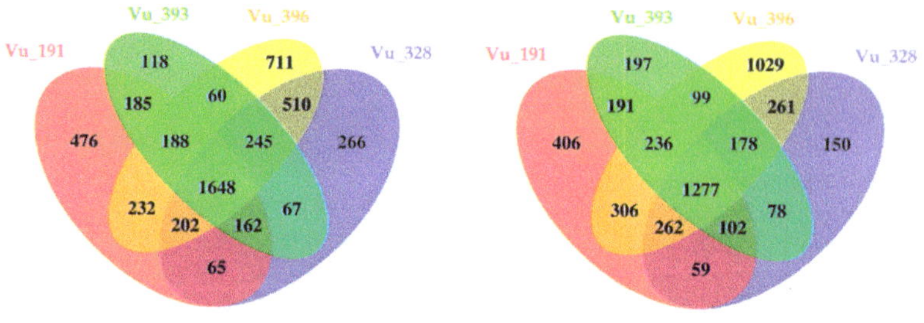

Figure 4. Venn diagrams showing the number of common and specific DEGs identified in cowpea germplasms with different levels of salt tolerance: (**a**) up-regulated and (**b**) down-regulated DEGs.

2.4. Clustering Analysis of the Identified DEGs

Hierarchical clustering analysis was conducted to confirm the gene expression patterns using information from the 9784 DEGs that were significantly expressed for each salt tolerance germplasm (Table S10). The identified DEGs were classified into six clusters containing 3710, 1762, 2961, 384, 276, and 691 genes, respectively (Figure 5).

For the C1 and C6 clusters, most of the DEGs were generally down-regulated while, for the C2 and C3 clusters, the DEGs were generally up-regulated across the four germplasms. Most of the relative expression levels were found to be similar in the four germplasms, but the C4 and C5 clusters exhibited distinct differences. The C4 cluster contained DEGs that were down-regulated in salt-resistant germplasm Vu_191 and those that were up-regulated in salt-sensitive germplasm Vu_396. This cluster contained DEG-encoding nodulin MtN21/EamA-like transporter family protein and NAD(P)-binding Rossmann-fold superfamily protein. On the other hand, the C5 cluster contained DEGs that were up-regulated in some salt-resistant germplasms and down-regulated in some salt-sensitive germplasms. This cluster contained DEG-encoding nitrate transporter 1.5 and the heavy-metal transport/detoxification superfamily protein.

GO and KEGG analysis was conducted to understand the functions of the DEGs included in each cluster (Tables S11 and S12). Overall, 384 DEGs in the C4 cluster were enriched for 29 GO terms, with 28 being independently classified into the MF category and 1 as a biological process (BP). In addition, many of the DEGs were associated with catalytic activity in the MF category, as is the case with the common DEGs, with a difference in that DEGs were detected for the response to oxidative stress in the BP category. However, in the C5 cluster, no GO terms were identified at a significant level ($p < 0.05$). As a result of the KEGG analysis of these genes, the C4 cluster was grouped into five major classifications and the C5 cluster into three (excluding genetic information and cell processing). In particular, metabolic terms were most common in both clusters, with sub-classifications dominated by global/outline maps related to pathways or metabolism, carbohydrate metabolism, and the biosynthesis of other secondary metabolites. However, this sub-classification had differences for six items, including membrane transport, transport, and catabolism.

qRT-PCR was employed to validate the expression of these DEGs (Figure S1). Of the DEGs with significant expression patterns, six were selected and analyzed further. The relative expression levels obtained from qRT-PCR exhibited trends similar to those of the Log2FC from RNA-seq.

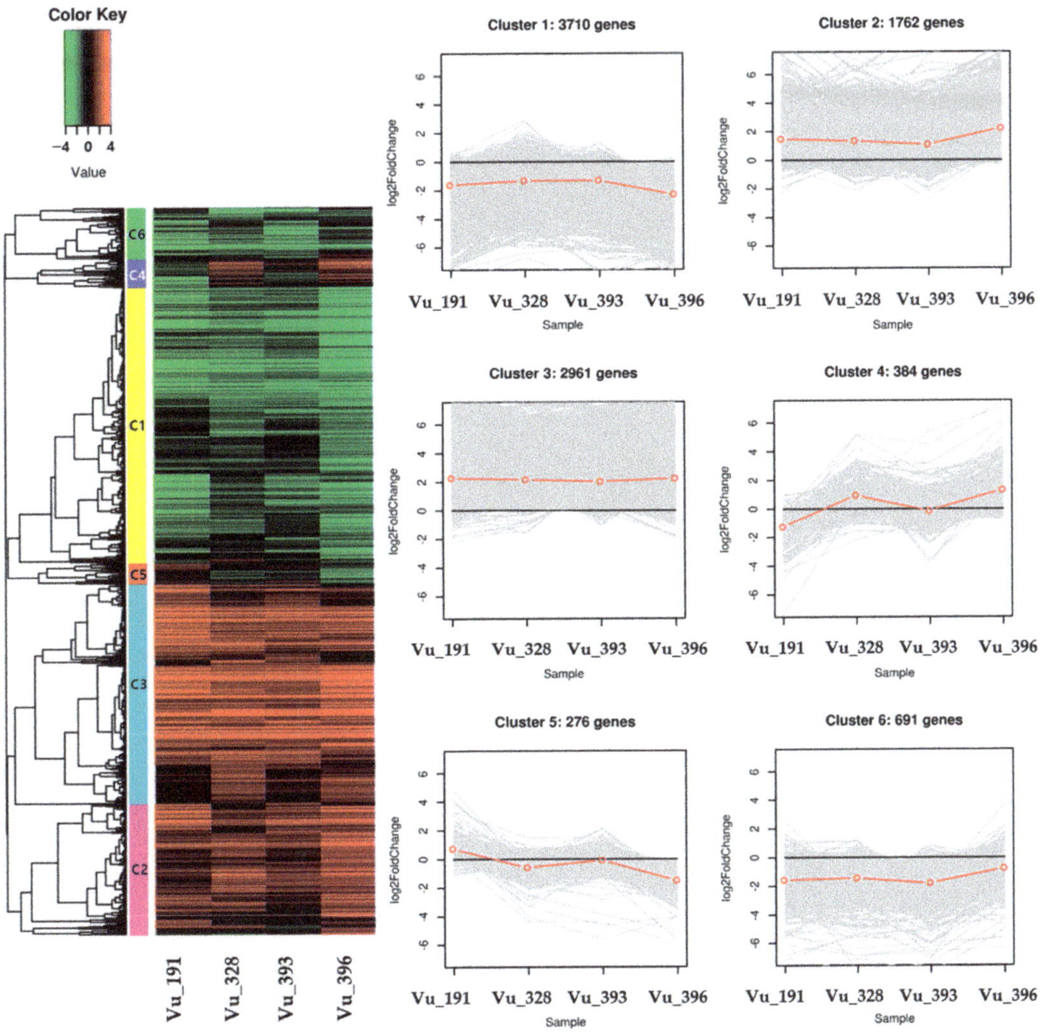

Figure 5. Heat map of the differential expression level of the genes in each cluster and line plots displaying the expressed clusters as patterns. The redline in line plots indicates the mean log_2foldchange (FC) value in gene expression.

2.5. Identification of Variations in the Target Gene

A total of 27 candidate genes were obtained based on the RNA-seq results, and two of these were not annotated (Table 1).

The 25 annotated genes included various genes related to salt stress, such as LATE EMBRYOGENESIS ABUNDANT PROTEIN 4–5, and POTASSIUM TRANSPORTER 6. Reference sequencing (re-seq) was conducted on the four germplasms to narrow down the candidate genes, and the results were integrated with those of RNA-seq. The whole genome re-seq results for each germplasm are presented in Table S13. Of the 27 candidate genes, two containing variations that may be associated with salt resistance and sensitivity (*Vigun_02G076100* and *Vigun08G125100*) were selected (Figure 6).

Table 1. Log_2foldchange (FC) and functional annotations for differentially expressed genes (DEGs) with significant expression patterns. The color of background indicates the gene expression level from red (up-regulated) to green (down-regulated).

Gene id	Position	Log_2FoldChange (FC) for Identified DEGs				Annotation
		Vu_191	Vu_328	Vu_393	Vu_396	
Vigun01g124200.1	30,107,016–30,107,917	3.71	2.51	−0.04	0.66	LATE EMBRYOGENESIS ABUNDANT PROTEIN 4–5
Vigun02g076100.1	22,812,209–22,818,056	3.35	−0.53	0.03	−0.91	POTASSIUM TRANSPORTER 6
Vigun02g087000.1	24,173,651–24,175,542	−2.55	−1.58	−1.40	1.73	ALCOHOL DEHYDROGENASE-RELATED
Vigun02g150700.1	29,780,314–29,782,539	3.70	0.77	0.71	−1.12	N-TERMINAL ACETYLTRANSFERASE
Vigun02g156800.1	30,300,350–30,306,429	4.74	−0.79	0.76	−2.85	OLIGOPEPTIDE TRANSPORTER-RELATED
Vigun03g036000.1	2,770,941–2,774,105	0.19	−0.98	−1.24	−2.64	RING-H2 FINGER PROTEIN ATL69-RELATED
Vigun03g195700.1	27,574,151–27,575,892	−0.49	−0.86	−1.45	−2.13	CYTOCHROME P450 89A2-RELATED
Vigun03g323700.1	51,942,392–51,948,451	3.10	−0.56	0.25	−2.84	ANION EXCHANGE PROTEIN
Vigun03g411400.1	61,890,298–61,892,363	−1.54	0.18	−0.65	1.31	PEROXIDASE 40
Vigun06g049900.1	17,502,581–17,504,135	1.89	1.00	0.25	−2.16	Hydroxycinnamate 4-beta-glucosyltransferase
Vigun06g206600.1	32,059,029–32,059,206	0.84	−0.07	−1.34	−1.61	Unknown
Vigun07g005500.1	439,864–440,793	3.34	1.49	1.55	−2.61	Hemopexin
Vigun07g044100.1	4,471,515–4,474,926	−2.91	0.52	0.32	2.81	3-oxoacyl-[acyl-carrier-protein] reductase
Vigun07g065400.1	7,663,218–7,665,263	−0.93	−0.34	2.22	1.80	Myb/SANT-like DNA-binding domain (Myb_DNA-bind_4)
Vigun07g164700.1	27,692,362–27,695,281	0.64	−3.29	−4.63	−4.02	ALUMINUM-ACTIVATED MALATE TRANSPORTER 10
Vigun07g217900.1	33,993,108–33,994,160	0.14	−4.35	−4.41	−5.92	Uncharacterized membrane protein
Vigun08g025300.1	2,190,472–2,192,518	−0.24	−1.85	−0.80	−3.35	HEAT STRESS TRANSCRIPTION FACTOR B-4
Vigun08g090000.1	20,808,945–20,813,724	4.14	−0.69	−0.46	−1.53	EamA-like transporter family (EamA)
Vigun08g116200.1	28,318,383–28,321,227	2.68	−1.19	−0.60	−2.22	COPPER TRANSPORT PROTEIN ATOX1-RELATED
Vigun08g125100.1	29,512,430–29,514,230	0.25	0.42	5.48	7.06	EXOCYST COMPLEX PROTEIN EXO70
Vigun09g086300.1	11,333,702–11,336,177	−2.42	1.31	2.73	3.33	INACTIVE POLY [ADP-RIBOSE] POLYMERASE SRO4-RELATED
Vigun10g015100.1	1,666,906–1,668,606	1.63	−0.20	−1.32	−1.96	CYCLIN-U4-1
Vigun10g180000.1	39,812,625–39,813,195	−0.79	0.97	1.66	2.02	Cotton fiber expressed protein (DUF761)
Vigun11g017700.1	2,206,680–2,207,648	−1.29	0.42	2.80	1.52	Unknown
Vigun11g018800.1	2,332,367–2,339,763	−0.10	−0.35	1.67	1.88	protein regulator of cytokinesis 1 (PRC1)
Vigun11g126800.1	33,386,966–33,389,552	1.96	−0.74	0.59	−2.28	MYB FAMILY TRANSCRIPTION FACTOR-RELATED
Vigun11g182400.1	38,574,316–38,575,114	0.20	−0.76	−1.17	−1.24	SAUR family protein (SAUR)

Vigun_02G076100 and *Vigun08G125100* were identified as encoding POTASSIUM TRANSPORTER 6 and EXOCYST COMPLEX PROTEIN EXO70, respectively. A total of eight coding SNPs (cSNPs) were found in the exons of these two genes. The five cSNPs found in *Vigun_02G076100* included three synonymous SNPs (sSNP) without amino acid substitutions and one synonymous variation without an amino acid substitution. However, the other SNP caused the substitution of lysine (Lys, K) in the positive amino acid with glutamic acid (Glu, E) in the negative amino acid when compared to the reference. This

SNP was found in the salt-resistant germplasm Vu_191. The three cSNPs found in the other candidate gene *Vigun_08G125100* included two non-synonymous SNPs (nsSNPs). When compared with the reference, one SNP led to the substitution of aspartic acid (Asp, D) in the negative amino acid with histidine (His, H) in the positive amino acid, while the other SNP led to the substitution of glycine (Gly, G) in the special case amino acid with Asp. These two SNPs were found in both salt-sensitive germplasms Vu_393 and Vu_396. Interestingly, the salt-resistant germplasms Vu_191 and Vu_328, for which no SNPs were found, had a missing allele rather than the reference allele.

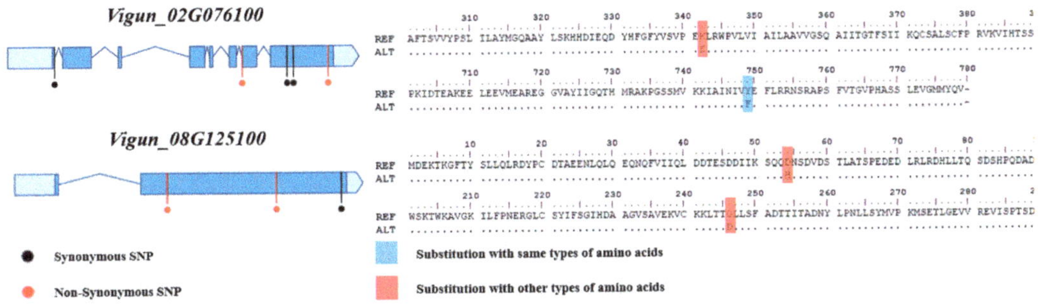

Figure 6. Gene model and variation information of two candidate genes related to salt-stress.

2.6. Validation of the Variations in the Candidate Genes

The variations in the two candidate genes were confirmed using DNA-seq and PCR analysis of the four cowpea germplasms. Of the cSNPs identified in *Vigun_02G076100*, one SNP, causing the substitution of another type of amino acid, was confirmed through DNA-seq. This was the same variation as observed in the four germplasms used for RNA-seq analysis, and the confirmed SNP was used to develop the KASP marker. In order to confirm that *Vigun_08G125100* was a missing allele, a primer producing a 1465 bp PCR product was designed (Table S14). Interestingly, PCR products were only generated for this gene with the salt-sensitive Vu_393 and Vu_396. The variations in the two candidate genes were validated using a total of 20 cowpea germplasms that included the 4 germplasms used for RNA-seq (Table 2).

Vigun_02G076100 was validated through the developed KASP marker. As a result, Vu_191 exhibited the same variation as seen in the re-seq analysis, and the SNP variation was observed in the salt-resistant germplasm Vu_111 (Figure S2). On the other hand, for *Vigun_08G125100*, PCR products were only generated for five salt-sensitive germplasms, including Vu_393 and Vu_396 (Figure S3), although this was not found in all salt-resistant cowpea germplasms.

Table 2. Validation and comparison of the variation in the two candidate genes using 20 cowpea germplasms.

Sample	Salt Tolerance Type	*Vigun_02G076100*	*Vigun_08G125100*
Vu_035	Sensitive	REF (AAA, K)	Existence
Vu_266	Sensitive	REF (AAA, K)	-
Vu_296	Sensitive	REF (AAA, K)	Existence
Vu_318	Sensitive	REF (AAA, K)	Existence
Vu_319	Sensitive	REF (AAA, K)	-
Vu_343	Sensitive	REF (AAA, K)	-
Vu_348	Sensitive	REF (AAA, K)	-
Vu_393 *	Sensitive	REF (AAA, K)	Existence
Vu_396 *	Sensitive	REF (AAA, K)	Existence
Vu_403	Sensitive	REF (AAA, K)	-
Vu_055	Resistant	REF (AAA, K)	-

Table 2. *Cont.*

Sample	Salt Tolerance Type	*Vigun_02G076100*	*Vigun_08G125100*
Vu_095	Resistant	REF (AAA, K)	-
Vu_111	Resistant	SNP (GAA, E)	-
Vu_129	Resistant	REF (AAA, K)	-
Vu_147	Resistant	REF (AAA, K)	-
Vu_166	Resistant	REF (AAA, K)	-
Vu_191 *	Resistant	SNP (GAA, E)	-
Vu_328 *	Resistant	REF (AAA, K)	-
Vu_336	Resistant	REF (AAA, K)	-
Vu_352	Resistant	REF (AAA, K)	-

* Four cowpea germplasms (Vu_191, Vu_328, Vu_393, and Vu_396) used for RNA-seq analysis.

3. Discussion

Cowpea is a legume crop that is widely grown in arid and semiarid regions because it is both heat- and drought-tolerant [1]. However, salt stress is becoming an increasingly serious issue for crops in these regions due to climate change [4]. In this study, we evaluated the ion accumulation response to salt stress using four cowpea germplasms with different levels of salt tolerance and conducted RNA-seq analysis (Figure 2). It was found that ion accumulation was significantly higher in the salt-sensitive germplasms rather than the salt-resistant germplasms. These results are in agreement with those reported by previous studies [19].

Of the four germplasms investigated in the present study, Vu_191 was classified as strongly salt-resistant, Vu_328 as weakly salt-resistant, Vu_393 as having a medium tolerance, and Vu_396 as a salt-sensitive germplasm. Based on these results, we screened candidate genes by focusing on DEGs with significant expression patterns in the comparison between Vu_191 and Vu_396.

The expression profiles of these DEGs were compared between control and treatment groups to identify genes associated with salt stress. Consequently, numerous DEGs related to salt stress were identified in the present study. For example, *Vigun_11G140800* encodes senescence-associated gene 12 (SAG 12), which is related to cysteine protease [35], and plays a role in plant aging and programmed cell death in response to biotic and abiotic stresses [36]. This was up-regulated in all four germplasms, suggesting that aging was accelerated by salt stress. In addition, *Vigun_09G159100* is a gene-encoding wall-associated kinase 3 (Wak3), which is associated with the pectin molecule in the cell wall and is essential for cell expansion [37]. It has been reported that a decrease in protein levels affects cell expansion and cell shape. In the present study, there was no significant expression value in the salt-resistant germplasms, but overexpression was observed in the salt-sensitive germplasms. This may be a form of plant defense to maintain homeostasis in response to stress such as excessive salt accumulation in salt-sensitive plants.

Another gene of note was *Vigun_03G195700*, which encodes CyP-89-A-5. The Cyp family is a large collection of proteins found in higher plants, and it has been assumed that they provide protection from various biotic and abiotic stresses. In particular, it has been reported that the suppression of CaCyP1 in pepper, which has a high homology with CyP-89-A-5 in *Arabidopsis*, increased the susceptibility to bacterial pathogens [38]. This gene is down-regulated in salt-sensitive germplasms, which is consistent with our results. Therefore, the Cyp gene found in cowpea is also assumed to affect salt tolerance via a similar mechanism.

We also conducted GO and KEGG analysis of the common DEGs and each cluster. The DEGs identified for the salt-resistant germplasms were generally related to catalytic activity (GO:0003824) and transferase activity (GO:0016740), while the DEGs corresponding to Cluster 4 were associated with small-molecule binding (GO:0036094), anion binding (GO:0043168), and ribonucleotide binding (GO:0032553). These GO terms play important roles in several salt tolerance mechanisms, including osmotic regulation. In particular, catalytic activity (GO:0003824) exhibited functions related to osmotic regulation and ionic change when exposed to salt stress [39]. These results also suggest that protein-coding genes, related to molecular structure and function, can be regulated in response to salt stress, and further indicate that anion reactions are associated with salt stress.

As a result of our KEGG analysis, most of the DEGs were associated with metabolism with four sub-classifications (global and overview maps, amino acid metabolism, carbohydrate metabolism, and biosynthesis of other secondary metabolites). This suggests that abiotic stress not only regulates metabolic processes via enzyme activity but also causes indirect or direct changes in proteins by affecting amino acids. Our results also suggest the involvement of the metabolism of various amino acids and the biosynthetic pathways of secondary metabolites such as phenylpropanoids. It has been reported that phenylpropanoids are activated under various abiotic stress conditions, including salt stress, to remove ROS [40]. These results thus help us to understand the molecular biological response to salt stress.

We subsequently selected 27 target genes, related to salt tolerance, based on the expression patterns and annotations for the DEGs in the RNA-seq analysis. One of these was *Vigun_01G124200*, which encodes LATE EMBRYOGENESIS ABUNDANT PROTEIN 4–5 (LEA 4–5). The LEA protein is a polypeptide that accumulates in later embryonic stages and is associated with the acquisition of desiccation tolerance [41]. It also increases resistance to osmotic and cold stress in various crops and is associated with water-deficient conditions [42]. This protein generally accumulates during periods of stress-induced growth arrest and is involved in stress recovery [43]. For example, AtLEA4–5 in *Arabidopsis* is known to be a member of the genes encoding the LEA protein involved in water deprivation tolerance [44]. This gene is usually suppressed by the repressor AtMYB44, but it has been reported that, when exposed to osmotic stress, the repressor is removed and normal expression occurs [45]. In the RNA-seq results, *Vigun_01G124200*, which encodes LEA4–5, was significantly up-regulated in the salt-resistant germplasms. However, the re-seq results did not detect significant variations. Interestingly, *Vigun_03G281700* encoding MYB44 was up-regulated in the salt-sensitive germplasms. This suggests that the expression of LEA4–5 in cowpea can be regulated by the same mechanism used in *Arabidopsis thaliana*, but that it is also regulated by an additional pathway.

Because interpreting the large volumes of data from the 27 target DEGs was difficult, we conducted re-sequencing to narrow down the range of the candidate genes. Most of the target genes had many SNPs in each salt-tolerant germplasm, but these SNPs were in the UTR or intron regions, which may not be involved in regulating gene expression. However, some of the SNPs in the two candidate genes, *Vigun_02G076100* and *Vigun_08G125100*, exhibited significant associations with salt tolerance. *Vigun_02G076100*, a gene-encoding POTASSIUM TRANSPORTER 6, was up-regulated in salt-resistant Vu_191. Potassium (K+) is an essential cation for plant growth and development and the regulation of enzyme activity, membrane potential, and turgor pressure [12]. High salinity is the result of the accumulation of excessive sodium (Na+) ions, which leads to ion stress. Plants are consequently unable to maintain K+ homeostasis, which ultimately adversely affects plant growth. Accordingly, one of the primary mechanisms associated with salt tolerance in plants is the maintenance of a balanced cation ratio in the cytoplasm [46]. In addition, the *Arabidopsis* KUP6 subfamily transporter is related to cell growth and potassium homeostasis and has been reported to be a major factor associated with osmotic control [47]. This suggests that the strong salt resistance of Vu_191 occurs as a result of the overexpression of potassium transporter 6. *Vigun_08G125100* encodes EXOCYST COMPLEX PROTEIN EXO70 and was up-regulated in both Vu_393 and Vu_396. The exocyst subunit EXO70 protein has been reported to be involved in anchoring and regulating membrane fusion and actin polarity in the plasma membrane of exocysts [48]. Some genes included in the exocyst gene family have been reported to be up-regulated with exposure to salt stress, but their exact functions have not been identified [49].

The variations in the two candidate genes were validated using KASP genotyping and PCR products. The cSNPs found in *Vigun_02G076100* were found in Vu_111 and Vu_191, both of which were salt resistant. This has been identified as a specific variation in some salt-resistant germplasms. Based on this, it can be assumed that Vu_111 had the same salt tolerance mechanism as Vu_191. Conversely, *Vigun_08G125100* was identified as a missing allele in NGS analysis. To validate these results, 20 cowpea germplasms were tested, with 75% classified as having the same salt tolerance type as before. Thus, it can be assumed that the loss of this gene has occurred as a result of the development of various salt resistance mechanisms, but the functional part has not been confirmed.

In summary, we identified two candidate genes related to salt tolerance that differed between cowpea germplasms with different levels of salt tolerance. These variations were developed as KASP and indel markers, respectively. The two developed markers thus have the potential to be useful molecular markers for the screening of germplasms in salt tolerance breeding programs.

4. Materials and Methods

4.1. Plant Material and Phenotyping of Salt Tolerance

In this study, 20 cowpea germplasms with different levels of salt tolerance (10 salt-resistant and 10 salt-sensitive) were used, and among them four showed distinct differences in salt tolerance under controlled conditions and were used for RNA-seq analysis (Table S15). The 20 cowpea germplasms were then used to verify the SNP variation. The germplasm seeds were obtained from the Rural Development Administration (RDA) Genebank at the National Agrobiodiversity Center, Republic of Korea. The four cowpea germplasm were treated with 250 mM NaCl for seedlings in the V2 stage with the same growth after germination. After three weeks of NaCl treatment, the entire plant was sampled to evaluate the accumulation of sodium and chloride ions. The ion content was extracted from dried and pulverized leaf samples (150 mg) using 30 mL of distilled water for 1 h and then filtered through Whatman filter paper. The sodium ion levels were determined using a Na+ measuring instrument (Horiba, Kyoto, Japan), while the chloride ion levels were determined using an ion-selective electrode (Mettler Toledo, Columbus, OH, USA).

4.2. Salt Treatment

One hundred seeds from each germplasm were sterilized with 70% ethanol for 1 min and then washed with sterile water. The sterilized seeds were germinated in a plant growth chamber under long-day conditions (16 h light and 8 h dark), and similar seedlings were selected and transplanted into 1/2 Hoagland Nutrient Solution for hydroponic use. After two weeks of salt treatment, seedlings with the same growth were treated with 250 mM NaCl, while the control seedlings were placed in a solution without NaCl. After 24 h of salt treatment, the roots of the NaCl-treated and control seedlings were sampled. Each treatment and control group had three biological replicates, which were randomly sampled at 10 points and mixed into a single sample. The samples were frozen using liquid nitrogen and stored at −80 °C for use in subsequent experiments. Overall, a total of 24 RNA library samples were analyzed.

4.3. RNA Extraction, Construction of cDNA Libraries and Short Read Sequencing

Total RNA was extracted using an RNeasy Plant Mini Kit (Qiagen, Hilden, Germany). The quality and integrity of the extracted RNA were determined using a 2100 Bioanalyzer RNA instrument (Agilent, Santa Clara, CA, USA). Poly-A+ libraries were prepared using an Illumina Truseq Stranded mRNA Library Prep Kit (Illumina, San Diego, CA, USA), and the generated libraries were sequenced using an Illumina NovaSeq6000 platform. Both RNA extraction and cDNA library construction were conducted according to the manufacturer's instructions.

4.4. Sequence Pre-Processing and Mapping of RNA-Seq Reads

In the sequenced transcriptome short reads, the adapter sequence was removed with cutadapt [50] and pre-processing was conducted using DynamicTrim and LengthSort in the SolexaQA package [51]. DynamicTrim removes low-quality bases at both ends of short reads to purify them, while LengthSort excludes trimmed reads of 25 bp or fewer from the analysis process. The clean trimmed reads were mapped onto the *Vigna unguiculata* (v1.2) reference genome from the Phytozome database (http: //phytozome.jgi.doe.gov/ (accessed on 1 December 2022)) using HISAT2 software [52]. HTSeq (v.0.11.0) [53] was used to measure expression as the total number of reads mapped to each gene. In order to avoid bias due to the germplasm in the sequencing numbers, normalization was conducted using the DEseq library [54].

4.5. Identification of Differentially Expressed Genes (DEGs)

DEGs were selected based on a twofold change in the number of mapped reads and an FDR of $\leq$0.01, with the adjusted p value calculated using Benjamini–Hochberg correction. Hierarchical clustering analysis was conducted using the amap [55] and gplot libraries [56] in R to determine gene expression patterns, which were calculated using Pearson's correlation, and grouping was conducted through the complete method. GO enrichment was analyzed using reference GO information [57]. The significance level was set at 0.05 and the GO terms were classified into biological process (BP), cellular component (CC), and molecular function (MF) categories. Functional annotation was conducted for an e-value of $\leq 1 \times 10^{-100}$ and best hits using amino acid sequences from the KEGG database [58] and BLASTP.

4.6. Quantitative Reverse Transcription PCR (qRT-PCR) for Validation of DEGs

First-strand cDNA was synthesized using SuperScript™ III First-Strand Synthesis SuperMix (Invitrogen, Waltham, MA, USA) following the manufacturer's instructions. qRT-PCR was conducted on a StepOne Real-Time PCR System (Applied Biosystems, Foster City, CA, USA) using a Bio-Rad iQ™ SYBR Green Supermix Kit (Invitrogen, CA). The reaction mixture, containing 20 ng of cDNA, was analyzed according to the manufacturer's instructions. The PCR conditions were as follows: holding, 1 cycle at 95 °C for 10 min; cycling, 40 cycles at 95 °C for 15 s and at 60 °C for 60 s. Then, the melting curve analysis was conducted to confirm the absence of a product and the dimer formation of the primers. The primers were designed using Primer3 software (v2.3.5) [59]. The CT values were normalized using the ubiquitin-conjugating enzyme E2 variant 1D (UE21D) gene stable under salt stress as a housekeeping gene [60] and gene expression was analyzed using the $2^{-\Delta\Delta CT}$ method [61]. Three biological replicates were analyzed using the average of two technical replicates.

4.7. Whole-Genome Resequencing and DNA Sequencing

Genomic DNA was extracted using a DNeasy Plant Mini Kit (Qiagen) following the manufacturer's instructions, and the integrity and purity of the extracted DNA samples were determined using 2.0% agarose gel and a Nanodrop ND 2000 spectrophotometer (Thermo Fisher Scientific, Waltham, MA, USA). The cDNA library was constructed and sequenced using the same NGS protocol as for RNA-seq. Paired-end reads were mapped onto the cowpea genomic reference genome and then entered into the nf-core/sarek's analysis pipeline [62]. The DNA was sequenced using the PCR products of the candidate gene on an ABI 3730XL analyzer (Applied Biosystems). The primers used to generate the PCR products were prepared in the same way as the primers used for qRT-PCR. More detailed information on this process is provided in Table S14.

4.8. Kompetitive Allele-Specific PCR(KASP) Primer Design and Validation

KASP primers were designed to detect SNP variation in the candidate genes according to the standard KASP protocol. Allele-specific primers included FAM (5′-

GCTATAACCAGAACAGGCCATCTCAATTT-3′) and HEX (5′-TAACCAGAACAGGCCATCTCAA-TTC). The KASP primers were used to genotype the 20 cowpea germplasms using StepOnePlus software (Applied Biosystems). Genotyping was conducted using a mixture consisting of 50 ng/5 uL of DNA, 0.14 uL of KASP assay mix, and 5 uL of KASP master mix. The KASP cycling conditions were as follows: pre-PCR reading, 1 cycle at 30 °C for 1 min; holding, 1 cycle at 94 °C for 15 min; cycling, 10 cycles at 94 °C for 20 s and 61–55 °C for 1 min (reduction of 0.6 °C per cycle), and 26 cycles at 94 °C for 20 s and 55 °C for 1 min; and post-PCR reading at 30 °C for 30 s.

4.9. Statistical Analysis

Statistical analysis was conducted using analysis of variance (ANOVA) and least significant difference (LSD) tests in SPSS 27 (IBM, Armonk, NY, USA), with $p < 0.05$ employed to determine statistically significant differences between groups.

5. Conclusions

Four cowpea germplasms with different levels of salt tolerance were used to investigate transcriptome variations in roots under salt stress. RNA-seq analysis of the salt treatment and control groups, with three biological replicates assessed for each germplasm, led to the selection of 27 candidate genes related to salt stress. Of these, two candidate genes with significant variation were investigated further in this study. The two candidate genes contained cSNPs in the exon region and represented a missing allele, respectively. The information provided on the two candidate cowpea genes in relations to salt stress and presented in the present study has the potential to be used for genetic improvements in cowpea breeding programs.

Supplementary Materials: The supporting information can be downloaded at: https://www.mdpi.com/article/10.3390/ijms24054762/s1.

Author Contributions: Writing—Original Draft Preparation, B.H.K.; Methodology, W.J.K., S.C., C.Y.M. and S.K.; Resources, S.-H.K.; Data Analysis, S.-H.J.; Funding Acquisition, T.-H.J., K.D.K. and B.-K.H.; Supervision, B.-K.H. and K.D.K. All authors have read and agreed to the published version of the manuscript.

Funding: This work was carried out with the support of the Cooperative Research Program for Agriculture Science and Technology Development (Project No. PJ015680012022), the Rural Development Administration, Republic of Korea.

Institutional Review Board Statement: Not applicable.

Informed Consent Statement: Not applicable.

Data Availability Statement: Not applicable.

Conflicts of Interest: Author (S.-H.J.) has been involved as a consultant and expert witness in Company (SEEDERS Inc.).

References

1. Ehlers, J.; Hall, A. Cowpea (*Vigna unguiculata* L. walp.). *Field Crops Res.* **1997**, *53*, 187–204. [CrossRef]
2. Rachie, K.; Roberts, L. Grain legumes of the lowland tropics. *Adv. Agron.* **1974**, *26*, 1–132.
3. Food and Agriculture Organization of the United Nations. FAOSTAT Database 2020. Available online: https://www.fao.org/faostat/en/#data/QCL (accessed on 16 December 2022).
4. Metternicht, G.I.; Zinck, J. Remote sensing of soil salinity: Potentials and constraints. *Remote Sens. Environ.* **2003**, *85*, 1–20. [CrossRef]
5. Sultan, B.; Gaetani, M. Agriculture in West Africa in the twenty-first century: Climate change and impacts scenarios, and potential for adaptation. *Front. Plant Sci.* **2016**, *7*, 1262. [CrossRef] [PubMed]
6. Sylla, M.B.; Nikiema, P.M.; Gibba, P.; Kebe, I.; Klutse, N.A.B. Climate change over West Africa: Recent trends and future projections. In *Adaptation to Climate Change and Variability in Rural West Africa*; Springer: Cham, Sitzerland, 2016; pp. 25–40.
7. Lacerda, C.F.; Assis Júnior, J.O.; Lemos Filho, L.C.; Oliveira, T.S.d.; Guimarães, F.V.; Gomes-Filho, E.; Prisco, J.T.; Bezerra, M.A. Morpho-physiological responses of cowpea leaves to salt stress. *Braz. J. Plant Physiol.* **2006**, *18*, 455–465. [CrossRef]

8. Moradi, F.; Ismail, A.M. Responses of photosynthesis, chlorophyll fluorescence and ROS-scavenging systems to salt stress during seedling and reproductive stages in rice. *Ann. Bot.* **2007**, *99*, 1161–1173. [CrossRef]
9. Zhu, J.-K. Plant salt stress. In *eLS*; University of California: Riverside, CA, USA, 2007.
10. Acosta-Motos, J.R.; Penella, C.; Hernández, J.A.; Díaz-Vivancos, P.; Sánchez-Blanco, M.J.; Navarro, J.M.; Gómez-Bellot, M.J.; Barba-Espín, G. Towards a sustainable agriculture: Strategies involving phytoprotectants against salt stress. *Agronomy* **2020**, *10*, 194. [CrossRef]
11. Roy, S.; Chowdhury, N. Salt stress in plants and amelioration strategies: A critical review. *Abiotic Stress Plants* **2020**. [CrossRef]
12. Arif, Y.; Singh, P.; Siddiqui, H.; Bajguz, A.; Hayat, S. Salinity induced physiological and biochemical changes in plants: An omic approach towards salt stress tolerance. *Plant Physiol. Biochem.* **2020**, *156*, 64–77. [CrossRef]
13. Chinnusamy, V.; Jagendorf, A.; Zhu, J.K. Understanding and improving salt tolerance in plants. *Crop Sci.* **2005**, *45*, 437–448. [CrossRef]
14. Farooq, M.; Gogoi, N.; Hussain, M.; Barthakur, S.; Paul, S.; Bharadwaj, N.; Migdadi, H.M.; Alghamdi, S.S.; Siddique, K.H. Effects, tolerance mechanisms and management of salt stress in grain legumes. *Plant Physiol. Biochem.* **2017**, *118*, 199–217. [CrossRef] [PubMed]
15. Roy, S.J.; Negrão, S.; Tester, M. Salt resistant crop plants. *Curr. Opin. Biotechnol.* **2014**, *26*, 115–124. [CrossRef] [PubMed]
16. Waśkiewicz, A.; Beszterda, M.; Goliński, P. ABA: Role in plant signaling under salt stress. In *Salt Stress in Plants*; Springer: New York, NY, USA, 2013; pp. 175–196.
17. Sudhir, P.; Murthy, S. Effects of salt stress on basic processes of photosynthesis. *Photosynthetica* **2004**, *42*, 481–486. [CrossRef]
18. Hasanuzzaman, M.; Oku, H.; Nahar, K.; Bhuyan, M.; Mahmud, J.A.; Baluska, F.; Fujita, M. Nitric oxide-induced salt stress tolerance in plants: ROS metabolism, signaling, and molecular interactions. *Plant Biotechnol. Rep.* **2018**, *12*, 77–92. [CrossRef]
19. Le, L.T.T.; Kotula, L.; Siddique, K.H.; Colmer, T.D. Na+ and/or Cl− toxicities determine salt sensitivity in soybean (*Glycine max* (L.) Merr.), mungbean (*Vigna radiata* (L.) R. Wilczek), cowpea (*Vigna unguiculata* (L.) Walp.), and common Bean (*Phaseolus vulgaris* L.). *Int. J. Mol. Sci.* **2021**, *22*, 1909. [CrossRef] [PubMed]
20. Flowers, T.J.; Galal, H.K.; Bromham, L. Evolution of halophytes: Multiple origins of salt tolerance in land plants. *Funct. Plant Biol.* **2010**, *37*, 604–612. [CrossRef]
21. Turner, N.C.; Colmer, T.D.; Quealy, J.; Pushpavalli, R.; Krishnamurthy, L.; Kaur, J.; Singh, G.; Siddique, K.H.; Vadez, V. Salinity tolerance and ion accumulation in chickpea (*Cicer arietinum* L.) subjected to salt stress. *Plant Soil* **2013**, *365*, 347–361. [CrossRef]
22. Maas, E.; Poss, J. Salt sensitivity of cowpea at various growth stages. *Irrig. Sci.* **1989**, *10*, 313–320. [CrossRef]
23. Ravelombola, W.S.; Shi, A.; Weng, Y.; Clark, J.; Motes, D.; Chen, P.; Srivastava, V. Evaluation of salt tolerance at germination stage in cowpea [*Vigna unguiculata* (L.) Walp]. *HortScience* **2017**, *52*, 1168–1176. [CrossRef]
24. Murillo-Amador, B.; Troyo-Diéguez, E.; García-Hernández, J.L.; López-Aguilar, R.; Avila-Serrano, N.Y.; Zamora-Salgado, S.; Rueda-Puente, E.O.; Kaya, C. Effect of NaCl salinity in the genotypic variation of cowpea (*Vigna unguiculata*) during early vegetative growth. *Sci. Hortic.* **2006**, *108*, 423–431. [CrossRef]
25. Al-Ashkar, I.; Alderfasi, A.; Ben Romdhane, W.; Seleiman, M.F.; El-Said, R.A.; Al-Doss, A. Morphological and genetic diversity within salt tolerance detection in eighteen wheat genotypes. *Plants* **2020**, *9*, 287. [CrossRef] [PubMed]
26. Zhang, J.; Chiodini, R.; Badr, A.; Zhang, G. The impact of next-generation sequencing on genomics. *J. Genet. Genom.* **2011**, *38*, 95–109. [CrossRef] [PubMed]
27. Malik, V.S. *RNA Sequencing as a Tool for Understanding Biological Complexity of Abiotic Stress in Plants*; Springer: Cham, Sitzerland, 2016; Volume 25, pp. 1–2.
28. Marguerat, S.; Bähler, J. RNA-seq: From technology to biology. *Cell. Mol. Life Sci.* **2010**, *67*, 569–579. [CrossRef]
29. Bahieldin, A.; Atef, A.; Sabir, J.S.; Gadalla, N.O.; Edris, S.; Alzohairy, A.M.; Radhwan, N.A.; Baeshen, M.N.; Ramadan, A.M.; Eissa, H.F. RNA-Seq analysis of the wild barley (*H. spontaneum*) leaf transcriptome under salt stress. *Comptes Rendus Biol.* **2015**, *338*, 285–297. [CrossRef] [PubMed]
30. Hu, L.; Li, H.; Chen, L.; Lou, Y.; Amombo, E.; Fu, J. RNA-seq for gene identification and transcript profiling in relation to root growth of bermudagrass (*Cynodon dactylon*) under salinity stress. *BMC Genom.* **2015**, *16*, 575. [CrossRef] [PubMed]
31. Xie, R.; Pan, X.; Zhang, J.; Ma, Y.; He, S.; Zheng, Y.; Ma, Y. Effect of salt-stress on gene expression in citrus roots revealed by RNA-seq. *Funct. Integr. Genom.* **2018**, *18*, 155–173. [CrossRef]
32. Zhang, F.; Zhu, G.; Du, L.; Shang, X.; Cheng, C.; Yang, B.; Hu, Y.; Cai, C.; Guo, W. Genetic regulation of salt stress tolerance revealed by RNA-Seq in cotton diploid wild species, *Gossypium davidsonii*. *Sci. Rep.* **2016**, *6*, 20582. [CrossRef]
33. Semagn, K.; Babu, R.; Hearne, S.; Olsen, M. Single nucleotide polymorphism genotyping using Kompetitive Allele Specific PCR (KASP): Overview of the technology and its application in crop improvement. *Mol. Breed.* **2014**, *33*, 1–14. [CrossRef]
34. Lei, L.; Zheng, H.; Bi, Y.; Yang, L.; Liu, H.; Wang, J.; Sun, J.; Zhao, H.; Li, X.; Li, J. Identification of a major QTL and candidate gene analysis of salt tolerance at the bud burst stage in rice (*Oryza sativa* L.) using QTL-Seq and RNA-Seq. *Rice* **2020**, *13*, 55. [CrossRef]
35. James, M.; Masclaux-Daubresse, C.; Marmagne, A.; Azzopardi, M.; Laîné, P.; Goux, D.; Etienne, P.; Trouverie, J. A new role for SAG12 cysteine protease in roots of *Arabidopsis thaliana*. *Front. Plant Sci.* **2019**, *9*, 1998. [CrossRef]
36. Grudkowska, M.; Zagdańska, B. Multifunctional role of plant cysteine proteinases. *Acta Biochim. Pol.* **2004**, *51*, 609–624. [CrossRef] [PubMed]
37. Zagorchev, L.; Kamenova, P.; Odjakova, M. The role of plant cell wall proteins in response to salt stress. *Sci. World J.* **2014**, *2014*, 764089. [CrossRef] [PubMed]

38. Kim, Y.-C.; Kim, S.-Y.; Paek, K.-H.; Choi, D.; Park, J.M. Suppression of CaCYP1, a novel cytochrome P450 gene, compromises the basal pathogen defense response of pepper plants. *Biochem. Biophys. Res. Commun.* **2006**, *345*, 638–645. [CrossRef] [PubMed]
39. Song, Q.; Joshi, M.; Wang, S.; Johnson, C.D.; Joshi, V. Comparative analysis of root transcriptome profiles of sesame (*Sesamum indicum* L.) in response to osmotic stress. *J. Plant Growth Regul.* **2021**, *40*, 1787–1801. [CrossRef]
40. Sharma, A.; Shahzad, B.; Rehman, A.; Bhardwaj, R.; Landi, M.; Zheng, B. Response of phenylpropanoid pathway and the role of polyphenols in plants under abiotic stress. *Molecules* **2019**, *24*, 2452. [CrossRef]
41. Pedrosa, A.M.; Martins, C.d.P.S.; Goncalves, L.P.; Costa, M.G.C. Late embryogenesis abundant (LEA) constitutes a large and diverse family of proteins involved in development and abiotic stress responses in sweet orange (*Citrus sinensis* L. Osb.). *PLoS ONE* **2015**, *10*, e0145785. [CrossRef]
42. Wise, M.J.; Tunnacliffe, A. POPP the question: What do LEA proteins do? *Trends Plant Sci.* **2004**, *9*, 13–17. [CrossRef]
43. Chourey, K.; Ramani, S.; Apte, S.K. Accumulation of LEA proteins in salt (NaCl) stressed young seedlings of rice (*Oryza sativa* L.) cultivar Bura Rata and their degradation during recovery from salinity stress. *J. Plant Physiol.* **2003**, *160*, 1165–1174. [CrossRef]
44. Cuevas-Velazquez, C.L.; Saab-Rincón, G.; Reyes, J.L.; Covarrubias, A.A. The unstructured N-terminal region of *Arabidopsis* group 4 late embryogenesis abundant (LEA) proteins is required for folding and for chaperone-like activity under water deficit. *J. Biol. Chem.* **2016**, *291*, 10893–10903. [CrossRef]
45. Nguyen, N.H.; Nguyen, C.T.T.; Jung, C.; Cheong, J.-J. AtMYB44 suppresses transcription of the late embryogenesis abundant protein gene AtLEA4-5. *Biochem. Biophys. Res. Commun.* **2019**, *511*, 931–934. [CrossRef]
46. Shabala, S.; Cuin, T.A. Potassium transport and plant salt tolerance. *Physiol. Plant.* **2008**, *133*, 651–669. [CrossRef] [PubMed]
47. Osakabe, Y.; Arinaga, N.; Umezawa, T.; Katsura, S.; Nagamachi, K.; Tanaka, H.; Ohiraki, H.; Yamada, K.; Seo, S.-U.; Abo, M. Osmotic stress responses and plant growth controlled by potassium transporters in *Arabidopsis*. *Plant Cell* **2013**, *25*, 609–624. [CrossRef]
48. Heider, M.R.; Munson, M. Exorcising the exocyst complex. *Traffic* **2012**, *13*, 898–907. [CrossRef] [PubMed]
49. Luo, Z.; Szczepanek, A.; Abdel-Haleem, H. Genome-wide association study (GWAS) analysis of camelina seedling germination under salt stress condition. *Agronomy* **2020**, *10*, 1444. [CrossRef]
50. Martin, M. Cutadapt removes adapter sequences from high-throughput sequencing reads. *EMBnet. J.* **2011**, *17*, 10–12. [CrossRef]
51. Cox, M.P.; Peterson, D.A.; Biggs, P.J. SolexaQA: At-a-glance quality assessment of Illumina second-generation sequencing data. *BMC Bioinform.* **2010**, *11*, 485. [CrossRef] [PubMed]
52. Kim, D.; Langmead, B.; Salzberg, S.L. HISAT: A fast spliced aligner with low memory requirements. *Nat. Methods* **2015**, *12*, 357–360. [CrossRef]
53. Anders, S.; Pyl, P.T.; Huber, W. HTSeq—A Python framework to work with high-throughput sequencing data. *Bioinformatics* **2015**, *31*, 166–169. [CrossRef]
54. Anders, S.; Huber, W. Differential expression analysis for sequence count data. *Nat. Preced.* **2010**, *11*, R106.
55. Lucas, A. Another Multidimensional Analysis Package. 2006. Available online: http://pbil.univ-lyon1.fr/R/pdf/tdr52.pdf (accessed on 16 December 2022).
56. Warnes, M.G.R.; Bolker, B.; Bonebakker, L.; Gentleman, R.; Huber, W. Package 'gplots'. *Var. R Program. Tools Plotting Data.* 2016. Available online: https://cran.microsoft.com/snapshot/2016-03-30/web/packages/gplots/gplots.pdf (accessed on 16 December 2022).
57. Ashburner, M.; Ball, C.A.; Blake, J.A.; Botstein, D.; Butler, H.; Cherry, J.M.; Davis, A.P.; Dolinski, K.; Dwight, S.S.; Eppig, J.T. Gene ontology: Tool for the unification of biology. *Nat. Genet.* **2000**, *25*, 25–29. [CrossRef] [PubMed]
58. Ogata, H.; Goto, S.; Sato, K.; Fujibuchi, W.; Bono, H.; Kanehisa, M. KEGG: Kyoto encyclopedia of genes and genomes. *Nucleic Acids Res.* **1999**, *27*, 29–34. [CrossRef] [PubMed]
59. Rozen, S.; Skaletsky, H. Primer3 on the WWW for general users and for biologist programmers. In *Bioinformatics Methods and Protocols*; Springer: Cham, Sitzerland, 2000; pp. 365–386.
60. Amorim, L.L.B.; Ferreira-Neto, J.R.C.; Bezerra-Neto, J.P.; Pandolfi, V.; de Araújo, F.T.; da Silva Matos, M.K.; Santos, M.G.; Kido, E.A.; Benko-Iseppon, A.M. Cowpea and abiotic stresses: Identification of reference genes for transcriptional profiling by qPCR. *Plant Methods* **2018**, *14*, 88. [CrossRef]
61. Schmittgen, T.D.; Livak, K.J. Analyzing real-time PCR data by the comparative CT method. *Nat. Protoc.* **2008**, *3*, 1101–1108. [CrossRef] [PubMed]
62. Garcia, M.; Juhos, S.; Larsson, M.; Olason, P.I.; Martin, M.; Eisfeldt, J.; DiLorenzo, S.; Sandgren, J.; De Ståhl, T.D.; Ewels, P. Sarek: A portable workflow for whole-genome sequencing analysis of germline and somatic variants. *F1000Research* **2020**, *9*, 63. [CrossRef] [PubMed]

MDPI AG
Grosspeteranlage 5
4052 Basel
Switzerland
Tel.: +41 61 683 77 34

International Journal of Molecular Sciences Editorial Office
E-mail: ijms@mdpi.com
www.mdpi.com/journal/ijms

www.ingramcontent.com/pod-product-compliance
Lightning Source LLC
Chambersburg PA
CBHW061622210926
51850CB00009B/2406
9783725861538